MELVIN SILVERMAN

THE TECHNICAL MANAGER'S SURVIVAL BOOK

McGRAW-HILL BOOK COMPANY
New York   St. Louis   San Francisco   Auckland
Bogotá   Hamburg   Johannesburg   London   Madrid
Mexico   Montreal   New Delhi   Panama   Paris
São Paulo   Singapore   Sydney   Tokyo   Toronto

HD
31
S535
1984

ABOUT THE AUTHOR

Melvin Silverman has a varied and extensive background as an engineer, operating executive, and university instructor and is a well-known consultant on project and engineering management. Dr. Silverman is the author of two books and numerous articles for the technical and management press. He is currently a Managing Partner in the consulting firm of Atrium Associates, Inc., in Cliffside Park, New Jersey. His academic background includes a B.S.M.E., an M.S.I.E., an M.B.A., an M.A. in clinical psychology, and a Ph.D. in industrial-social psychology. He is a registered Professional Engineer. With over twenty years of direct technical experience as a manager in, and as a consultant to, a broad spectrum of industrial organizations, he is able to combine an extensive training in applicable management theory with the seasoned approach of a practicing professional.

Library of Congress Cataloging in Publication Data
Silverman, Melvin.
The technical manager's survival book.

Includes bibliographies.
1. Industrial management.   I. Title.
HD31.S535   1984   658.4   83-13555
ISBN 0-07-057515-0

   2 3 4 5 6 7 8 9 0   DOC/DOC   8 9 8 7 6 5

ISBN 0-07-057515-0

The editors for this book were Harold B. Crawford and
Stephan O. Parnes, the designer was Riverside Graphic Studio,
and the production supervisor was Reiko F. Okamura.
It was set in Caledonia by J. M. Post Graphics, Corp.

Printed and bound by R. R. Donnelley & Sons Company.

CONTENTS

PREFACE

People make choices during their lives. They make choices about the kind of work they do, the companies they will do it in, and how long they will stay at it. Sometimes these choices appear to be limited and at other times almost unlimited; but in all cases, there are always choices. Often, the choice that is made depends upon the normative or predictive expectancy of the person(s) making it. This means that there is an expectation that the alternative chosen (from those alternatives perceived to be possible at the time of the choosing) will result in the best future situation. Therefore, the choosing process depends upon the personal theory of the chooser, and that personal theory in turn is based to a great extent upon the chooser's past (education, experience, etc.) and how that past is coupled with a subjective evaluation of the future. It's not a completely logical process; emotion is involved. There is no such thing as complete objectivity where people are involved; personal values and assumptions determine which "facts" that one sees.

Much of our education and training as technical managers, engineers and scientists attempts to minimize this subjectivity by proposing a relatively value-free, objective view of the world. Those attempts can be only partially successful, since it is obvious that this education process too has a hidden value: value-free data and decisions are best. That "best" approach can work only if no humans are involved. Values, ethics, and prejudices are vital attributes of each individual's personality. These attributes are often prime contributors to our success or failure as managers, yet they are often overlooked in management books.

That is not the case here. This book is the result of many choices that I have made over the years in dealing with human attributes in technical organizations. These choices were based partly on extensive management experience in industry and academia and partly on the same kind of "value-free" training that all engineers are supposed to receive. Therefore, they included partially biased selections of various theoretical points of view, literature reviewed, analyses made and recommendations offered. The major bias behind these choices is my belief, based on experience, that managers of technical operations are potentially among the more influential managers in any organization. Another bias is that these technical managers are often poorly equipped to handle that influence because their training provides insufficient background about the complex (and less-than-objective) human interactions within all organizations. Technical success is usually based on observation and implementation of the relatively fixed, logical relationships found in nature. Management success requires this and the more elusive ability to respond to and use the relatively flexible human relationships of people. This book, therefore, results from my choices about these ideas. As you read it, compare your own choices and theories with those that I suggest, and be aware of the similarities and differences between them.

Before we begin I would like to stress one important point: this book is intended to benefit you. It is intended to help you improve the choices that all managers must make. You and I have a common goal here: to help you survive as a technical manager and to optimize your position now and in the future. This survive and optimize approach is almost unprecedented in industrial organizations. Technical people now can choose the quality and quantity of work to be done and can therefore directly affect the future of their organization. This kind of individual power or influence at work is relatively new.

WHAT IS THIS BOOK ABOUT?

In our modern, complex organizations producing technical products and services, I believe that the decision makers (managers, technical staff, etc.) determine the direction of the company through the decisions that they make. This decision making is often decentralized. It is the technical workers who design and develop innovative products who keep the company alive and growing. Typically, these workers operate in small groups or teams that actually do the work. The groups may be organized by project or by function (e.g., by Blue Turbine project or Hydropack project or by engineering design, new products development, or quality assurance). However organized, the outputs of these groups are vital to the company. They are managed by highly skilled, independent, technically qualified leaders, and it is really these leader-managers who direct the productive and growth capacity of the company, and who are responsible for technical achievements and continued organizational good health.

This book is about the power of managers to make decisions, the power that they (and you, as one of them) inherently possess because of their technical competency, and the control over company growth that this power gains for managers. This book is intended to help managers optimize this power. First we will review recently developed management techniques that strengthen often underemphasized management areas. These areas include psychology, sociology, anthropology, information science, economics, and finance, topics that often seem to be missing in technical manager's backgrounds. We deal with these areas and more in order to develop and improve the quality of your decision making as a technical manager. This development and improvement should result in at least an equivalent improvement for the company. Being better equipped to deal with uncertainties in decision making in both technical and human problems can only result in a general improvement for everyone concerned. There are many ways, however, to improve. Since we all have different strengths and ways of learning, I have tried to make the approach eclectic. We will review appropriate management readings, compare them with applicable practice and theory, and prescribe some uses for them. Your responsibility is to select the prescriptions that you think will best fit your situation. You will be making the final choices; you will develop your own management methods.

It is necessary to develop your own methods because technical managers in different organizations are involved in different problems that are situationally dependent. These managers rarely have the luxury of solving problems in uniform ways from organization to organization or even within different departments of the same organization. They are required to be as creative and flexible as the people they manage, and different amounts of creativity and flexibility are required in different situations. When nonrepetitive problems must be solved, creativity and flexibility are strengths. Conversely, in repetitive situations, they could be weaknesses. A problem that is solved once should not be handled again if it reappears. That would be a waste of your own invaluable thinking assets. Solved problems are placed in books or in organizational policy manuals. Then, if the problem happens again, the solution is at hand.

Another reason managerial responses or choices should be unique is that technical groups occasionally must respond to unusual crosscurrents in directions received from others: e.g., "We need some new creative ideas around here; but be sure that everyone adheres to company dress codes and working-hour standards in addition to keeping time sheets. That's company policy and everyone has to do it." Responding to that kind of direction requires the adroitness of a management Houdini, and certainly requires a unique management theory. However, even unique theories do not preclude consistent personal guidelines. New situations generally have some old, familiar parts that have been faced before. Without some consistent personal management guidelines, decisions might just as well be made at random.

This book consequently addresses the problem of developing a consistent personal theory of technical operations management (e.g. treating people as individuals). It begins by showing you, the technical manager, the inherent advantages you have in your organization. It describes some of the better methods you can use in improving your personal management techniques and assists you in optimizing your position. Your decisions as a technical manager affect the growth of your organization, and when your decisions improve, both you and the organization gain. But first and foremost, this book is for you, to help you improve your knowledge and abilities. The organization will automatically profit if you do.

FOR WHOM IS THIS BOOK MEANT?

This book is intended for the manager or would-be manager who makes the decisions in the technical departments of the company—those decisions that eventually determine the company's future viability. Although I have tried to make this book easy to read, I have also tried to avoid simplification of the material to the point where it becomes just another "management cookbook." I believe that you, as the reader, want more than that. Therefore, I have tried to keep my simplifications and interpretations of the materials to the minimum needed to coordinate and explain them.

These materials are among the most complex that we have; they involve human beings and their behaviors. The book, therefore, might be a bit difficult to read and absorb in one sitting. I suggest a more reflective pace. Read one part or chapter at a time, and integrate it thoroughly in your mind before moving on to the next part. Examine it well and ask yourself as you go along, "How can I use this in my job?." In my opinion, management in general, and technical management in particular, is not learned in "ten easy lessons." The process involves understanding difficult situations and people, learning about what others have done in similar settings, setting up the theories that are intended to work in your own situation, and then testing them on the job. The process never ends, since management improvement never stops.

As a technical manager, you are among the more influential managers of the modern technical organization. When you started your professional life as an engineer, technician, or scientist, the variables you considered in problem solving and decision making were relatively limited, since the laws of physics, chemistry, metallurgy, and similar fields seemed to be quite stable. They were very predictable and in general, applicable to all the technical situations you encountered. However, as a manager you find that some rules no longer hold uniformly and the number of variables has increased geometrically. The complex psychological, social, and economic relationships are not as predictable as relationships in physical laws, and applicability across different situations is quite limited. For example, you can consider the politics that affect the use of your company's products and service. There are many more political variables in the product user's environment than the relatively few variables of pressure, temperature, and so on in the product designer's environment. You, as the technical manager, must now consider more sets of variables when new products are on the drawing board. Additionally, the situations that involve human variables are not always predictable; therefore, the more structured thinking of technicians, engineers, or scientists—who used to be concerned primarily with the product's function and cost—now is forced to deal with social variables that might have been irrelevant not-too-many years ago. These new management situations require new understanding, different modes of analysis, and the development of unique management responses. They must be unique since you (a unique creature) are not only responsible for managing the situation but are also part of it!

Your success in technical management therefore depends upon a multivariable, very involved process. That process of management includes both the variables of the technical background that you learned in the past and a new, more flexible, set of variables. When these new variables (and their changing relationships) are mastered, your personal opportunities are greatly increased. If you can survive (and then optimize) in your own situation, your organization will gain by being able to respond successfully to the changing demands of its environment. And when it does, some of that success should redound to you and your further advantage.

WHAT MAKES THIS BOOK DIFFERENT?

In the past, many texts dealt with management as if it were a static activity, and the recommendations in the texts seemed to apply to a generalized "one best way" to plan, to organize, and to control. These "best ways" have followed styles in management just like the styles in manners or clothing. For example, earlier in this century scientific management theory was popular. This theory relied on a triangular, classically hierarchical, organization structure in which the manager was to have a limited span of control (six to eight people), report to one boss, and send information up, with directions coming down the organizational structure. I've worked in that kind of an organization and I know that *sometimes it works*. On the other hand, *sometimes it doesn't*.

But styles change. There was a period during which decentralization was a popular recommendation for organization design, and the model of the large multidivisional corporation was used to show how well this particular concept worked. For quite a while that also seemed to work. However, times changed. More recently, foreign auto builders have used quite different management models and seem to be doing very well, thank you.

Obviously, other factors, even in the limited area of organizational design, helped determine organizational growth, and these were still not defined. Within very recent times, the development of a relatively new branch of management theory and practice that is *situationally* determined seems to have produced ideas and practices that include many of the positive aspects of prior theories and yet is able to overcome many of their deficiencies. This relatively new set of theories provides the flexibility to organize and manage differently to meet the different needs of complex technical organizations. As an example, it seems to be able to explain why and when both rigid, centralized structures and flexible, decentralized structures are appropriate. It can also handle both functional and project-matrix management structures. It can even resolve why all these structures might be necessary within the same company—because different situations may give rise to different problems at different times and places. We will explore those alternatives. Since few, if any, books apply these relatively new and powerful set of theories to the unique problems and opportunities of technical managers (from their own viewpoints), this book could be a very practical tool for managers in technical organizations who wish to optimize their situations.

HOW IS THIS BOOK ORGANIZED?

This book is organized into three parts, each of which is intended to follow a familiar sequence of explanation, analysis, and synthesis. Part One is concerned with developing backgrounds, explanations, and a general model or operating hypothesis for the organization against which you can test your perceptions of your own organization. It deals with applicable theory in two general ways: first by description of applicable concepts and then by synthesis or prescriptions for the use of these concepts in your own situation. Part Two disassembles the general model into its major components of people, structures, and technology and shows how you can modify and use these components in building your best management "style." Part Three deals with the special problems of communications systems and leadership in technical organizations and shows how to develop systems that provide you with the data that you need and how to respond to the changing leadership needs. It also summarizes some current data on developing change processes needed to implement your theories.

WHY IS IT NECESSARY?

For many years, first as an operating technical manager and then as an educator and consultant, I have felt that there was a need for a flexible and yet effective set of management concepts that could be adapted to fit most situations in technical management. Existing texts and systems seemed to be oriented either top-down, with some mythical "top management" making major decisions that everyone else was supposed to implement, or bottom-up, with the "all-knowing" lower managers sending up their needs and requirements to be satisfied by a cooperative company. Neither of these extremes reflected the typically varied behaviors of successful managers in the fast-changing economic environments associated with technical products and services. There might be an occasional similarity, but situations changed and successful behaviors, systems, and products always seemed to be in a state of response to this change.

This did not match the static approach many management texts seemed to recommend. An example is the often-found recommendation for constant leadership behavior (which is usually to be both supportive and participative). If one followed this advice, the successful supportive, nondirective "theory Y" leader of a technical group would invariably fail when external economic conditions became difficult. A different and more flexible leadership style would then be needed. Conversely, the successful directive, results-oriented "theory X" leader of a similar technical group would fail if changed economic conditions require advancement in the state of the art. Leadership needs change with changed situations, and the technical manager who is aware of these changed needs may be able to respond better if he or she is also aware of some of the various responses that can be used. The manager may not be able to respond completely (it's very difficult, if not impossible, to change our personalities), but he or she will be able to recognize the changes in the situation and will therefore have an increased ability to modify some situationally dysfunctional behaviors.

In addition to leadership needs, there are other management needs and responses that are similarly variable. These also depend in part upon the situation. If we can make an obvious assumption that managers, as individuals, are unique, it should follow that the manager-situation interaction is also unique. The ideas in this book attempt to assist you in handling these unique situations. Many of the ideas have been developed over many years and have been tested against the perceptions of hundreds of successful technical managers in the diverse organizations with which I have worked. These ideas do work, but they also require cooperation on the part of the user. This is a different kind of "how-to" book; it requires you to take the components and concepts offered here and build your own management framework, a framework that can be modified as the situation changes.

ACKNOWLEDGMENTS

As an engineer, a manager, and a consultant, I was always aware that few of us accomplish major achievements alone. Interactions with others are vital elements, and I could not have written this book without those supportive interactions. My appreciation and thanks are therefore extended to all who have been involved in it. I particularly wish to thank those who were kind enough to read innumerable (and deadening) drafts of the manuscript and offer the constructive, supportive criticism that extended beyond the call of duty. They include Drs. William Engbretson, Michael Grimes, Leslie Kanuk, Harvey Levine, Michael O'Neal, Frank Riessman, A. E. Russo, Herb Shepard, and Alfons Van Wijk. The continuing moral support of my family (especially Liz) helped me get over many "writer's dry-up" days. Finally, the many professional associates and attendees at my continuing education seminars who thoroughly tested the book's concepts and case studies before they were included were invaluable.

Now a word about the case studies. Those for the first eight chapters are fictionalized accounts based upon both my experiences and those of professional associates. They are not intended to represent any existing people or organizations.

Having mentioned existing people and organizations, I would like to note here that the contributions of women are becoming increasingly important in technical management. I therefore hope that where terms such as "draftsman" are used, you will understand that this is merely the common terminology used in our industries.

A final word before you begin this book: I freely acknowledge any errors, and I would like to hear from you about any that you find. I'm sure that they exist. I would also like to receive any suggestions for improving the ideas that I present here or anything else that you feel is important. McGraw-Hill will forward your comments to me. I hope that you enjoy reading the book.

Melvin Silverman

PART ONE

DEFINITION AND BACKGROUND

Introduction
TECHNICAL MANAGEMENT AND WHAT IT'S ALL ABOUT

UNIQUE PROBLEMS OF TECHNICAL MANAGEMENT

You, the technical manager, are a key executive in the rapidly changing modern industrial organization, since the responsibility for the innovation supporting the growth of that organization rests in the technical groups you manage. Satisfying that responsibility is often difficult. It depends on both the cold logic of high technology and the often warm creativity of human beings. These two inputs may not always be related to each other, which makes optimizing the combination of science and psychology a core problem in managing technical operations. It is the major problem addressed in this book and requires human management art as well as technological logic. Therefore, this book provides summaries of useful research and practical applications to help you, as a manager, improve your management "arts."

Managing (or optimizing) the use of science is easier and probably more familiar to most of us than managing (or optimizing) the use of people in organizations. Much of the content of engineering and technical curriculums of universities and other teaching institutions is oriented toward very objective and measurable goals. For example, we were taught how to increase production of high-quality goods and services and then to deliver this output efficiently to waiting customers. The results would seem to be quite effective if we lived in a logical world with a sensible orderliness about it, an orderliness based on the relatively fixed and predictable laws of nature. And after all, the laws of nature always work consistently when applied to the design and production of goods and services.

But they don't always work in other areas. Our world is a combination of predictability and seeming irrationality. Although many products are intended to be designed according to the logic based on the laws of nature, both the designing process itself and the purchasing behavior of customers expected to buy those products are affected by human behavior that responds to less predictable laws dealing with emotion. Human behavior is the most difficult part of management. In technical operations human behavior is not only difficult, it is the crucial variable that determines success as a manager. A good knowledge and background in human behavior seldom gets enough attention in the training of technical managers.

For example, the technical student quickly learns that the current in an electrical circuit flows according to immutable, predictable rules; when the wires are connected correctly in the lab, the circuit performs its task. If not, the circuit breakers interrupt the current (if the student is lucky) and a prompt learning process is completed. That is exactly as it should be, since Mother Nature is a strict and attentive instructor who exacts retribution immediately when excessive "creativity" intrudes. If the circuit does not agree with the rules she has established, there is an immediate negative feedback. The rules are clear and unchanging and the attentive student can understand them quite easily. Mother Nature has a general rule: Repeated inputs of some specific type of variable generally result in the same type of repeated output.

PEOPLE AND IRRATIONALITY

But when the technically trained, logically oriented person becomes a manager, this rational thinking framework, which contributed to success in the past, no longer works uniformly. The repeated inputs can now lead to alternative outputs, such as repeated outputs, different outputs, no outputs, or any combination of the above. Human beings do not always respond as predictably as machines do. The effective management of people in organizations, either as individuals or in groups, requires different kinds of inputs or knowledge, because the variables of emotions and human thinking are different and are greater in number than the variables considered in designing machines. The relatively simplistic rationality of the design of the machine or the electrical circuit no longer applies. We now must deal with the more complex (occasionally irrational?) thinking processes and emotions of humans and with the intricate interpersonal relationships of groups of people who work in any organization. Mother Nature is also very complex. The more we learn about her rules, the more we appreciate how intricate the world really is. However, her intricacies are comparatively quite easily understood using the relatively logical frameworks of the "hard" or natural sciences to which we were first exposed as students. Typical academic subjects such as physics and chemistry follow straightforward rules. In fact we can predict some of the elements of human behavior on the basis of those sciences since our bodies are, in effect, mobile chemical factories. For example, it is obvious that starving for extended periods will cause people to be very concerned about food and not at all concerned about meaningful jobs. Therefore, the motivations (and management) of marginal workers are not going to be the same as those of highly qualified technical workers.

However, human beings do other things that are less predictable. Obviously, they are more than consuming, moving, reproducing, machinelike creatures whose actions are completely described in biology texts. Human beings do things that are not as easily explainable. These behaviors are the ones of utmost concern to us both as technical managers and as coworkers. They include the results of creativity or innovation, cooperation or conflict, and, in general, any of the large repertoire of other behaviors that are exhibited in all work situations. They do not always match the logical, predictable frameworks that we learned to use in first exploring our technical specialties in the university. However, as technical managers, we must deal with both frameworks: those of the rational hard sciences and those of the "soft" sciences of psychology, sociology, anthropology, and the other ologies (less rational, surely). Technical management is one of the few organizational departments that depends on both areas. Objectivity and creativity have to exist together here.

Creativity and emotion are not necessarily rational, but they are as much the basis for innovation as the physical laws of Mother Nature. While most of us who have had some technical training are more or less familiar with the more predictable laws, the equally important aspects of technical management that deal with psychology often require strengthening. This book is directed to that strengthening.

I intend to use a personal approach to teaching technical management here, to help you to succeed as a manager. There will be few references to the company, the group, or the worker unless that reference centers on benefits to you. I will include suggestions about how you can optimize or improve your management capabilities, even though improving those capabilities is not going to be easy. I have no panaceas, but I do have some suggestions that can improve your management skills. Those with some experience in managing others must know that no panaceas that can be applied uniformly really exist. Instead, you must develop the style of management that best fits your situation. I will assist by providing description, analysis, and other inputs, but the successful use of those inputs depends on how you modify and apply them.

I start by developing a *process model* for learning intended to change thinking. That process model (or learning how to change) will include another, more tangible, model (or what actually exists) of the organization itself. The *organizational model* is intended to help you to visualize how the parts of the technical organization operate (visualization helps the learning process). Therefore, the process model will include an analysis of the organizational model's components (both from a human-emotional and a physical-rational viewpoint) and an attempted synthesis of those components into a new tool that you should be able to modify or change to suit your situation. The sequence to be followed in this process model of analysis and synthesis is:

1. *Description and explanation:* What have others done? How do their actions relate to technical management, and what do they mean to me?

2. *Prediction and testing:* How can I use this? When it is used, what happens, and how effective is it when applied to my situation?

That sequence is part of the much larger process model of learning. The larger process model involves both *cognition* (thinking) and *behavior* (managing or doing). Both your cognitive and your behavioral changes are interrelated in the learning process. But since we'll be dealing first with attempted changes in cognition, a word of caution is in order. Those changes depend to some extent upon how the new data (such as in this book) are interpreted, both by the authors I refer to and by you. When these authors or the other writers of the research findings noted in this book present their results, I will always be in the background cautioning you about the uses of the research data described. Psychological findings often do not have the same objectivity as findings in the hard sciences. In all sciences data are viewed differently by different people, but physics and mathematics, for example, have achieved a more objective frame of reference than, say, psychology and sociology. However, even that objectivity is relative. Occasionally, in the hard sciences the same information has been interpreted in very different ways (Gould, 1981, p. 8).

For example, Galileo insisted that planetary orbits were circular, even after Kepler had provided meticulous observations showing that they were not. (As you know, they are ellipses.) And we all know about the trouble that occurred when Galileo stated that the earth was not at the center of the universe. So the interpretations of data may often match the researcher's concepts. The same is true of your own concepts. If the data reported do not match your concepts, you have two basic alternatives: disregard or discard the data (e.g., Galileo's discovery that the earth was not the center of the universe was data that were "discarded," as was Galileo himself when the Church imprisoned him for a while) or recognize that you have an incomplete set of concepts and that those concepts have to be changed. (For instance, most of us now agree that the earth is *not* the center of the universe.)

To help you to achieve that change in cognition or concepts as we deal primarily with the human side of managing technical organizations, I'll use a tool that is fairly familiar. That tool is the *scientific method*.

1. Define the terms and develop the theory.

2. Set up the hypothesis that tests the theory.

3. Gather data and compare results against the predictions of the hypothesis.

4. When the variance between predictions and results is equal to or less than expected, accept the hypothesis. Otherwise, go back to step 1 and reevaluate.

DEFINING THE TERMS

We therefore start by defining our terms. The first definition can be of *theory*. There is more than one definition, but I like the one that follows because it refers to the individual and describes how he or she organizes his or her thinking.

> . . . a theory is a set of related concepts or best guesses about what is going on in a given area. . . . By beginning with theory we develop the basis for processes of—
>
> 1. Inference or conclusion based on a guess or hypothesis;
>
> 2. Test of the inference (what is the right guess?); and
>
> 3. Correction, based on feedback from behavior.
>
> A theory, therefore, is chiefly a mode of organizing one's thinking. (Levinson, 1976, p. 3)

The point is not always self-evident, but it is always true that the ways in which we organize our thinking directs the consequent thought processes. This becomes more obvious as we consider management problems, since any potential solutions are usually limited and directed by the knowledge and experience that we use in our personal theories. The selection of the appropriate definition of variables, parameters, and constants, as part of a theory or philosophy followed by a hypothesis, prepares the mind to learn to see things in a different way. Data that are received are evaluated with respect to some personal theory that we have developed. We see what we are prepared to see, and this seems to occur both *before* and *after* receiving and interpreting the data or, in more formal terms, the results of tests of hypotheses. Just the act of defining the terms and developing the plan (i.e., theory and hypothesis development) prepares us for learning in a predetermined way. Without this kind of preparation, the world can be a random pattern. The following quotation says this a bit more elegantly.

> . . . reason has insight only into that which it produces after a plan of its own, and that it must not allow itself to be kept, as it were, on nature's leading-strings, but must itself show the way with principles of judgement based upon fixed laws, constraining nature to give answers to questions of reason's own determining. . . . Out of the profusion of words uttered by nature, the scientist selects those that he can understand. (Kant, 1958, preface)

Being immersed in our own organizations, we can hear the "profusion of words" uttered, but we try to limit that profusion by defining our terms, building theory, and limiting our hypotheses closely. This book can assist in those tasks by providing analyses of selected research findings that seem to be applicable. However, you are the only one who can really select those that are applicable, since you are the theory builder or (in Kant's words) the scientist who selects (those words) that he can understand.

The research findings presented here are based on two general data sources: academic research summaries and personal experience collected over the years. Both sources are used in the first step of limiting that profusion as we define and analyze the environment around the organization. It is a seemingly simple step but an important one that is needed to separate the organization from the environment. Without separation, we would have no place to start our study of the various aspects of the organization and our place within it. With no starting points, there is no beginning or end and everything is related. But since our time and efforts must be limited, we start with a limitation or definition of the place in which we find the technical organization: the environment.

DEFINITION: THE ORGANIZATION'S ENVIRONMENT

The environment is defined as important interactions outside the organizational structure itself. This definition is not exact (in fact, it's a bit circular, but nonetheless quite useful), but that inexactness happens because there is little agreement, either in the research literature or in my experience in technical operations, on exactly where the organization stops and the environment begins. (Psychologists call this type of definition an *operational definition;* see Skinner, 1953.) For example, do the organizational boundaries include stockholders and customers? What about unions and regulatory bodies, or families of employees? This general definition is not as exact as it could be if we were dealing with one specific organization and could make some assumptions about that organization and its multiple internal and external relationships, but inexact measurements do not preclude useful results.

Consider the builders of the great pyramid at Gizeh. They lived before the development of the accurate and precise surveying techniques that could achieve the absolute flatness of the foundation needed to carry the millions of limestone blocks which were to be placed without mortar. Any unevenness in the foundation would probably result in an eventual sideward thrust that would cause the blocks to move and the pyramid to collapse. In spite of this lack of exact definitions of flatness, they developed a pyramid that, after thousands of years, still stands square and vertical: perfect enough not to need cement to hold the blocks together. The squareness and verticality were not exact, but they proved to be entirely adequate for these several thousand years and that's good enough. With our "good enough" definition of the limits of the organizational environment completed, we'll move on to a similar definition of the *people* in the organization and determine why these people in the technical groups have become so important within recent times.

DEFINITION: THE ORGANIZATION'S PEOPLE

The people in the organization are those who interact *regularly* with it, draw economic and/or social payments from it, contribute physical and mental inputs to it, and are *primarily* employed in the direct creation or production of the company's goods and services. Their tasks within the organization are affected and defined by changes in the organization's economic environment. One change in the economic environment is the recent general movement from a manufacturing to a service economy. We are no longer primarily a manufacturing economy, if that type of economy is defined as one in which production of physical goods is the controlling force. Our economy is now concerned mainly with the production of services. Our labor force has moved rapidly from the manual work of prior years to the knowledge work of today.

Also, there has been an explosion of education and knowledge that is propelling rapid improvements and expansions in technology. These changes have not been matched with equivalent investments in machines and plants but rather with investments in human beings. The individual technical worker has become the human capital that supports economic growth, thereby displacing machines and plant facilities, which were the capital of yesterday. Now the innovation that fuels the increase in real wealth comes from the skills and minds of professional and technical workers, instead of from the physical capital of equipment, as in the past.

The change from a primarily goods-producing to a services-producing economy is fairly conclusive. Recent demographic research concerning producer services, which includes engineering and technical functions, indicates that the value being added by producer services now approximates that added by manufacturing in the U.S. economy (Ginzberg & Vojta, 1981).

In other words, the technical (or knowledge) workers are contributing as much wealth as or more than both the manual workers and physical capital put together in manufacturing. Since this trend appears to be continuing, it seems obvious that our primary production source has become this knowledge worker ". . . who puts to work what he has learned in systematic education, that is, concepts, ideas and theories, rather than the man who puts to work manual skills or muscle" (Drucker, 1973, p. 32).

This knowledge worker is not the same type of person who used to be in technical organizations. The present-day technical professional has a higher level of education upon entering the work force than his predecessors of the 1950s and 1960s. In some superficial ways, he resembles the artisan with highly developed skills, but there are major differences. Artisans apply their skills in a fairly logical and repetitive fashion within an extensive but still limited repertoire, while the technical professional must add unique cognitive contributions to those craftsmanlike skills to develop the new products and innovations upon which the organization depends.

The definition of the people, or the knowledge workers, in our modern organizations has also been affected by the changes in the size of the organizations themselves. The growth and proliferation of large, diversified industrial organizations has affected the use of technical professionals in those organizations, thereby impacting the definition of the way in which they will work. Although the multinational conglomerate is not a new phenomenon (within the last two centuries, Hudson's Bay Company controlled much of North America, while the East India Company ruled most of India), the increase in the numbers of these organizations within the recent past has affected the lives of thousands of employees, vendors, and customers, and even the course of governments. Some of these organizations have grown into multinational conglomerates that have the power to redistribute work and resources in ways that many nations of the last century were unable to do. For example, the Krupp companies that produced steel and coal in Germany directly assisted in the growth of militarism in Europe during the late 1930s.

In this country, we have placed many legal restraints upon our industrial organizations that are designed to preclude this type of interference in our political processes, but other effects of this organizational growth can still define our economic processes. These large organizations require many technically qualified people. The result is often a competition for the services of the knowledge worker that has begun to equalize the previously unequal power relationship between the organization and these workers. Technical professionals are recognized as the source of innovation and organizational growth. With increased competition for their services, they have become quite mobile and have easily transferable skills (and equally transferable benefit plans). They have fewer loyalties or economic ties to any one organization. They move to other jobs very easily if dissatisfied with or pressured in their present jobs, which naturally leads to a loosened pressure for behavioral conformity.

THE TECHNICAL MANAGER AND HIS CONCERNS

In addition to dealing with problems concerning highly mobile, more independent knowledge workers, and larger, more complex organizations, the technical manager faces other problems trying to manage the special blend of logic and creativity the workers' jobs require. One element of a creative mind is an independent capacity for judgment and a resistance to the conformity of group pressure. The latter mental element is important to the technical manager since the manager, by definition, supervises creative people. We know that both as individuals and in groups people do not function as consistently and predictably as the laws of nature do. Creative people are, if possible, even less consistent than the general population. They do, however, have a major asset in their ability to see novel (or, if you will, heretofore illogical) patterns in data that others did not see before. That ability often results in innovative products and services.

On the other hand, the laws of Mother Nature that provide the raw data used by creative people often seem to be irresistibly logical, if the person looking at the data has been trained to understand them. The laws may not seem to be very flexible, but they are just as important in developing innovations as the creative (illogical?) thinking of the innovator. Innovations that violate obvious physical laws fail. Therefore, logic and illogic (or emotion, creativity, etc.) need to be matched.

It is unfortunate that the consistent, logical thinking (based on the models of nature's laws) that is taught very well in the technical and engineering schools, is not supplemented by an equal emphasis on the flexible thinking of the creative innovator, which provides personal and organizational growth. This flexible thinking is often learned on the job if, indeed, it is ever learned at all. It is rarely, if ever, taught in academia. In many respects, managing technical functions or organizations is closer to the innovative process than it is to the disciplined, logical process. The data being used by the technical manager are extremely variable, including, as they do, the behaviors of very independent people, changing organizational structures, and expansions of technology.

This management is more an art than a science and art is primarily a subjective rather than an objective process, always involving human creative and emotional elements. With rapid product and environmental changes affecting the technical departments, the management part that is art is even more important. Only this part of the human perception and interpretation process supports adequate organizational response to these rapid changes.

As examples we have the two changes discussed before: the growth in the importance of the knowledge worker (which increases his independence) and the proliferation of multinational organizations, with their ability to move all kinds of work, including knowledge work, among company divisions or even countries (which increases worker dependence).

These changes are both very new. The adequacy of the manager's art and the speed of response in dealing with changes such as these often will determine if the company survives and grows or dies. On the other hand, when dealing with the "science" part of management, actions are generally slower and more deliberate because one proposes questions to nature very carefully. Mother Nature's responses to wrong assumptions are often very painful. With rapid change in the economic environment or in the organization, response time becomes critical and a management reaction that in more limited technical questions was quite slow must now be fast and almost instinctive—in many ways like the artist who "knows" what is good or bad in an artistic situation.

But developing artistic capabilities requires more than reading books (even this one). Although this book (and others) presents a general process model of learning and personal change (e.g., the scientific method noted before) and a general tangible or static model of the organization (that we will discuss in Chapter 2), both the method of learning that *you* finally use to become a successful management artist and your organizational model may be quite different from those suggested here. That happens because the most vital part of this learning process, *you*, is outside the scope of this book. Try to recognize these differences, which occur because of the difference in *you* as the manager. Test the hypotheses presented, use the parts of the models presented that you can, and retain the rest for possible future use. Build your own theory by selecting the parts that fit your perception of your situation to support your own artful and successful management style, since no two situations or organizations (or managers, for that matter) are exactly alike.

Of course, no style can ever be static; the world and you are always changing. Therefore, you must provide for an iterative testing of your style and consequent modification based on new data received. This is not a simple process, and under the best conditions might require you to learn in a calm and reasonable environment, which the working organization may not have.

Research shows that the activities of managers during a "normal" day rarely reflect any description of a calm, reasonable environment in which to learn about technical management. There are many descriptions of what a manager does and many myths about managers that, in general, do *not* reflect their real situation. One myth is that the manager is a careful, reflective planner and if he is managing effectively, he has no regular duties to perform. According to this fantasy, management is, or at least is quickly becoming, a science and not a profession. Many management courses in academia seem to be taught this way.

According to research findings, the real-life situation is a bit different.

> . . . the job of managing does not breed reflective planners; the manager is a real-time responder to stimuli. . . . The managers' programs to schedule time, process information, make decisions, and so on remain locked deep within their brains. Thus, to describe these programs, we rely on words like judgment and intuition, seldom stopping to realize that they are merely labels for our own ignorance. (Mintzberg, 1979, p. 110)

Building the model of the art (or the management style) that works for you is a process that involves off-the-job thinking (there's that calm, reasonable environment), hypothesizing, and on-the-job testing of these hypotheses. Simply reading about models of managing and using what others have done as a basis for changing your manager's mental programs is insufficient by itself. In other words,

> Cognitive learning is detached and informational. . . . But cognitive learning no more makes a manager than it does a swimmer. The latter will drown the first time he jumps into the water if his coach never takes him out of the lecture hall, gets him wet and and gives him feedback on his performance. (Mintzberg, 1979, p. 122)

On the other hand, there are those who suggest that on-the-job learning is best and the "art" of management cannot be obtained cognitively. Therefore, instead of taking the time to study the various models of management, why not just go out there and try to swim by observing how others do it? This might work some of the time, even when the "other" being observed is your boss: At least he (or she, as the case may be) can't then complain about your style of management.

But many times it may not result in an optimum management style *for you*. For example, assuming that the samples of behavior that are observed are entirely adequate for your present situation, they may not prepare you for any surprising changes that may come from other parts of the company or from the outside environment. On-the-job training is inadequate by itself because it has several implicit assumptions: that the trainer has the best model, that you can duplicate what you see, that only action triggers mental programs, and that you're ready for anything once you've finished your course of training.

The fallacies of these assumptions become obvious once they have become explicit. Management, as art superimposed upon logic, can be learned through practice, but that practice has to be *corrected and predicting* practice, based on solidly researched theory and the successful experiences of others. Your trainer (and the training situation) has to be able to provide this type of practice; it is also obviously important to recognize that your trainer's perceptions and behaviors are not yours. (Only Picasso painted like Picasso; the imitators could never duplicate his work exactly.)

Fortunately, duplication is neither necessary nor desirable, because situations change rapidly. The art is in learning the flexibility and creativity of being able to set up your own successful solutions to situations that have not occurred before. In addition to the variables in your state of mind and in environmental change, other variables affecting the situation include the organizational structures and the technical processes used. Consider the possibility that a "Picasso" of management in a well-structured organization producing standard electrical parts may not function adequately in the relatively unstructured matrix of a defense plant. In that unstructured organization, the "Picasso" would probably fail if he behaved the way he had in the closely structured company. The structure is different, the technology is different, and the people are probably different. To summarize, it would seem that both introspection (i.e., cognitive change) and *corrected* testing (i.e., behavioral change involving on-the-job training either in a formal or in a self-taught manner) are required to learn the art of successful technical management.

PRACTICING MANAGEMENT AS A PROFESSIONAL

The part of the process that involves testing (behavior) should be concerned with development of a management style that reflects the new duties and responsibilities of the technical management job and not with the duties and responsibilities of the job from which the manager was promoted. Most of the positive behaviors of technical personnel that result in their promotion into management can become negative once they have gotten those promotions. This is generally true no matter how high up in the organizational structure a person rises. A successful equipment designer who still likes to check the drawings after he is promoted to chief engineer is not performing as a chief engineer. He has not recognized that his new job is no longer concerned with the adequacy of the design; it is now concerned primarily with the adequacy of the designer (possibly the one who replaced the engineer who was promoted) and the design group itself.

The job is entirely different. The superior technical expertise of the designer promoted into management is no longer a major consideration; now only a basic familiarity with technical considerations is required. The expertise requirements have been changed to include less structured people and organizational requirements, such as those needed for personnel, finance, marketing, sales, organization design, industrial relations, and a myriad of other areas that have relatively little to do with the technical operations that the designer promoted into management had to contend with before the promotion. Those areas are the ones that have to be "tested," using new management styles until the manager finds one that is effective for him or her. If this is understood, the problem of "How can I manage people who are more qualified technically than I am?" becomes unimportant. The job is no longer that of managing the relatively fixed (and more consistent) technical rules of Mother Nature; it is now one of managing the more complex, varying, and interrelated rules affecting people, organizational structures, and the impact of the environment on the company.

Therefore, the first task of that new job is to develop a different set of cognitive processes and behaviors—a new theory that will work for you in your new situation. That theory and the hypotheses that make it operational are intended to help you to develop your own management style. The second task is to modify and sharpen problem-solving methods, since the number of variables to be considered has increased dramatically, and the amount of data to be processed through the "human computer" has also increased.

Processing those data accurately and rapidly to extract the relevant facts upon which to act is part of the manager's technical expertise. It is very different from the technical expertise of the engineer or technician because of the greater number of variables and the incomplete (and greater quantity of seemingly irrelevant) data to be processed mentally. Management never gets enough data to make an absolutely correct decision every time. There comes a time when it costs more to wait for corroborating or more data than would be gained from making a better decision later. The general who waits to be *absolutely* sure of the enemy's strength before attacking will probably lose the battle. By the time that his forces are committed, the enemy may have changed his position, making the general's data obsolete. Therefore, if this analogy holds, the data upon which most decisions rest are usually incomplete.

Developing your own theory and decision-making talents is really not that difficult since you, as a technical manager, have quite a bit of freedom to experiment with various behaviors and to develop your own management style. You are one of the most valuable assets that the organization has, and if the general value of the technical professional is increasing, your value as a manager of these technical professionals is increasing even faster. Part of your responsibilities is to direct and coordinate their efforts.

In addition, there has always been an unsatisfied demand for competent, technically qualified managers to manage the "human capital" of the organization. That demand has increased even more within recent times because of the difficulty of obtaining creative managers who can handle the increased rates of change in the human capital, the economic environment, and the organization itself. It requires more art than ever before.

Art is built on innate human capabilities, and not everyone has equal abilities to mix the colors the artist uses; but sometimes innate capabilities can be improved. Therefore, both your increasing value and the unsatisfied demand for effective technical managers should provide you with the opportunities to try new (and possibly better) concepts to improve your decision-making and management abilities. However, learning and experimenting (or testing) occasionally lead to setbacks. In the present organizational situation, these setbacks should not be considered failures; they are chances to revise and modify your management model and to be used as guides for obtaining better solutions to the problems raised by rapid changes.

SUMMARY AND REVIEW

The technical manager is faced with an environment and an organization in rapid change. There are opportunities in these changes for the technical manager to learn and grow. Some of these changes and the opportunities for growth follow.

Interpreting the environment or the world has never been an entirely objective process. Therefore, you need a personal theory to aid you in subjectively organizing your thinking. Defining terms and selecting hypotheses based upon some theory helps you to find potential answers.

The technical professionals in the organization have become its human capital, and that human capital is responsible for the production of more income than the physical capital. There has been an increase in technical education of "knowledge" professionals. These professionals have become more mobile and are less dependent upon any particular company for continued employment.

There has been a change from a manufacturing to a service-oriented society.

There are many ways to approach the problem of learning how to manage technical operations, since one must consider both the participants and the situation; neither are fixed for long.

Successful technical management requires creativity and logic or, in other words, art combined with science.

Understanding what others have done in management is an initial step in developing your own art (why start at the beginning again?), but you are cautioned to test that understanding in your own situation. If the results of your tests meet the goals you selected, use the hypothesis behind the test as part of your behavioral repertoire; otherwise, go back to the starting point and take another look at the research of others for improvements. This testing, with on-the-job feedback from peers, subordinates, and superiors, is my suggestion for a guided learning pattern that can produce the positive cognitive changes you need.

In other words, developing a successful management style requires combining creativity with logic, or art with science. While the situation on the job generally does not support the quiet, introspective learning that is usually associated with cognitive change; there is still the demand for rapid, almost instinctive, but accurate responses to difficult management situations. This suggests that the learning be done very well, until your responses are almost automatic. This type of response requires careful on-the-job training in addition to introspective learning. Consequently, a recommended process would start with introspection (i.e., reading and study), development of a theory and hypothesis for action (i.e., a plan), and testing that hypothesis on the job (i.e., on-the-job behavior). Then, success as a technical manager will be yours only after the results of those tests are used to modify your future behavior.

You, as the technical manager, are a tremendously valuable asset to your organization, and that knowledge should provide you with the confidence to test your hypotheses. (Organizations don't discard valuable assets if there is an occasional loss, and you can't make progress if you don't stick your neck out once in a while.) Those hypotheses that do not work out as well as expected may have to be modified to cope with the changing situation. (We also learn from experiments that fail.) Finally, when you are able to develop effective and useful management hypotheses (as measured by your success as a manager), you have learned the basics of combining the logic of technical requirements with the art of human requirements. In the next chapters, we will explore more of the "how to" that might be used to further and expand that success.

Case Study
THE CASE OF THE QUALITY PROGRAM

CAST

Tom Berry: President of Thorax Medical Products. His father had started the business. Tom had taken over ten years ago after working in various training positions throughout the company following his graduation from Alpha Technical University.

George Beardsley: Vice president, sales. Had been with the company for fifteen years and was primarily responsible for the growth of sales over that time.

Marvin Loren: Chief engineer. Had been with the company for more than twenty years. He had started as the first draftsman the company ever had and had been promoted as the company grew.

Bob Spelvin: Chief, quality control. With the company six months. Had been recruited from an aerospace company to start a formal quality program.

Thorax Medical Products had been manufacturing high-quality stethoscopes and other mechanical instruments for hospitals and physicians for more than fifty years. Its products were considered to be the best in the field. Within the last several years, sales had begun to drop when one of their competitors, Atlas Medical, came out with a series of remote-reading electronic stethoscopes.

Tom Berry, the president of Thorax, had immediately instituted a product redesign program, and that program had been in progress for two years. During that time, the engineering design group had developed several concepts that seemed to be ahead of anything else on the market. Some of these products included a microcomputer that could take vital measurements at any bed in a hospital automatically, and a diagnostic chair for physician's offices that could take a patient's blood pressure, temperature, heart signs, and other data automatically when the patient sat in it.

George Beardsley, vice president, sales, had been pressing Tom to get the factory going because his salesmen had been out with preliminary brochures describing the new products for several weeks and tremendous interest had been shown by key customers, even though delivery was not promised for another six months. Marvin Loren, chief engineer, had assured Tom that these products were well-designed and even though the company had not done this before, he felt that the products could be manufactured by subcontracting out major parts to several computer manufacturers. Mechanical parts and assembly were to be completed in Thorax's own plant. Bob Spelvin, chief, quality control, had sent a memo to Tom a week before, with copies to George and Marvin, flatly disagreeing with Marvin. He had recommended that an extensive program of vendor quality inspections and in-house quality audits be instituted. Tom didn't even understand some of Bob's language. What was a quality audit, anyhow? With pressures building up because of dropping sales, Tom called a meeting in his office the following Monday morning. At 8:30 A.M., everyone was there.

Tom: Well, you're all aware of our situation. This meeting was called so that we could thoroughly air our differences of opinion and decide what to do next. Who wants to start?

George: Well, I'm sure everyone knows what our position in the market is. It's deteriorating, and unless we do something fast, we'll be in serious trouble. Our field representatives have been screaming for something to beat the competition with and we have to move on it or we'll lose our market and the best group of field reps that any company ever had. Those people are independent and they won't stick with a loser, especially since they work on commission.

Marvin: I couldn't agree with you more. Some of the ideas that we have come up with are really fantastic. Why, I personally spent my vacation right here checking the detail drawings and assemblies for our new microcomputer products. And Bob here has been a great help in developing life tests for the equipment, getting it past the Underwriter's tests, and making sure that every component going in was thoroughly checked out. Frankly, I don't understand the reason for this meeting. We have great products and they work beautifully, both here in our labs and in the field tests.

Bob: Well, Mr. Berry, as I see it, the computer chips that we are buying and the other electronic components are unlike anything this company has handled before. Incoming inspection cannot do an effective job of checking them and we cannot afford to have any field failures of our equipment. Some of those components may deteriorate over time and I'm trying to prevent a field repair program that could ruin our reputation. As far as our own plants are concerned, I'm not sure that quality control is the answer anymore. I'd like to place the responsibility for quality with the plant operations office and start a quality assurance program that would prevent problems before they happened, rather than quality control, which would find the problems afterward.

Tom: So far, Bob, I haven't heard anything to disagree with, but why should this affect our new-product introduction schedule? Engineering and sales have agreed, so why not go ahead and produce?

Marvin: Well, I've heard some things that I really don't understand. Bob, why can't incoming inspection check those components? The prints are very clear about the specifications, and our plants have been running under quality-control systems for years. Why should we change now? It is just an additional frill. Why raise our costs for something that could happen but has never happened before?

Bob: You're right, we can check components being received, but that is not effective enough. I suggest that we develop a vendor quality evaluation program in which we check the way our vendors design and manufacture our parts. In that way, we'll have a higher confidence level that we are getting what we are supposed to get. As far as our plants are concerned, they should be working to the same procedures as our vendors. Let them produce quality parts and we, in the quality department, will just be checking their methods and procedures. If those methods are right, the product will be right.

George: Bob, how long will this take and how much will it cost?

Bob: I don't have the final details but I would guess about six months and an increased budget to provide four more quality engineers.

Marvin: Bob, you're going overboard on this. We're not Amalgamated Aerospace, where you came from. We can always fix things and there are always a few bugs in any new product. Everybody knows that and the customers have always accepted it. I vote for getting into production now.

Tom: Fellows, the decision to move ahead with production has to be looked at very carefully. I think that we should sleep on it. Let's meet again tomorrow.

QUESTIONS

1. What actions did Marvin take, and what should he have done?

2. If you were Bob Spelvin how would you have been prepared before this meeting?

3. What should George have done before the meeting? What should his recommendations be tomorrow?

4. What should Tom Berry do before the next meeting?

5. What are the implicit theories and hypotheses that each of them is using, and how are they different?

SUGGESTED ANSWERS TO CASE QUESTIONS

1. It seems to me that Marvin had not determined that his role as a technical manager is different from that of a designer. His personal theory is out of date since it applies to his previous job with the company. He mentioned that he had spent time on the drawing board determining that all the components would work well. That behavior is more appropriate for the designer that he used to be. To set up his theory, he should understand that there have been changes in technology and in the company, which is moving into an entirely different kind of market and manufacturing process. He should be concerned with whether the product design goals can be met in the marketplace, since the product has been drastically modified from a mechanical to an electronic device. The product operations and controls are different. Because the product works differently, it might not be as easy to define product acceptance tests.

2. Without objective data to back them up, a person's statements can only be evaluated subjectively, and Bob Spelvin had not been with the company long enough for that to work for him. Therefore, data were required. For example, in a worst-case scenario, what would the company's liability be if one of the new diagnostic chairs were to fail? What would the costs (i.e., of the quality assurance program) be to minimize that potential field failure? In effect, Bob should have come in with a cost-benefit set of data.

3. George seemed to be as unprepared for this meeting as the rest of the group. He certainly was aware of the problems (and opportunities) in producing the new products. George should have supplied data for the meeting that were obtained from his field representatives showing how serious (in a quantifiable measurement such as lost sales) they thought the delay would be. Or he could have offered an experimental hypothesis with one obvious cost-effective alternative that could have prevented possible trouble in the field. For example, after development of final acceptance tests, the company could provide the diagnostic devices to several teaching hospitals on an experimental basis through selected field representatives. Data obtained from the devices would then be checked at the hospitals by more conventional means, giving some indication of product effectiveness. Meanwhile, the field representatives would be able to maintain contact with their markets (i.e., the hospitals) and would be able to get the sales credit when the superiority of the new products was established.

4. Define the terms. (Set up an agenda and require each attendee at the meeting to provide specific recommendations with justifications for them before the meeting so that they can all understand each other's position.) Propose a theory. (The company needs new products because. . . .) Develop a hypothesis to be tested at the meeting. (Take either Bob or Marvin's position and, independently, determine what the potential effect would be.) Then use that working hypothesis as a standard against which to measure the recommendations of others.

5. The implicit or mental theories of the participants seem to be:

Marvin: The future is almost an extension of the past. If that's the way we did it in the past, that's the best way to do it in the future.

George: Sales is where it's at. It doesn't matter if the technology has changed. Selling complex products is just the same as selling simple ones. Just get out there and sell.

Bob: The solution to a similar problem in my last organization was to set up a vendor quality program. This organization (and, implicitly, the situation) is the same, so the answer should be the same.

Tom: Have no theories or hypotheses of your own. Depend upon the advice of the "experts." How can you manage people who are more qualified technically than you are?

Do you agree or disagree with these interpretations? Why? Do you see how they are related to the chapter? Is this situation similar to those in which you have been involved? What happened?

REFERENCES

Drucker, Peter F. *Management: tasks, responsibilities, practices.* New York: Harper & Row, 1973.

Ginzberg, Eli, & Vojta, George J. The service sector of the U.S. economy. *Scientific American,* March 1981, *244*(3), 48–55.

Gould, Peter. So human a science. *The Sciences,* May–June 1981, 6–30.

Kant, Immanuel. *Critique of pure reason* (2d ed.), (N. K. Smith, trans.). New York: Modern Library, 1958.

Levinson, Harry. *Psychological man.* Cambridge, Mass.: The Levinson Institute, 1976.

Mintzberg, Henry. The manager's job: folklore and fact. In *On human relations,* New York: Harper & Row, 1979, pp. 104–124. Reprinted by permission of the Harvard Business Review. Excerpt from "The Manager's Job: Folklore and Fact," by Henry Mintzberg (HBR, January–February 1975). Copyright © 1975 by the President and Fellows of Harvard College; all rights reserved.

Skinner, B. F. The operational analysis of psychological terms. In Herbert Feigl & May Brodbeck (Eds.), *Readings in the philosophy of science,* New York: Appleton-Century Crofts, 1953, pp. 104–124.

FURTHER READINGS

Adams, Brooks. *The theory of social revolutions.* New York: Macmillan, 1913.

America's restructured economy. *Business Week,* June 1, 1981, 55–100.

Bell, Daniel. *The cultural contradictions of capitalism.* New York: Basic Books, 1976.

Cyert, Richard M., & March, J. G. *A behavioral theory of the firm.* Englewood Cliffs, N.J.: Prentice-Hall, 1963.

Drucker, Peter F. *The future of industrial man,* New York: Mentor Books, 1965.

Gartner, Alan, & Riessman, Frank. The culture of the service society. In Roy P. Fairfield (Ed.), *Humanizing the workplace.* Buffalo, N.Y.: Prometheus Books, 1974, pp. 99–111.

Hunt, Pearson. Fallacy of one big brain. In *On human relations,* New York: Harper & Row, 1979, pp. 19–29.

Kanter, Rosabeth Moss. The politicization of organizational life: skills for critical issue management in a changing organizational environment. *O.D. Practitioner,* September 1981, *13*(3).

Kuhn, Thomas S. *The structure of scientific revolutions* (2d ed.). Chicago: Univ. of Chicago Press, 1970.

Lewin, Kurt. *A dynamic theory of personality*. New York: McGraw-Hill, 1935.

Lombard, George F. F. Relativism in organizations. *Harvard Business Review*, March–April 1971, 55–65.

McCall, George J., & Simmons, J. L. *Issues in participant observation*, Reading, Mass.: Addison-Wesley, 1969.

Maccoby, Michael. Who creates new technology and why? In Roy P. Fairfield (Ed.), *Humanizing the workplace*. Buffalo, N.Y.: Prometheus Books, 1974, pp. 87–97.

Merchant, M. E. The factory of the future: technological aspects. In *Toward a factory of the future*, (PED Vol. 1). New York: ASME, Winter Annual Meeting, November 16–21, 1980, pp. 71–82.

Shannon, Robert E. *Engineering management*. New York: Wiley, 1980.

Tawney, R. H. *Religion and the rise of capitalism*. New York: Mentor Books, 1954.

Yankelovich, D. The meaning of work. In J. M. Rosow (Ed.), *The worker and the job*. Englewood Cliffs, N.J.: Prentice-Hall, 1980, pp. 19–47.

Zukav, Gary. *The dancing Wu li masters*, New York: William Morrow, 1979.

1
THE MANAGER

Case Study

THE CASE OF THE ORDINARY DAY

CAST

Leona Russo: Chief engineer

Bill Watson: Head, quality assurance

Sam Snith: Controller

Arnold Mitch: Design engineer

Leona kept the car well within the speed limits as she drove toward the plant on a sunny Wednesday morning. There was still plenty of time to get to her desk, review the capital budget for next year, and make some last-minute adjustments before her presentation at the board meeting scheduled for 10:00 A.M. that day. She went over her morning's schedule in her mind:

1. Review the budget.

2. Approve the final design review notes on the latest turbine pump.

3. Find out what happened during the last interview that Arnold Mitch had with the graduating engineer they had recruited from State University.

4. Check out the vendor assessment that Bill Watson had promised her on the potential hydraulic motor vendor.

She had everything all set up in her mind by the time she wheeled into the parking lot. As she approached her office, promptly at 8:30 A.M., her secretary was waiting for her with a worried expression.

Secretary: Good morning, Leona, the board's secretary just called and said that the meeting had to be moved up to 9:00 A.M. Can you be ready? Here are some other phone messages that just came in.

She quickly checked over the phone messages. One bothered her. It was from Sam Snith. She decided to return his call first because he rarely called unless there was a major problem. She picked up the phone and dialed.

Leona: Hello Sam, what's up?

Sam: Leona, I think that you're in for a bit of a problem at the board meeting. George Wishley asked me for a summary review of project costs on the last three capital expenditures and I had to give it to him. You remember, those were the installations that finally ended up costing about twice what the appropriation had allowed. I know that there were extenuating circumstances, but you might have a little explaining to do this morning. I called you late yesterday afternoon but you had left for the day.

Leona: Thanks Sam. Yes, I have the data on those. I'll talk to you later.

As she hung up, Bill Watson poked his head in the door.

Bill: Hi Leona, got a minute? (Without waiting for an answer, he draped himself over one of the office chairs and continued.) The material review board has been approving rejected materials from Appleby Valves for the last three months. Your engineers don't want to change the drawing and we've "bought" ostensibly "unacceptable" materials every time that my inspectors rejected them at incoming inspection. Look, we either have to reject that stuff for real the next time or, if it's really OK, let's change the tolerances on the drawing and maybe negotiate a better price with Appleby Valves. We just can't go on wasting time like this. You know that I'm shorthanded and those board review meetings take up a lot of time. I'm sure that your review engineers have other things to do too.

Just then, Arnold Mitch walked by; he saw Bill in Leona's office and waved at them both. Bill got up, walked over to the door, and said:

Bill: Hi Arnold. If you have a second, I'd like to get your opinion on the latest lot of Appleby Valves.

Arnold: Well, if you ask me, those guys at Appleby are getting away with murder. They swore up and down, when we did the vendor review at their plant, that they could meet our tolerances, and our evaluations tend to support that. Now they ship late, and we always have to waive our requirements because we need their valves, and another thing. . . .

Leona interrupted him:

Leona: Look guys, we've had these discussions time and time again. Both the design team here at the plant and the vendor review team that visited Appleby agree that we need those valves and it was cheaper and more effective to have Appleby supply them. There doesn't seem to be too much that can be gained by going over the same old ground. What would you like to do today? This problem doesn't seem to be going away. I'll work on it and lay out some policy by the end of the week. Meanwhile, Mitch, what happened at the design review meeting on the turbine pump?

As Mitch launched into a rambling discourse on the meeting, she thought about the rest of her schedule that morning. He was interrupted by a call from Leona's secretary.

Secretary: The data processing center just called and said that the central processing unit was down and would be unavailable for at least two hours.

Leona was visibly annoyed. She had requested her own central processing unit (CPU) for the engineering department for the past two years, but had never been able to justify spending the money to the board of directors. The central CPU was down on the average of once a month and always when there seemed to be an impending disaster. At that point Leona noticed the time. It was five minutes to nine. She quickly shooed Bill and Arnold out of the office, gathered up her notes from the previous night, and headed down the hall toward the board meeting.

QUESTIONS

1. What management techniques should Leona have used that would have helped her to handle the situation?

2. How should she resolve the engineering-quality dispute on the Appleby valves?

3. How is she resolving the problem of handling uncertainty, and what would you do?

4. How should she handle any problems about the capital overruns if George Wishley brings them up at the board meeting?

My suggested answers for these questions will be found at the end of this chapter.

A QUICK REVIEW OF SOME MAJOR POINTS

The introduction described new and recent changes in the various organizational environments, such as those caused by differences in the work force and the sizes or types of the organizations themselves. Those changes are rapid and increase the need for equally rapid solutions to the problems raised in these novel situations. The standard solutions that used to be applied uniformly, for example, when workers were considered to be interchangeable, don't work anymore now that these workers have to produce innovative products and services. Those products and services can only be created when the worker (in this case, the knowledge worker) is free to create. This means that the worker is treated as an individual, not as part of a uniform group, because creativity is initially based in some one person's unique thoughts, not in a group's collective mind. It's very difficult to follow some uniform group or company policy and be creative, since the act of creating is, by definition, inimical to standardization.

The changes in the work force have also been affected by the relatively recent spread of multinational organizations. These large organizations have extensive resources and an ability to move them almost anywhere in the world if they wish. However, the advantages of increased organizational size and relative economic freedom have a price—the loss of the speed of response to changes in economic, political, or other environments. Therefore, there are always smaller, more agile companies that are able to move into and take over specialized parts of the market, and this agility is often determined by these smaller companies' technical function.

Technical functions provide two general outputs for their companies: controlled extensions of existing profitable products and innovation, upon which each company's future depends. Both of these come from the knowledge worker, who is a relative newcomer to the industrial scene. That knowledge worker, in turn, needs two tightly interrelated inputs: adequate physical resources and the personal freedom to create. In very general terms, the larger organizations usually have the resources and the smaller ones, the organizational freedom. Your job, as a technical manager in any size organization, requires a blend of these inputs, similar to the combination of the logic of physical resources and the art of human emotion.

Speculatively, larger companies could be more logical in their development of overall corporate policies intended to control or standardize human behaviors. But they seem to be less agile in their response to changed environments. Innovation and fast response are usually easier to achieve when dealing with individuals or very small groups because (speculatively speaking, of course) of the lower human inertia involved, There are fewer cultural blocks that have developed in the smaller organization, since that organization usually has had a shorter history during which to develop them.

Therefore, successful innovation seems to be greater with small companies or smaller, independent divisions within larger companies.

The underlying idea is that there are advantages to being structured for consistency and logic, but there are at least equal advantages to innovation and emotion. Needs for either of these positions are not fixed; they change as the environment does. It is this more or less ongoing change that we must deal with as technical managers. The comparison between small and large and policy and innovation might even be related to the comparison between the logic of Mother Nature and the art of human thinking and behavior. It's almost a type of continuous balancing act between opposite needs.

LEARNING HOW TO LEARN

Logic as a required course in the natural sciences is often justifiably emphasized in technical schools and universities, since the output is expected to be scientists and engineers, not technical managers. Logic is needed to be a technical person. Art is needed to manage those technical persons. But the manager, in order to understand his or her people, must also understand the logic with which they work. However, it is art (in my opinion) that determines the success of scientists, engineers, and technicians who become technical managers, since art determines the quality and direction of management's innovative responses to changing conditions.

This art, like creativity, is not gained easily, but it can be learned to some extent. The introduction described one method for learning it: learning from *corrected* practice derived from a typical process model called the scientific method and using the results from hypotheses tested in on-the-job training. Since each individual practices art differently, that process is intended to assist you, as the individual (the technical manager), to improve and develop optimal practices for managing people and situations. Even though people are all different, the introduction managed to give general suggestions on developing the optimal and unique theories of management that will help you to understand these differences. With respect to the situation, there is a more tangible model (see Chapter 2) for you to use in personal theory building. It is a model of the organizational situation. We now continue with the process model to help develop specifics to be used in your personal management theories.

The process model includes directions for both cognitive change (such as that represented by reading this book) and behavioral change (represented by testing your hypotheses in your organization). We begin with definitions of the terms to be used in your cognitive change. Then we describe the methods of theory building (a theory is a general explanation for a class of events), develop testable hypotheses, gather test data, and, if necessary, restructure the theory to fit the data received more closely.

That description seems clear enough, but this obviously logical and rational process is not always the clear and supportive guide that we need in our management responses. It doesn't equip us to deal with those theories that we carried with us before we began the process. Those implicit theories we had before are not always easy to change, especially when the new data received seem to conflict radically with our previous notions. Reluctance to change theories can occur even when the new data and the results seem to be irrefutable. One example of this occurred in the theoretical world of physics when Albert Michelson and Edward Morley obtained experimental results that conflicted with Newtonian mechanics. Zukav describes it well.

> The problem is that no matter what the circumstances of the measurement, no matter what the motion of the observer, the speed of light *always* measures 186,000 miles per second (in a vacuum). . . . Suppose the light bulb is standing still, and . . . we are moving away from it at 100,000 miles per second. What will the velocity of the photons measure now? . . . 86,000 miles per second? . . . the speed of light minus our speed as we move away from the approaching photons? Wrong. . . ! it should, but it doesn't. The speed of the photons still measures 186,000 miles per second. . . . Two American physicists, Albert Michelson and Edward Morley . . . completed an experiment (in the late 1800's) which seems to show that the speed of light is constant, regardless of the state of motion of the observer. (Zukav, 1979, p. 149 –151)

This crucial experiment contradicted most of the previously closely held theories of physics of the day, but for those who accepted the results, it eventually led to important modifications to all our theories about the universe. One of those modifications resulted in the mixed blessing of atomic energy. However, since the original Newtonian theories still seemed to hold on a macro level, they then became a theoretical subset of the larger theories of modern physics that this experiment showed existed. In another example, there was a tremendous uproar caused by Darwin's theories of evolution, since they conflicted with closely held religious theories of his day. (The uproar has still not died away.) The unwillingness to change those closely held theories that we carry into the learning process can be a difficult psychological stumbling block for anyone to overcome.

It is especially difficult in technical management because the technical part was based in apparent logic and consistency, and even if that logic is occasionally disrupted, the many past technical successes which were based on the theories we held before we became managers were probably the reason for our promotion into management. These were usually rational, linear thinking kinds of successes. But now the job has changed radically from being primarily technically oriented to being people oriented. The "management" part of the new title is successful only as it minimizes the past sources of success and becomes less structured (or rational, if you will) in order to deal with the emotional inputs of people. When that happens, the closely held theories that may have seemed to be right for you because they led to your present position as a manager become more fluid as you become more attuned to people. This current theory is just as important as the past one. It has, however, a larger framework that now includes people and Mother Nature.

I suggest that the ability to change one's personal theories as new data are received is one of the bases of learning the art of technical management. A word of caution is needed here, however. I am not implying a complete discard of your prior personal views, but only an increased ability to modify and consider other viewpoints, possibly less rational, than your own. Michelson and Morley showed that Newton's ideas were a subset of larger theories, but they also showed that larger theories existed. Newton's laws still worked, but within a more limited context. Changing closely held personal theories is therefore recommended as you learn or as the situation changes, but it doesn't always require discarding previous successes. Maybe those successes were only a subset of a new and larger theory. Conversely, you should not lack any personal theories of your own because of possible future changes. With no theory of your own, any answer may be as correct as any other, and rarely does a random answer work as well as a working hypothesis answer. Random answers are as good as any other answer when you have no theory at all.

For example, I'm sure that you have heard the statement "But that won't work here because we're different." That statement is possibly quite true at the time because the "difference" may be due to a very limited theory held by the person making the statement. With a more general theory that supports some common understanding upon which we could all agree, at least at the beginning, we could never learn to apply management lessons from one situation to another. If that were true, every situation would be completely unique in all respects. Very small children seem to think that way since they have little experience or personal theory to work from. Everything is entirely new to them. We, on the other hand, always have some minimal common ground from which we can gather data to expand our personal theory.

Maybe the differences that were supposed to be the reason for something not working here are really too great and no experiment by Michelson and Morley will show us how to adapt. I'm not minimizing these differences (after all, they are what we need to know to operate in a particular situation) but I am saying that they are probably not as great as we think they are and they might really define a subset of the larger theories that we, as technical managers, have to build to manage our changing technical functions.

MODIFYING THEORY: HOW TO MAKE DECISIONS

Our process model begins with definition. Without definition, there is no starting point to build personal theory (or anything else) describing your brand of technical management. We can assume that you have been successful in the technical part of the job, so we'll start our defining process with the less familiar part of our title: the management part. Some of the definitions of management that appear in the texts are not very specific.

> Management is defined here as the accomplishment of desired objectives by establishing an environment favorable to performance by people operating in organized groups. Each of the managerial functions . . . is analyzed and described. (Koontz & O'Donnell, 1964, p. 1)
>
> . . . management can be defined as a process, that is, a series of actions, activities or operations which can lead to some end. (Gibson, Ivancevitch, & Donnelly, 1976, p. 30)
>
> Management can be defined as involving the coordination and integration of all resources (both human and technical) to accomplish various specific results. (Scanlon & Keys, 1979, p. 7)

The common ideas seem to be that management is a process resulting in some positive accomplishment and that process is associated with a group of people or an organization of some kind. The process appears to operate by itself as if it were in an organizational vacuum. These definitions do not show a direct relationship between the management process itself and specific organizations. One might assume from them that management is independent of the situation.

These definitions provide little guidance to anyone who wishes to manage since there are no directions for personal cognitive or behavioral adaptation. Therefore, they could be universally applicable, except that we now know that can't really be so. Obviously we need better guidance for our useful definition to begin the development of a personal theory. We need a definition that connects the manager with the situation.

I suggest that an adequate definition of management is that it is the process of *making decisions* that are applied to people in organizations, i.e., people who are defined as being in the situation. That definition is the "what is it?" and the "how is it done?" of management, as decision making infers a mental process called *absorbing uncertainty*. That process is inferred because it cannot be observed, but can be understood from observations of applicable behavior.

For example, consider the managing processes of the members of the "perfect" organization. Each day the chief technical officer arrives at the office and, instead of attending to the many political and social tasks that are usually associated with that position in other less-than-perfect organizations, the chief technical officer spends the day in behavior which indicates that he or she is thinking. For many months, that is what he or she does. Then, based upon all the data available from the rest of the organization and the results of the thinking, he or she makes a decision and announces it: "The technical group will set up a separate organization consisting of project operations personnel and facilities, thereby ensuring a stream of new and improved "widgets." He requests the rest of the organization to produce the detailed development plan to get that decision implemented.

In this "perfect" organization, the chief technical officer makes very few of these decisions in any year, but when they are made, they shake the entire organization. These decisions are generally unique and concerned with future events. While there may be inputs such as recommendations from his staff, the buck stops at his desk. More specifically, *his* management process of decision-making "buck" stops there. The uniqueness of the problem to be solved, its magnitude, and the period of time in the future with which it is concerned seem to be almost directly related to the level of management in the organization. Even though the method of decision making (i.e., absorbing uncertainty) is the same throughout the company, the amount of uncertainty absorbed is supposed to vary directly with the management level. The higher the level, the higher the amount of uncertainty absorbed.

On the other hand, consider the decision making of the first-line technical supervisor in the shop for that same "perfect" organization. He or she also arrives at the office each morning and also spends the time thinking. But that thinking period is usually quite short; decisions are made quickly. They are concerned with less unique situations and shorter time periods in the future and have less magnitude. For example, there is a problem with some components and the decision must be made rather promptly to reject, repair, or replace them. Otherwise shortages will develop in production. And so it goes all day long, decision after decision. In both examples, we have eliminated the repetitive decisions that plague managers in less than perfect organizations and allowed both the chief technical officer and the first-line technical supervisor to concentrate on the peculiarly human skill they both have, that of making decisions by absorbing uncertainty.

It is now appropriate to define the term *uncertainty*, since I have proposed that it is at the core of decision making (which, in turn, is the reason for management to exist). I suggest that rational decisions can be classified under three general categories (modified from March & Simon, 1958, p. 137).

1. *Certainty*. There is complete and accurate knowledge of the consequences that will follow on each alternative available to the decision maker.

2. *Risk*. There is correct knowledge of a probability distribution of the consequences of each alternative and the decision maker has that knowledge.

3. *Uncertainty*. There is a set of consequences for each alternative which belongs to some lesser set within all the possible consequences, but it is impossible for the decision maker to assign definite probabilities to the occurrences of the particular consequences. The probability distribution is there but the decision maker cannot use it. If there were no probability distribution, decisions could be made at random, since they all would be equally acceptable.

I have not suggested that managers make decisions under random conditions because people are not really needed then. With any decision being equally acceptable, there is no personal or even any other kind of theory needed: just a set of honest dice.

While one major difference between the two examples above of the chief technical officer and the first-line technical supervisor was the amount of uncertainty that each absorbed in order to make decisions, I suggest that another was the frequency of those decisions. An inverse relationship between uncertainty and the frequency of decisions might even be possible. However, it would probably be too simplistic to suggest that this relationship be linear. Organizations are much too complex for that.

In my opinion, the amount of uncertainty absorbed by our "perfect" chief technical officer is much greater than any linear relationship defined by the number of levels of management, the number of employees supervised, or any other measurement of management responsibility that I know of. Therefore, a curve of uncertainty (vertical axis) versus frequency of decisions (horizontal axis) would probably decrease exponentially from left to right (i.e., minimum decisions at the maximum uncertainty to minimum uncertainty at the maximum number of decisions).

As noted before, maximum uncertainty could be treated as randomness, in which case almost any decision would be equally correct, so there is an assumption that randomness is not in our type of decision making (i.e., when $x = 0$, $y = $ infinity), and complete certainty is not, either (i.e., $x = $ infinity and $y = 0$). If randomness were to occur, you could make decisions with any set of dice, and if you had complete certainty, you could use a computer. In both cases, human thought would not be necessary. (See the curve in Figure 1-1.)

Personal theory is intended to assist you in choosing alternatives or making decisions that will optimize your position (first) and the efforts of the organization (second). Without you (as the representative manager), there is no organization. The process of absorbing uncertainty supports the primary

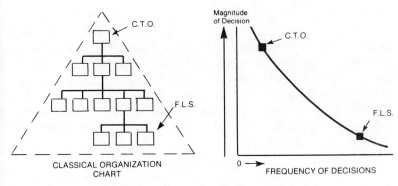

Figure 1-1   Organizational position and decision making.

management task of decision making (within your organization). Therefore, the next step in developing this inferred process would be to learn how other existing management processes could apply. In effect, use definitions first, modify as needed, develop hypotheses, test them in situations, and determine how closely the data received from the tests match the predictions of the hypotheses.

Then, if necessary, you can go back and modify your management model (developing new hypotheses). When you modify that model, you are including experience in your personal theory, and your theory can only be improved. Of course, so will your decision making (and, possibly, your position). Since the situation is in almost continual flux, your theory will probably always be in a somewhat equivalent change and improvement process. The development process is fairly straightforward to describe, but the challenges of improving personal theories could make the description very difficult to follow. You need discipline and concentration. This is how you can apply the scientific process model:

> *First the general theory:* Management is a process of optimizing the efforts of an organization.
>
> *Then the definition:* After evaluating applicable management definitions, we selected a process defined as effective decision making about people in organizations that is based on the inferred ability to absorb uncertainty. (These first two steps of theory and definition are interrelated; their sequence, therefore, is not important.)
>
> *Now the selection of the first hypothesis:* Evaluate applicable research results and practices against your special needs as a technical manager. Those are generally described in this and other books. Select the hypothesis that fits both *you* and your perception of the *situation*.

Then the testing of the hypothesis: Apply selected techniques within your limited areas of responsibility: Compare the data that result against your predictions. If the data do not match the predictions of the hypothesis (within predetermined limits that you had set), change your theory or disregard the data as unacceptable, erroneous, or in conflict with your preconceived notions. (I would hope that most of the data we look at in this book will match your hypotheses.) Books like this help in the beginning of cognitive development, but feedback from associates at work guide the corrected practice of improved management behaviors.

This generalized definition and theory building, cognitive change, and behavioral practice is a learning process that never really ends. There will always be a need for updating your management skills as new findings in both research and practice become available and the organization itself changes. We are all becoming lifelong students, irrespective of the management position we have in the organizational structure. Knowing that can be an asset, especially if we are confident enough to handle it by recognizing our own inadequacies. And competent managers do know this. Real problems require suspending one's status or organizational position and assuming the humble status of a student under competent teachers. This process is a problem for insecure, second-rate people, but those who find no difficulty in learning are probably not even aware that it might be a problem (Boettinger, 1979, p. 199).

We have many competent teachers. Those that I quote from or are listed in the bibliographies are just a small sample of those available. There are others; the libraries are full of them. To be an effective manager is to be a student for life.

WHICH MANAGEMENT MODELS DO WE CHOOSE?

There often seems to be just too much information in the world. Therefore, some type of classification scheme must be used to select applicable data. Otherwise we select at random, but that method provides few guidelines for future use. The classification scheme that we will use is a *model* and the learning process (i.e., selection and testing of ideas) is known as *modeling*. A model is some representation, qualitative and/or quantitative, of a process or situation that shows what we think the significant factors are for the purposes we have chosen. Scientists and engineers use models as exploration plans to demystify nature. With no plans, all facts would be equally relevant. However, these plans can never be completely free from some degree of subjectivity.

Observation and experience can and must drastically restrict the range of admissible scientific belief, else there would be no science. But they cannot alone determine a particular body of such belief. An apparently arbitrary element, compounded of personal and historical accident, is always a formative ingredient of the beliefs espoused by a given scientific community at a given time (Kuhn, 1970, p. 4).

Our technical community has subjectively accepted many different models of management over the past century. Many still exist, but they are changing. Both the models, as exploration plans, and the ideas of the researchers who support them are continually being modified as new data and new beliefs are brought forward. Therefore, models are based on both objective (research data) and subjective (researchers' opinions) entities.

As a simplified example to illustrate the point, consider all the alternative designs or models available for cap screws that are available to the machine designer and must be considered before an assembly drawing can be completed. If the screws were evaluated at random, the designer would be unlikely to select the best screw. Similarly, if all of them were evaluated each time, the design time would be excessive. To minimize uncertainty, *someone* in the organization develops standards—which are really repetitive design decisions—for a small number of sizes and styles of cap screws. Some novel and creative cap screws are probably eliminated, but the major goal of focusing the attention of the designer on the important alternatives concerning the product's functioning and cost is achieved. The model served its purpose of limiting alternatives and directing thinking, but it also incorporated the subjective beliefs of the original standards designer. And the models selected for this book, for example, include many of my subjective beliefs.

While this example is almost trivial, the process of building and using models is hardly that. It requires some careful thinking about what the important variables are (of course, assuming that we are even aware of them, which is another subjective consideration) and determining how those variables can be manipulated to gain some desired result. Also, the models themselves often change during the time that the models are in use. Typically, "knowledge about the process being modeled starts fairly low, then increases as understanding is obtained and tapers off to a high value at the end" (Chestnut, 1965, p. 130).

USING MODELS: ADVANTAGES AND DISADVANTAGES

When we assume that we have selected the relevant variables of whatever we wish to understand and have been able to develop the relationships among them, we can develop a relatively consistent and powerful thinking process that allows us to hold parts of the model (or the situation, in this case) constant while we change the other parts. For example, in our model of the organization, if we hold the organizational structure constant, we might be able to predict the effects on total organizational effectiveness when we change a relationship- or person-oriented manager for one who is task or results oriented. The prediction might not be perfect and, in fact, rarely is because of the self-limiting nature of all models but at the least, we could build a testable (i.e., predictable) hypothesis before the change was made and the organizational situation irrevocably altered. Therefore, model building assists in understanding the central processes of the situation that are supposed to be represented. It involves abstraction and is intended to represent the essential properties of whatever we are studying.

A disadvantage includes the obvious limiting of thinking to the variables and relationships that were selected. Some variables that are not considered may be quite vital. Conversely, some that are considered might have justifiably been neglected. Obviously, the former would be quite a problem, while the latter would not be as important. This disadvantage of limiting thinking could be particularly important to you as a technical manager since, as I have noted before, the training that we all received prior to becoming managers was often quite structured and provided little reward for creativity. It supported limited, structured thinking.

Engineering and technical curricula stress logic and consistency. This is quite reasonable, since it is the basics of natural science that must be learned first. Structural engineers, for example, do not design at random. They learn to develop foundations and then the supporting framework for the building before even attempting to lay out the beautiful outer building shell that will be so universally admired by the editors of architectural magazines. Creativity often does not stand alone in technical operations, but is married to a foundation of logic. However, it is rarely taught as well, if at all. We can all learn the foundation work easily but learning how to create the beautiful building shell is a much more difficult learning process.

SELECTING OUR FIRST MANAGEMENT MODELS

Our management models are also married or joined in their parts, since they deal with both the natural and the social sciences. These social sciences do not have the developed, more objective, methodology of their predecessor, the natural sciences. Therefore, it is not as easy for new data to alter old social theories as it is for new data to alter theories in the natural sciences. For example, when new research in physics disproves old theory (Remember Newtonian physics and the Michelson and Morley experiments?), that theory is generally modified or discarded in today's scientific communities. However, new data in some of the social sciences may not mean either the acceptance or the discarding of old theories. The social sciences are too young to have been able to develop the integrating processes or the superordinate theories that the more mature natural sciences have. Therefore, early ideas such as hierarchical organizational structures may live quite comfortably in the same organization as more modern structures, such as project and matrix management. Old social theories have been discarded and new superordinate theories in management have been developed [such as in the example of using technology as a variable (see Chapter 5) to categorize hierarchical and project organizations], but instances are still quite rare.

Therefore, when measured by the standards of the natural sciences, most general management models are relatively inexact, if not inaccurate. In addition to inexactness, management models dealing with technical operations have some special problems of their own. Technical operations have the two diametrically opposed organizational requirements placed upon them (described in the introduction), which doesn't seem to be true for any other organizational function. It is almost as if there were two opposing lists of needs that had to be satisfied simultaneously. One list might simply be labeled logic, consistency, and control. These are the logical left-brain organizational needs that typically stress design standards, drawing controls, growth through product improvements, and fixed procedures that promote organizational stability. An opposing list might be labeled innovation, uniqueness, and creativity. This list exists at the same time and contains equally insistent intuitive right-brain organizational needs that typically stress innovation or radically new products and processes and change that generally reacts to organizational instability.

If these differences were organizationally separated, they could have different labels such as the research and development function as opposed to the sustaining engineering function. However, in many organizations the responsibilities are combined into one technical area, and the technical manager not only has to wear the proverbial many hats of general management but also has to maintain the two general psychological heads (or brains, if you prefer) of both logic and intuition. And that is done through an obviously difficult process of developing decision-making processes that are adaptable to both areas. In other words, in their personal management theories, technical managers must develop two related but distinct processes and structures that can suit the differing needs of the overall organization. That task is more difficult than developing many other management skills, and involves developing an equivalent bilateral model decision-making process to fit.

WHICH MANAGEMENT MODEL? SITUATIONAL THEORY

The model I believe to be most applicable fits within a general theory set called *situational theory*. It defines management as being contingent on or related to forces in three interdependent areas: the manager (this chapter), the situation (Chapters 2 through 6), and the interaction with the people being supervised (leadership, in Chapter 7). This can be contrasted with the management which is concerned primarily with modifying the physical factors of production. Situational theory is concerned primarily with the psychological factors in the organization.

These psychological factors are not as limiting in developing our own theories as they might seem. Psychology can even expand our options as managers. Although most of us are more comfortable with more concrete theories such as those in operations management because they're more objective and tangible and they fit in with the technical background we all have, we recognize that operations don't move by themselves. People move them and people (not only others, but ourselves as well) are the major concern of technical managers. Situational theory includes the people and all the other, more objective, areas that are more familiar (they are all part of the situation). It typically includes topics such as organization design, management information systems, motivation, and small-group theory, as well as individual psychologies. Since situational theory is being discussed by a group of research contributors with supplementing and slightly overlapping views, it has the promise to become the type of overall explanation that was discussed earlier in this chapter. It often coordinates information that might seem to conflict otherwise. However, as a body of knowledge, it is still relatively young. It has not matured enough to provide all the answers—just some of them.

Even though the theory deals with three areas—manager, people, and situation—it is apparent that those three areas are pieces of each other. It is difficult to separate them in order to understand and, eventually, manipulate them. Therefore, we shall try to simplify things by separating our analysis into two major, and more easily handled, divisions—the manager and the situation. Later, we will cover manager-subordinate interactions in discussions about leadership. In the context that we shall use here, situational theory (or contingency theory) deals with predicting actions to be taken under described contingencies in a particular situation. Those actions might involve the manager and what he or she does to achieve the best fit or congruence among the various components of the organization or between the organization and its environment. In this example, we will consider the manager's evaluations and actions in dealing with the contingencies of the situation.

We start our study of situational theory with the first major area, *you* as the technical manager. This approach follows those of the physical sciences, since we start with the most independent variable, you. In the following chapters we will deal with the more dependent or malleable variable, the situation.

THE MANAGER AS THE INDEPENDENT VARIABLE

Understanding human behavior is a special concern of psychology and is much more challenging than understanding the physical factors in technical operations. Among other things, psychology is concerned with apparently nonrational (but very human) aspects such as creativity, aspects that are obviously not as predictable as many of those of the natural sciences. In some ways, psychology and engineering are analogous to you and the situation. You are more complex than the situation, since there are many more variables to deal with when we deal with you as an object of investigation. However, when you first become part of a group of many, as in a group of people, and the group then becomes part of a larger organization, human behavior, *on the average*, becomes more predictable.

It is easier to make predictions about working groups than about individuals. This is one of the reasons that life insurance companies make money. Their actuaries can predict with great certainty how many people will die and how many will live in a group of people, but they can't predict at all on an individual level. Putting it another way, it seems to be a safe assumption that you are more complex and less easily modified than the situation is because you are likely to contain fewer predictable (rational?) elements than the situation. We all have many nonpredictable aspects, such as love, drive, loyalty, and creativity, and they make us what we are: individuals. We therefore start our theory building by dealing with the more independent variable, you. The less independent variable, which is the situation, is dealt with in the next chapter.

A good beginning would be to describe and understand some of those inferred mental processes that individuals seem to follow in managing. However, we will be less concerned with grand ideas about individual philosophies and more concerned with relatively limited ideas that involve the specifically human management process of decision making. After describing some general ideas about decision making, we'll develop a prescriptive action-oriented model that should support your unique decision-making hypotheses by *absorbing uncertainty*. That general process, description followed by prescription, is an appropriate method for learning new ideas.

THEORY CLASSIFICATION: DESCRIPTIVE AND PRESCRIPTIVE

There seem to be two approaches that most researchers take to the study of management and organization. One approach may be positive; that is, descriptive or explanatory. The other is usually normative, which means that it is prescriptive and evaluative. We will take the descriptive or explanatory concepts first, and then move to the more complicated normative or prescriptive approach. This is a reasonable sequence, since we can then use the descriptive ideas that we have first defined as a basis for the prescriptions for our own personal theories. Most of us accept prescriptions better when we understand the reasons behind them, then the potential benefits others have achieved using them, and finally the problems in using them. We will use generic theories, then narrow down to more specific and directly applicable theories or models (which are really subsets of theories). We will develop hypotheses to be tested.

The data collected from the tests are the stuff of learning experiences or empirical information that improves us as managers. If the theory (or model), hypotheses, and data fit your situation, I would prescribe using it as is. But since it is quite rare to find an exact fit, a more useful process might be for us to develop the specific models and tools from the generic models and then use these specific prescriptions as further raw materials to be adapted as required, in a repetitive or iterative way. Our first prescriptions are concerned with the primary responsibility of managers: that of making decisions.

DESCRIPTIVE THEORY: DECISION MAKING

Management decision making, as we have discussed before, can take place under conditions of certainty, risk, and uncertainty. Making decisions under conditions of certainty is a fairly straightforward process and, if it were possible to follow a completely rational path in that process, the path would be:

> In the certainty decision model the decision between alternate strategies is relatively obvious—namely select the strategy whose payoff is largest or smallest depending upon whether the decision-maker is maximizing or minimizing. (Archer, 1967, p. 455)

This is self-evident, since all the decision maker has to do is to evaluate the available information and select the optimum alternative. It's quite logical and consistent, similar to a computer program or the inevitable (and only acceptable) answer when you solved mathematics problems correctly in school. But this is not that kind of a school, and decision making under conditions of certainty and rationality is not always possible.

By definition, neither the organizations nor the people who work in them are rational. They depend on human emotions, and not all the information about these human emotions is accessible to you or amenable to easy maximizing-minimizing calculations. How do you know what a maximum or minimum payoff is, when each person has different values for these things? Even your own value ideas are not fixed; they can change with your circumstances. Therefore, I suggest that the certainty process of decision making is quite a rarity in most operating organizations.

The next process of decision making, deciding under conditions of risk, is less rare. "Under risk . . . review the expected value or the sum of the payoff each multiplied by its respective probability of occurrence. . . . Select the strategy that optimizes expected values" (Archer, 1967, p. 455). This process apparently could be a bit more complicated than making decisions under conditions of certainty, since we have to gather more information about expected frequencies in addition to expected values. It is, however, easy to describe: Just determine what the frequency of the occurrence or answers will be, multiply each by its expected value to the decision maker, and add the products together to get your answer.

But this process is based specifically on human emotions and judgments. It is therefore easier, since the required objectivity of certainty is diminished, if not eliminated entirely. It is now acceptable to define the expected values and the frequencies of their occurrence subjectively. Calculating those expected values of the decision under conditions of risk might appear to be quite a bit more complicated than determining a single value under conditions of certainty, but we have now legitimized the nonrational human evaluation processes themselves. That makes it easier, because there is less need to come up with the only "correct" answer. It's "This is what I think the probabilities are," rather than "This is the answer."

Although there are tools that can be applied in these two modes of decision making, they are only the rational foundations that support the nonrational, peculiarly human, thought processes that are superimposed upon them. Both of these are vitally needed to produce solutions. Decision making in organizations is rarely, if ever, based completely on the powerful mathematical tools that provide optimum alternatives in a coldly analytic, logical evaluation of all the facts. Some of these tools have been described in the literature on operations research. For example,

> Algorithms and heuristic processes; An algorithm is a process for solving a problem which guarantees a solution in a finite number of steps if the problem has a solution. (Example—finding the maximum of a function for which the equation is known: Take the first derivative, set it to zero, solve for X, etc.)
>
> A heuristic is a process for solving a problem which may aid in the solution, but offers no guarantee of doing so. . . . A familiar and widely employed heuristic is the use of analogy: Look for an analogy between the situation with which you are attempting to deal and some other situation with which you have successfully dealt in the past.

> Heuristics are helpful but they are not guaranteed to work, they are not algorithms.

> A Means-End Analysis: Compare what you have with what you wish to obtain; identify a difference between the two; find and carry out an operation which may reduce the difference; repeat the procedure until the problem is solved. (Taylor, 1965, p. 73)

Algorithms are extremely useful in relatively low-level organizational decision making, where the rules are clear and unchanging. In those very rare instances when this does occur, decisions are made under certainty conditions. "Is George entitled to participate in the corporate profit-sharing plan or is he not?" The profit-sharing algorithm has everything that is needed for this decision, such as the regulations on length of service, compensation, and the other relevant considerations. These types of decisions, with very limited circumstances, use algorithms best. The other tools noted above require the skill and undefinable input of a human being to work; even then, the answers do not spring forth in their complete and full beauty.

The reasons for this limitation lie in the fallibility of the human thought processes. We do not have the mental ability to search our mental and physical environment totally and provide the complete data that exist. We just cannot perceive it all. Without this optimum data input, it is impossible to get an optimum output. In other words, there is always another and probably better answer out there somewhere, but we rarely find it since we don't know where it is, and even if we knew where, we probably would not have the time to look for it. Since our decision making is usually based on the data provided through human relationships (and they are somewhat nonrational), our search for the mathematical perfection of the definite and optimum answer is bound to fail.

Some other supporting reasons for this are intuitively obvious. For example, we can never know exactly what others are thinking and therefore can never include their processes in our decision making when that decision concerns them. About the best that we can do is observe behavior and (depending upon our own criteria) infer what that behavior means in terms of that other person's thinking processes. One obvious problem with this type of inference is that we don't know how much of the behavior results from the present circumstances and how much from those remembered by the person from the past. Present behavior is to some extent an extrapolation of past behaviors. The person's expectation of some future occurrence also affects present behavior. A valued future reward can spur working behavior, just as no reward can slow it down. Observations do not supply all the reasons, and our inferences, therefore, are limited.

Another reason (which was briefly touched on before) is that the limitations of our own senses to receive and process data prevent our being aware of all the factors that could be relevant to our particular problem of the moment. As an example, consider the oil company forecasters of the early 1970s who were aware of the repeated demands of the raw petroleum producers in the mideast for increased crude oil prices but were unable (or unwilling?) to think about the consequences of a possible concerted action to raise prices on a worldwide basis. The problem may be that some managers don't optimize, but instead are satisfied with a less than best answer that will temporarily solve the problem. This type of behavior is called "satisficing" rather than optimizing. There is a part of situational theory, called administrative theory, that deals with this. For example,

> . . . The theory of intended and bounded rationality, of the behavior of human beings who satisfice because they have not the wits to maximize (p. 13). . . . The central concern of administrative theory is with the boundary between the rational and the non-rational aspects of human social behavior (Simon, 1965, p. XXIV).

There are many other examples of this kind of thinking about decision making in the literature. Drucker (1973) gives us these two:

> A decision is a judgment. It is a choice between alternatives. It is rarely a choice between "right" and "wrong." It is at best a choice between "almost right" and "almost wrong." (p. 470)

> People do not start out with a search for facts. They start out with an opinion. (p. 471)

These are not the rational processes of the physical sciences. But why do these descriptions of decision making, which include both nonrational behavior and "satisficing" descriptions, seem to be simultaneously correct and incorrect? Because they refer to a type of decision making that is really the central concern that we have as technical managers: that of uncertainty in which no one answer or even range of answers can be evaluated, analyzed, and used objectively to optimize the decision-making process.

Even though we have discussed decision making under conditions of both certainty and risk, or as "satisficing" as if they were one process, they cover many different processes (perhaps as many as there are managers?), and untangling them to select the best process for ourselves is a very difficult task. Even more interesting, we know that if we could untangle the best process, what is best for me may not be best for you. There is no one decision-making process, there are many processes, but fortunately those ". . . . processes important in problem solving are also important in decision making and creative thinking" (Taylor, 1965, p. 48).

This complicates the selection of decision-making processes even further since we are now dealing with a multidimensional definition that involves creativity. Yet, as we all know, managers do make decisions, they do solve problems, and they do create. Therefore, although the process may be difficult to understand and prescribe because of its multidimensionality, it is being used continually. We know that it includes rational and nonrational (or, if you feel better about it, undefinable) processes, and one of those processes is very familiar to us: intuition.

Using one's intuition means making a choice that relies on unconscious processes. These processes (which are probably nonrational since they are unconscious) cannot be used easily without an established frame of reference (Levinson, 1981, p. 18). If it is possible to develop this frame of reference for intuition, that same frame of reference may let us handle other, less accessible, processes as well. One goal of this book is to help the reader build an established frame of reference. This can be done by going through cognitive change in learning, observing successful examples in other situations, selecting the more useful parts of those examples, and attempting to duplicate them within your own special situation. That is similar to using a heuristic methodology. If you take the following steps, you will probably be able to develop the disciplined, decision-making processes that will work for you:

1. Look for an analogy between the situation with which you are attempting to deal and some other situation with which you have dealt successfully in the past.

2. Set up a self-developed framework that will support repetition of your past success in the present and in the future.

3. Continually revise the framework as the situation (and, much more slowly, you) changes.

This may sound like practicing until you get it right, but it's not intended that way. Practice does not make perfect: *corrected* practice does. Corrected practice begins with some type of ongoing development of the unique, personal theory that matches perceptions of and expectations for (or hypotheses for) the situation with the responses that the situation produces. Developing a continuously modified theory based on changing sets of these hypotheses is a better way to make optimum decisions. Define the elements of the situation that you can, but even if you can't get a complete definition, you can still use partial elements. With these as bases, develop and modify your theories as the test data come in. That adaptation should be in accordance with successful practice, but the world being what it is, it might also be modified by some less than successful practice. The task is a repetitive one. In other words,

A creative executive must be able to plan a program, try it out, tear it apart and then start over again before settling on a system that works. Many executives feel such pressures to get things done that few examine closely what they are actually doing. They do not see the implications of their actions and often prematurely close on an incomplete system which must be redone later. (Levinson, 1981, p. 221).

Therefore, assuming that these conditions apply:

- You have been able to avoid the trap of being blind to the implications of your choice and actions before you act.

- You have been able to build a preliminary theory framework against which you will measure the results of your actions.

- You are aware that you cannot emulate others exactly but you can use your observations of others to contribute to the construction of your own framework.

- You have to evaluate the data you receive from testing your hypotheses and modify your theory. In effect, all the data received from such things as on-the-job experiences, nonrational thought processes, and learning through cognitive change (e.g., reading books) can become just another part of personal development and growth in your management job.

But you have an advantage in developing this process that many of the other managers in different functional specialties may not have. Since you have succeeded in managing technical functions, you are probably quite familiar with the left-brain logical thinking that training in the natural sciences usually required. The bilateral organizational design of the technical function requires both logical, consistent (left-brain) processes that reflect the natural sciences very well and creative, changing (right-brain) processes that reflect the social sciences quite well. Since our academic background before becoming technical managers has included fairly comprehensive training in handling rationality and logic, our major tasks now include concentrating on the less familiar aspects of decision making that involve nonrationality and creativity. Half the job requirements seem to have been satisfied. The special responsibilities of managing changing technical operations can now be concentrated on the right-brain, human side of the organization. And since both making decisions and solving problems are related to creative thinking, both you and the organization gain as you become better at it: you as a manager and the organization as the user of human creativity.

Usually we, as scientists, engineers, or technicians, enter the organization in jobs that include fairly rigorous tasks and therefore use our training in objective, logical thinking. I remember that one of the first jobs I had as a neophyte engineer in training was to check material lists to be sure that nothing was missing and all the parts were appropriately listed. While this sort of job is almost obsolete today (the computers do it so much better than human beings), it emphasized objectivity, logic, and painstaking accuracy. (I hated it.)

Becoming a successful manager requires a different approach. There is less need for the mind-deadening logic of repetitive technical jobs. The approach now is primarily intended to support innovation, improvement, and change. Our new organizational tasks require the individuality that each of us has. It is therefore more emotional and creative (i.e., nonrational) and we, in turn, must develop personal management models that include those aspects if we are to optimize our management decision making.

PRESCRIPTIVE THEORY: DECISION MAKING BY ABSORBING UNCERTAINTY

We finally come to the actual process of making decisions and the mechanisms of absorbing uncertainty that are the central part of that process. In the "perfect" organization I described before, uncertainty was a major variable that distinguished the various levels of organizational responsibility. Maximum uncertainty resided at the top and diminishing levels of uncertainty followed downward with the decreases in organizational responsibility. Uncertainty occurred on every level because the decision maker at every level could not assign a probability to all the alternatives that existed. He or she could not perceive them all since the information needed for evaluation might not even be available. In some cases, decisions were made using data on hand because the decision maker accepted a nonrational choice to stop searching for more data and was "satisficed" with what he or she had. We cannot deal here with any specific individual's reasoning in decision making because of the complexity of the subject and the limitations of this book, but I will suggest (and prescribe) a less than perfect but still usable general behavioral approach that you can adapt to learn how to optimize your decisions under conditions of certainty, risk, and/or uncertainty.

In this prescription, I continue to assume that the continuing major rationale for management to exist is the capability to make decisions under conditions of uncertainty. Therefore, one initial step is to categorize all problems requiring decisions into repetitive and nonrepetitive classifications. The repetitive problems can be handled under conditions of certainty. With repetitive problems there should be some type of organizational decision matrix that produces an answer *every time*, as with decision making under conditions of certainty using algorithms. These matrixes exist in every company, although not in the same detail. Typically, they might include such decisions as employee benefit and insurance programs, design standards, and corporate policies. Creating them is one of management's first responsibilities, since they represent major decisions that have dealt with prior uncertainties. They have now become decisions under conditions of certainty. Sometimes they're called *company policy*. These repetitive decisions are the answers that can be produced by algorithms (e.g., a computer program).

The next task of management is handling nonrepetitive problems under all other conditions. In this model, both risk and uncertainty are subsumed under the heading of uncertainty because the assigning of an array of possible values in decision making under risk seems to involve many of the same nonrational choices as making decisions under conditions of uncertainty. I believe that the dividing line between risk and uncertainty rarely exists outside of textbooks.

Therefore, my prescriptive model of decision making for a technical manager to use as a beginning tool is as follows:

1. Determine if you have (or can or even wish to assume) the responsibility to make a decision. If it has not been made before, it seems to be a major decision, and you do not or cannot make it, then push it upward to the next level of responsibility or determine if the problem is a symptom of a potentially negative situation that you believe can be solved or avoided without your intervention. If that is likely, delegate it to someone else within your area. If the problem cannot be delegated and seems to require your intervention, go to step 3. However, if you can or wish to accept the responsibility, go to step 2, since you have now arrived at the point in the decision-making process that requires your efforts.

2. Next, determine if it is a repetitive or nonrepetitive problem. If it is repetitive, design a matrix (e.g., some type of system) or delegate the design task that contains all the solutions. If this is done, *stop* and let the matrix provide the answers *(don't keep solving repetitive problems)*. If the problem is nonrepetitive, go to step 3.

3. At this point, if you have had the discipline (or the success) to follow the preceding steps, you are ready to absorb the uncertainty required at your level of responsibility. Now is the time to test the problem against the theory that is your general strategy for solutions. Establish your hypothesis, make your decision (logically, intuitively, or any other consistent way that you feel fits the situation), and determine how closely the results of that decision matched your predictions. If the variance between predicted and actual data is within your tolerances,

stop. Why modify something that seems to be working well enough? Paraphrasing an old New England saying, "If it's not broken, don't fix it." On the other hand, if the variance exceeds your tolerances and those variance limits are still valid, modify your theory and hypotheses, *beginning with the largest variances first*, and try again.

This decision-making process assumes that there is no one better equipped to solve that particular problem than the person to whom it is finally assigned in the organization. In other words, the amount of uncertainty depends in part on the decision maker's organizational level and how much uncertainty he is willing, or is able, to absorb. The initial decision was to determine within whose area of uncertainty the problem rightly belonged. That obvious statement does not provide you with objective guidelines for absorbing uncertainty, but there are interesting, and equally obvious, statements that might assist you in setting these guidelines.

For example, no one in the organization knows more about your responsibilities and your job than you do. You're the expert and you always will be! Organization charts and job descriptions don't cover more than an insignificant part of any management job, and in many organizations, the basic reason the structure even functions at all is because managers (and, in fact, all participants) do things they feel need to be done. These things are not always covered by the job description.

Interestingly enough, the formal organization is rarely aware of these actions, but it would probably come to a grinding halt very quickly if participants did only what they were told to do. Therefore, no person in the organization is better qualified than the manager to determine when, and if, the organizational level requires her or him to make a decision. If the decisions are made in accordance with your own prescriptive model, problems will tend to be placed at the lowest organizational level for solution. And that's the optimum.

For example, if a lower-level manager finds that he or she does not have the resources, or the authority, to solve a problem at that level, he or she pushes it one level upward. "I can't solve this because it involves areas beyond my responsibilities." The next upward level manager must either solve it by absorbing uncertainty and make a decision or decide that it is also beyond his or her responsibilities, in which case it is sent upward another level, and so on until it is finally solved. If the problem can be delegated downward, that should be done, but the delegation must include the resources or authority for a successful implementation of the eventual solution. The process can be illustrated by the flow diagram in Figure 1-2.

There are, of course, difficulties in learning how to use this prescriptive model of decision making, since it includes many nonrational processes that are not easily described. Intuition may be one of those processes since you

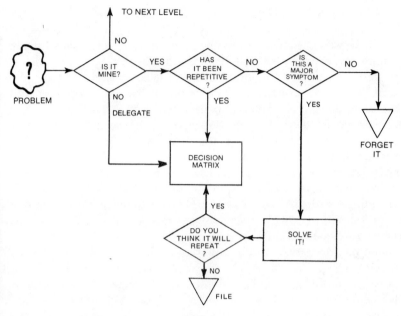

Figure 1-2 Suggested decision process.

have developed a method of corrected practice that allows you to improve your hunches. For example, it might be useful to make time available during the decision-making process to jot down all the reasons for supporting a particular decision (i.e., documenting the data behind your limited hypothesis). When the results come back, check them against your predictions and correct your mental model for the next time. If this sounds like the previous prescription for overall personal model building, you're right. As noted before, it is not practice that makes perfect, but *corrected* practice. In many cases, you can improve your intuition and other nonrational processes. When that happens, intuition may become practical experience. The name of the nonrational process is not important. What is important is that it can be useful to you as a manager.

Returning to the decision-making model and the process of absorbing uncertainty, there are several operational questions that may come up. For example:

How do I know when and how much to delegate?

Delegation is measured by the *amount of error* that can be allowed. That error is a function of time and potential loss. It is similar to the insurance premium that increases with a greater potential loss. The greater the amount of time between your act of delegation and the feedback or response of the individual to whom you delegated the task, the greater the amount of delegation (assuming all other things equal, of course). The cost of the delegation process itself (which is measured by the time you took to explain what you want) must be less than the potential value of the job delegated. This is similar to the premium on an insurance policy which must be less than the potential loss (the destructive fire or other peril that the policy covers) if the task delegated is not taken care of as expected.

In other words, there must be a relationship between the amount of delegation and the potential loss if the delegated problem is not solved quickly and effectively. Defining this relationship in terms of time or cost is an excellent process to follow in order to develop your nonrational delegation hypotheses. The definition process doesn't have to be formalized. Documenting it with notes on a scratch pad can be quite effective. Learning may be achieved just as well from your notes in a daily personal log as from a formally developed corporate policy.

Assume that the problem to be resolved involves accepting or rejecting a new vendor as qualified to supply your organization. If you recognize this as a repetitive problem, you follow step 1 and determine if there is a matrix for this kind of decision. If a decision matrix exists, follow it (remember, this problem is repetitive) and the decision will be automatic. If no matrix exists and the problem is yours to solve, now is a good time to consider designing a new decision matrix for vendor evaluations, using this vendor's qualifications to test it. If you are not responsible for this problem being solved (and remember, you are the only one who can really make this decision) but the problem does have to be solved, push it upward to the next level of management *with appropriate suggestions for action*.

Conversely, if this problem has not been solved before and you decide that it can be resolved within your area, the next step is to determine who is to handle it. Do you want to tackle it yourself or delegate it? If you decide to resolve it, you must determine how it fits into your management theory about this situation. That fitting must be done either by you or by some upper level of management (e.g., "All major proprietary assembly suppliers are to be approved by the engineering manager before being placed on an approved vendor's list because these assemblies are critical for correct product operation.").

If it is to be delegated, a hypothesis within that theory helps you to determine the magnitude of the potential loss and thereby assists you in determining whether to assign this particular task to the neophyte quality engineer in your department or the departmental veteran who has handled many similar jobs well. Since you used to be the expert on this vendor's products, you can easily determine errors in judgment and therefore might give the problem to the neophyte as a training exercise. On the other hand, if you have no expertise on this product, you give it to the veteran, but you follow up closely because you are facing greater uncertainty and therefore a greater potential loss (and you also want to learn something from observing how the veteran does it).

Some might argue that the veteran to whom you delegated it would be able to decrease the potential loss in delegation, but that is not relevant here unless your expertise is at least equal to that of the veteran. I am concerned here with the potential loss as measured by your own *level of technical knowledge,* not someone else's. This method eventually teaches you about the areas with which you are less familiar. It is part of the management education process. A really straightforward procedure, isn't it? Of course, this example is deliberately simplified to illustrate the decision-making (or, if you will, the decision-delegating or -placing) process that is correctly matched to the level of uncertainty; but simplified or not, most of the time spent in managing can be wasted if this type of procedure is not developed, corrected, and practiced continuously.

To emphasize the other point about delegation, it is not often intuitively obvious but it is logical to delegate the greatest amount about the things you know best. For example, the chief hydraulics designer should be able to delegate the hydraulics designs very easily when he or she is promoted to head of the mechanical design group. With his or her expertise in hydraulics, the potential loss in delegation is minimized for this area, since he or she can spot potential errors quickly. He or she therefore has a lower potential loss in hydraulics if he or she spends time elsewhere. It also follows that the less you know about a subject, the less should be delegated until you learn enough to be comfortable with it. Many of us tend to carry our expertise with us when we change management jobs, and that could be an error. The sales manager who can outsell many of her or his salespeople and does is not doing the job of the sales manager. The new management job requires delegating almost all of the actual selling, concentrating instead on less familiar areas, such as developing better administrative procedures or better communications between customers and internal order clerks.

> How can problems be pushed upward? That might involve a risk to my position as a manager.

Pushing problems upward because they are beyond your capacity to absorb the uncertainty can be relatively straightforward provided the rules for doing so have been fairly well-defined beforehand. Those rules should be set up using heuristic methods and should be discussed with upper management levels before they are implemented. These discussions could be a foundation upon which to build your management theory or framework. You might start by looking for similarities between particular past kinds of problems and their solutions and the present situation. How were the problems solved successfully in the past? What kinds of extrapolations or modifications could be accepted? As the expert in your particular management job, you are best qualified to determine your level of uncertainty and to suggest when that level has been exceeded. This should result in uncertainty guidelines. Developing these guidelines with upper management beforehand tends to minimize surprises, and surprises are generally regarded unfavorably.

As noted before, if appropriate suggestions for decision making accompany a problem upward, only two alternatives are possible: acceptance or rejection. If your suggestions are accepted, you have either explicitly or implicitly been given the responsibility to make similar decisions. In that case, your level of uncertainty has been raised. If they are rejected or returned to you for modification, you have a better concept of the limits of uncertainty (and the kinds of decisions that will be made) at the next upper level of management. Since you will be modifying your personal management theory as you receive new data (such as these data on the level of uncertainty), your theory will be that much better next time. The process is not one of taking a risk but of learning how to manage both yourself and others, including the boss.

> How do I know which repetitive problems to work on first in building a decision matrix?

One mechanism is to keep a histogram or frequency distribution of problems on a regular basis over a specific period of time. The x axis should note each separate class of problem or decision and the y axis the frequency of that problem or decision. Using a type of risk analysis, you should subjectively assign a potential loss factor (i.e., how much that problem cost you each time it occurred). Following the classical description of risk analysis, the highest product of frequency and loss defines the highest priority, the next highest product determines the next priority, and so on. With this priority sequence in hand, you can determine if the problem is repetitive and what the cost would be to develop a decision matrix (e.g., design an algorithm for it). That cost would depend on who is to develop the algorithm, you or someone to whom you delegate it.

> I've managed to handle repetitive problems by developing decision matrixes and building them in sequence of frequency of appearance and to pass major nonrepetitive problems upward and minor nonrepetitive problems downward. Now how do I handle the nonrepetitive problems that are left?

If you have gotten this far, you have probably found that you are making fewer but more important decisions. For example, if the cost of developing an algorithm is less than the cost of the product of the problem frequency and its potential loss, solve it. If not, forget it, since the economics are really not worth it. Some problems just go away by themselves and you don't want to make a major contribution to a minor problem.

Problems that don't go away are relatively important. But since you are the expert in your job and are best qualified to absorb uncertainty at your level, now is the time (and I'll say it again) for combining your evaluations of successful management patterns of others in your organization with your reading and continuing education, developing your theory, setting up your hypotheses that forecast specific results, and solving the problem or making decisions in accordance with those hypotheses. Then measure the variance from your prediction, and if acceptable, stop. If not, you have some data that do not agree with your hypotheses, so you either have to discard those data or modify the hypotheses for the next time. In any event, since there was no one better qualified to make the decision at the time, your existential management framework was the best that existed, and that framework can only get better as you modify it.

SUMMARY

The problems associated with technical management have some similarities but many differences from those faced by other managers in the organization. There are few other operations within the organization that are continually and simultaneously responsible for both logic and creativity. The technical groups as the suppliers of innovation are major organizational assets, since they create the organization's future, and optimum management of these technical operations is necessary if maximum effectiveness is to be achieved.

Organizational management theory, modified and adapted to fit the situation, is a basis for the technical manager to develop the unique theory he or she believes will fit the situation. Within that large body of theory, the more limited situational theory defines the relationships between people and the situation. It seems most useful. The similarities and differences between repetitive, logical operations and nonrepetitive, creative ones seem to support a multiple input-output type of management theory that optimizes the fit between the manager and the situation.

Situational theory is both descriptive and prescriptive. It deals with the variable of the manager and the situation and the interactions of these two general areas. The more predictive elements of this theory deal with the larger groups of people in the situation. Therefore, you, as the manager, are the more independent variable, and we dealt first with that independent variable, you, and how to do your job of managing, which is primarily that of a problem solver and decision maker. That job is based on the inferred general mental process called *absorbing uncertainty* and includes emotional processes, such as intuition. Several prescriptions that should assist in building your personal management theory were discussed. The following chapters deal with the other part of the theory, the situation. These chapters are intended to help you select the sections of situational theory (and other theories, too) that you think apply best.

Questions

These questions are not intended to test your retention of the contents of this chapter (after all, if you have missed something, you can always go back and read it again), but they are expected to help you to assimilate the applicable ideas and apply them to your own situation.

1. Which of the conditions—certainty, risk, or uncertainty—dominated your decisions before you became a manager? Did anything change when you became a manager? Has anything changed recently? What is or was responsible for that change? What would you like to do about it? How? When? Why?

2. How many times have you made the same decision in the past six months? What was it? Why did you do that? How can you change that, considering your evaluation of your situation?

3. In terms of your management strategy in your present situation, have you received any data that are intuitively reasonable and yet seem to conflict with your own and/or company policy? What did you do? Why? What should you do now when data do not agree with existing frameworks? How?

4. Can you really refuse to make a decision because it is beyond the limits of your uncertainty? If so, how would you do it? If not, what are the constrictions upon you and should you do anything about them?

5. Can we define the limits of our uncertainty? How? Isn't that a contradiction, defining what we don't know.

Since your answers will always be the right ones for you, why not write them out? It's reasonable to suggest that behavior can change thinking, and just the taking of notes does help many of us to learn. Try putting the notes aside for a few days to age and then take another look at them. Would you want to change some of those notes? Why? How? What happened?

SUGGESTED ANSWERS TO CASE QUESTIONS

1. She started out well by reviewing the things that had to be done within the near future, but she did not stick to her plans. There was no valuing of the many problems that were dumped on her that day. She had not categorized the problems into those that had to be resolved immediately (for the meeting) and those that could be put off (such as the problem of the Appleby valves). She had not calculated her potential losses if she had delegated properly, or set up her own decision process. See the first decision box, "Is it mine?" in Figure 1-2.

2. This seems to be a repetitive problem requiring a general answer. Possibilities include:

   a. When engineering has accepted a product that fails to meet specifications twice, the prints shall automatically be changed through the issuance of an engineering change order and purchasing shall be required to negotiate a different price with the vendor.

   b. All products that are rejected twice in sequence shall be removed from the accepted vendor list and a new source obtained by purchasing.

   c. Fill in your own answer. The repetitive answer is not as relevant as the fact that an answer has to be given. See the decision box, "Has it been repetitive?" in Figure 1-2.

3. She is not handling uncertainty at all. On a more limited scale, she could have listed all her tasks of the day and categorized them as *A* (to be done first), *B* (to be done this week), and *C* (to be done when I have time). There are other ways. Have you any of your own?

4. In line with my suggestions for handling uncertainty, I would advise Leona to point out to George that the item was not on the agenda and (diplomatically, if possible) suggest that it be put on the next agenda so that everyone attending the meeting would be prepared. My feeling is that you do not have to know everything, and a functional or problem-solving meeting should be run under predetermined rules in order to minimize uncertainty.

Do you agree with these answers? Do you think that you could use them in your own situation? Why and how?

REFERENCES

Archer, Stephen H. The structure of management theory. In Walter A. Hill & Douglas Egan (Eds.), *Readings in organization theory*. Boston: Allyn & Bacon, 1967, pp. 448–466.

Boettinger, Henry. Is management really an art? *Harvard business review, on human relations*. New York: Harper & Row, 1979, pp. 195–206.

Chestnut, Harold. *Systems engineering tools*. New York: Wiley, 1965.

Drucker, Peter F. *Management: tasks, responsibilities, practices*. New York: Harper & Row, 1973.

Gibson, James L., Ivancevitch, John M., & Donnelly, James H., Jr. *Organizations*. Dallas, Tex.: Business Publications, 1976.

Koontz, Harold, & O'Donnell, Cyril. *Principles of management* (3d ed.). New York: McGraw-Hill, 1964.

Kuhn, Thomas S. *The structure of scientific revolutions* (2d ed.). Chicago: Univ. of Chicago Press, 1970.

Levinson, Harry. *Executive*. Cambridge, Mass.: Harvard Univ. Press, 1981.

March, James G., & Simon, Herbert A. *Organizations*. New York: Wiley, 1958.

Scanlon, Burt, & Keys, J. Bernard. *Management and organizational behavior*. New York: Wiley, 1979.

Simon, Herbert A. *Administrative behavior* (2d ed.). New York: Free Press, 1965.

Taylor, Donald W. Decision-making and problem solving. In James G. March (Ed.), *Handbook of organizations*. Chicago: Rand McNally, 1965, pp. 48–86.

Zukav, Gary. *The dancing Wu li masters*. New York: William Morrow, 1979.

FURTHER READINGS

Barnard, Chester I. *The functions of an executive*. Cambridge, Mass.: Harvard Univ. Press, 1938.

Control Data Corporation. *Employee development manual*. Minneapolis, Minn.: Control Data Corp., 1981.

Cummings, Thomas G. *Systems theory for organization development*. New York: Wiley, 1980.

Cyert, Richard M., & March, J. G. *Behavioral theory of the firm*. Englewood Cliffs, N.J.: Prentice-Hall, 1963.

Fiedler, Fred E. *A theory of leadership effectiveness*. New York: McGraw-Hill, 1967.

Filley, Allan C., & House, Robert J. *Managerial process and organizational behavior*. Glenview, Ill.: Scott Foresman, 1969.

Follett, Mary Parker. *Dynamic administration: The collected works of Mary Parker Follett* (Henry C. Metcalf & Lionel Urwick, Eds.). New York: Harper, 1942, pp. 185–199.

Kahneman, Daniel, & Tversky, Amos. The psychology of preference. *Scientific American,* January 1982, *246*(1), 160–173.

Pelz, Donald C. Conditions for innovation. In Walter A. Hill & Douglas Egan (Eds.), *Readings in organization theory*. Boston, Mass.: Allyn & Bacon, 1967.

Taylor, Donald W. Age and experience as determinants of managerial information processing and decision-making performance. *Academy of Management Journal,* March 1975, 74–81.

Weiner, Norbert. *The human use of human beings* (2d ed.). Garden City, N.Y.: Doubleday Anchor Books, 1954.

2
THE MODEL

Case Study
THE CASE OF THE RELUCTANT ORGANIZATION

CAST

Paul Sliffer: Vice president, marketing and administration

Walter Medlock: President and general manager

Maude Finch: Controller and vice president, systems

Michael Moriarity: Manufacturing manager

The background: Maximum Motors Incorporated had a spotty history of cyclical profits. For the past several years, there had been a tremendous growth in its markets because of the increase in world demand for small powered equipment. However, Maximum never seemed to be able to deliver as fast, or as well, as its competition. Its reputation in the field for product quality and integrity was excellent but deliveries were almost always late. Therefore, any growth in sales occurred after competing companies had completely filled their order books, leaving the remainder to Maximum.

A project management system was intended to act as a buffer and expediter between the company and its customers, but deliveries were getting longer and there was more conflict within the organization than ever before. It appeared that the company's systems were not working. Therefore, Walter decided to call a staff meeting of his top people to try to come up with an answer that would work. The meeting started promptly at 10:00 A.M.

Walter: Good morning. You have all received comparisons of actual motor deliveries versus contract delivery dates for this year and last. They seem to be getting further and further apart. What's going on? Every quotation is thoroughly reviewed by everyone in this room and approved before it is sent out. Each of you agrees that the work can be done on time, and yet it never is. Paul, you're our main contact with our customers. What's your opinion?

Paul: Well, as you pointed out, Walter, every contract is signed off before we formally submit it. The budgets are made up and there is another coordination meeting after the contracts are signed, so we seem to be starting off right. It's what happens next that is frustrating. About two or three months into the project, my project managers begin to lose control. Engineering seems to take forever to answer our customers' questions, and after we get those answers, the cost estimates that are provided by accounting if there is a contractual change take too long to be of any use. We have to give the customers answers promptly. If we don't, they send their expediters in, and that often really interferes with our work. Our customers claim that we're not responsive to their needs. Then when they agree to pay for changes, there are additional charges because of rework that our delays caused and that never should have happened in the first place. In fact, one of my best project managers just quit because of frustration.

Michael: Look, Paul, we are trying to work with your project managers, but they seem to have a million questions about every little detail and they insist on written answers to everything. Your people are the real causes of the delays. My engineers are trying to cooperate, but we only have so many hours in the day and if you involve them in answering questions, they can't concentrate on engineering, which is their real job.

Walter: Well, Paul?

Paul: Walter, Michael knows that our business is based on supplying specific motors for specific applications. We don't sell from a catalog. Customers always have had questions, and a lot of our reputation has been built on producing to meet those customers' needs. The past few years have seen a tremendous rise in competition. Our customers are demanding more service and getting it from our competition, but the engineering department still treats customer inquiries as annoyances rather than as opportunities to provide service. Time has moved on and it is no longer adequate to propose a design, negotiate a price, sign a contract, and deliver exactly as proposed about a year later, as we used to. Things change between contract signing and delivery of the hardware, and we have to adjust our organization to respond to those changes or our problems are going to get a lot bigger.

Michael: Just a minute, Paul. My people just follow company procedure. They answer questions, and one of the reasons that we have maintained the reputation that we have is that every change is thoroughly documented and checked before being implemented. We've worked that way for years and we haven't had some of the failures that a couple of our competitors have had. We may be slow, but we're sure. Perhaps, Paul, some of your people can answer some of the obvious questions that have no major effect on safety or function themselves. That would save time.

Paul: We tried that, but your project engineers then refused to sign off on the changes that we accepted. For example, they insisted on running new load tests on some hydraulic motor piping changes that the project manager had approved, even though everybody knows that those pipes are overdesigned. It was just a waste of time, and then the accounting staff insisted on adding the extra costs of the tests to the contract instead of charging them to quality control or overhead.

Walter: Well, it looks like this attempted solution isn't working either. Does anybody have any suggestions? Maude, you're in charge of costing and general systems. What can we do to allow more flexibility?

Maude: We've tried to change our systems before with very little success. Perhaps things are bad enough now that we can get something done. Perhaps a study committee can handle it. Our auditors insist that all financial changes go through the company books, and they always want to know who made the decisions about a change of any scope. Another thing—our engineering and accounting systems have kept us out of trouble. Remember the problems that the pump division had with the government several years ago because of inadequate engineering and cost data? Walter, you've been here about two years now and you know how difficult it is to change the way that Maximum Motors does things. We have familiar ways of doing things and it's tough to get any changes made around here.

Walter: OK. I think the study committee idea is great—and you three are it! I suggest that you get started by picking your own committee chairman and deciding on an agenda fairly quickly. We'll get together next Thursday. This meeting is over.

QUESTIONS

  1. a. How would you determine the major reasons for Maximum Motors' problems? What would you say they were?

    b. Do you think that Walter can effect a change in the organizational culture? How? Why do you think so?

2. a. How would you write the agenda for the meeting if you were selected as the committee chairman?

    b. What would the problems look like from each of the participants' viewpoint? How do you explain the differences?

3. Why do you think these three managers were selected to be on the committee? Would you have changed the committee's composition? Why? How?

4. What kind of decision-making technology is being used here?

5. How does certainty or uncertainty affect decision-making here?

6. What examples from your own experience seem to apply to this situation? How were they resolved? Do you believe the results would have been the same if you had been an active participant, knowing what you know now? Why?

INTRODUCTION AND REVIEW

In Chapter 1 we learned about situational theory and described how it could be used to help solve problems in managing technical operations. However, that chapter was primarily about you, as the manager of those operations. This chapter is about the second part of that theory, the situation, but you, the technical manager, are not forgotten since you are also a part of the situation.

To review briefly, I suggested some basic tools to use in the development of your personal management micro theory within the situational macro theory. The first step was definition and started with a search of existing, applicable management research. That search resulted in the selection of appropriate foundation methods for making decisions by absorbing uncertainty. The methods were prescriptive, but they were based on descriptive findings dealing with both the logical and the creative parts of technical operations.

THE MANAGER IN THE SITUATION

We started the development of personal theory using the scientific method as a preliminary development tool. That method is:

1. Define the terms and set up a general theory or explanation.

2. Develop appropriate hypotheses.

3. Test them in your situation (similar to on-the-job experimentation).

4. Depending on the amount of variance between the results received and those you expected, either revise the hypothesis and possibly the theory or, if the variance is within tolerable limits, allow them to stand.

There were two alternatives in that evaluation and revision section (see items 3 and 4 above) of the process when the data did not match the hypothesis. One alternative was to discard the data. ("That will never happen again." "The data are no good because the test tubes were dirty." "The boss won't like these answers so I'd better try again.") The other was to revise the hypothesis and possibly the theory. ("I guess I was wrong when I expected everybody to work harder because we gave them an across-the-board raise." "Boss, making them show up at a stand-up meeting at 7:00 A.M. Monday morning doesn't seem to be supporting your requirements for greater innovation in the design group. It's just causing resentment.")

The second alternative is often dangerous because it can disturb the existing status of both you and the organization. However, revising personal theory is probably the most important ongoing management task, since these revisions and our behavioral interactions (or, if you will, implicit hypotheses) on the job enable us to learn through corrected practices. They teach us how to handle different or expanded job responsibilities.

For example, the personal theory that you used when you were a draftsman must change when you become head of the drafting department. The duties and responsibilities have been changed and the resulting behaviors must also change. This applies even if there is no obvious change in formal organizational status such as an announced promotion. It's an ongoing process because situations in organizations are always in a state of change. Our perception of that change defines the potential modifications that must take place in our personal theory if we want to respond to the best of our abilities.

Changes occur as we modify our cognition and/or our behaviors. It's an interaction similar to a two-way street. When our attitude or cognition changes about something, our behavior toward it also changes. Similarly, when we change our behaviors, our thinking is also modified. There are both logical and emotional elements in this two-way modification process and the logic is not always separable from our emotional thinking. For example, it is very logical to read the management or technical literature (Why reinvent the wheel?), and selectively to add the new logical data we have learned to those acquired by watching others on the job. We know and can avoid the problems with on-the-job training, so we use only those parts that apply specifically to us. But logic is fallible, because human beings often use other criteria when they decide to stop searching. It is almost an emotional decision at that time. We know that it is emotionally harder to keep searching for answers when we have found one that is satisfactory, even if we know that it is not optimal; i.e., "satisficing," by Simon's description (Simon, 1957, p. xxiv). So human beings do stop. It is similarly logical to make only those decisions that are in our area of responsibility (i.e., to make decisions under our limits of uncertainty), but we can also decide emotionally to extend the situation and temporarily take charge of other organizational responsibilities. ("This part has a crack in it, so even though I have nothing to do with quality control in my company, I'll bring it to the attention of that group.")

These two obvious examples involving logic and emotion that can either limit (i.e., by satisficing) or expand our mental processes indicate how tightly both logic and emotion are intertwined in the learning and adaptation processes that we use to improve or modify our personal theory. To try to separate them won't produce distinct elements for analysis; it produces nothing, since these elements do not stand alone. As human beings, we are total systems with each part of our system contributing to and being modified by the other parts. The parts themselves are incomplete when examined separately, since the interactions among them are not then defined.

For example, when the person is "disassembled," where is the mind? Does it reside in the skull? How do you find it? How is it related to the body? Is there a real difference? (This, as you can imagine, is not a particularly novel set of questions. Philosophers and psychologists have worked on it for many years.) Personal theory is a type of gestalt or whole theory that integrates all the parts into something else. It "starts from empirically observed general principles of phenomena . . . and deduces from them results of such a kind that they apply to every case that presents itself" (Helson, 1973, p. 79). Personal theory, therefore, must be modifiable by new data, and this modification process contains both cognitive and behavioral variables. It is a part of your gestalt, the whole you. It guides your actions and thoughts when you are acting as a manager.

In this book, I suggested that the major rationale for management itself was decision making. I believe that this must come before the usual management goals of achieving results, continuing growth, or anything else. And those decisions are expected to be nonrepetitive and involve modifying people's behaviors. Otherwise you are no better than a computer using machinelike algorithms which do not take advantage of the positive emotional elements that you alone (as a gestalt) would use in solving problems that had not arisen before.

Those algorithms, with their inevitable answers, are to be set up whenever possible so that the tasks of the managers would be to make decisions under conditions of risk or uncertainty. Making decisions under risk was defined as multiplying some subjective distribution of potential answers (absorbing uncertainty in defining the distribution?) by the proposed value of those answers. Finally, this human decision-making process moved onto handling uncertainty, which was defined as absorbing some unknown but existing frequency distribution of potential answers. And, finally, the amount of uncertainty to be absorbed is supposed to match the level of responsibility within the organization.

DO WE EVEN NEED A THEORY? CHALLENGE AND RESPONSE

I have suggested that the theory- and model-building process is one of the better ways to learn how to manage well. An opposing viewpoint seems to consider every situational challenge as almost novel, requiring novel responses that are rarely based on any predetermined management theory. It is a very pragmatic approach. Therefore, the need for a scientific or any other kind of predetermined approach based on one's own theory does not seem to be justified. It suggests that you should not be limited by any one set of concepts; just look for the answers to specific problems whenever they occur.

That viewpoint could be useful if we were all the well-qualified, creative, responsive managers we would like to be. However, many of us are not, and since some situations do repeat themselves—even if they don't do it exactly the same way—they have common elements from which we can learn. I therefore find it prudent to use some personal mental framework that has worked before or appears to be able to work in the future in problem solving and decision making. It's always useful to have some point to start from. In other words,

> The challenge and response approach takes a pragmatic outlook on management. It does not attempt to build a consistent framework of thought or knowledge but stresses that management is a practice which should employ all ideas developed by sciences and arts to increase achievements in business performance. The manager is challenged by the situations in which he finds himself and must seek answers to his particular problems unrestricted by any single conceptual framework. This approach . . . does not itself seek to discover generalizations; its orientation is not toward the theoretical or scientific increase in general knowledge, but toward the answers to specific problems faced by managers. (Massie, 1965, p. 417)

An example of this type of challenge-and-response approach is typified by a remark by Drucker. "One great mistake management makes is to look things up in a book. Management is a diagnostic discipline. One does not first read a book. One first looks at the situation" (Drucker, 1981, p. 12). I agree, but with some modifications. This approach seems to assume that management, in general, should not use the documented generalizations of others in the past as a basis for future actions. Even pragmatism has its limits. Why repeat the mistakes of others when you can learn all about them beforehand? (We're all creative enough to make our own mistakes.) This approach might even be justified if management were all art and little science. But even art is rarely totally dependent upon the person's innate abilities. More often, it depends upon abilities *and* extensive, successful experience. Let's not exclude the genius artist entirely. One *might* be born—but even Michelangelo spent time as an apprentice learning basic stone-cutting techniques from others.

I believe that managing technical functions requires even more training, because of the obvious requirement for a basic knowledge of the physical sciences and of human psychology. It is another example of the bilateral design that requires the manager to know both logic (the physical sciences) and art (the social sciences). Logic uses plans; and plans (by definition) require some predetermined theory behind them. The challenge-and-response approach may have more validity in functional management areas that are not concerned with technical specialties because there might be fewer requirements for the primarily logical, rational (and theory-based) management style in those other areas.

Another reason for personal theory building, briefly noted before, is that even though the situation may be unique, there are always parts of it that have occurred before. For example, new-product development is generally based on some established product line and/or market base. Therefore, neither the challenge nor the response can ever be *completely* unique, and in technical operations, we make few major leaps of faith. Usually, each step is tested thoroughly before the next one is taken. Nothing is completely new. Therefore, I suggest that iterative theory building and testing—logic first and then emotion, in a never-ending iteration—is the way to succeed as a technical manager, and we should expect continually to develop and modify our theories as we gain more data and more experience.

THE SITUATION

How It Originates

We now begin to define and develop concepts that apply to the less independent group of variables in the double set including you, as the manager, and the situation: We'll do this by outlining a total organizational model of the situation from the inside out. Most organizations are not built this way, but instead depend upon the limited viewpoints of very few people looking at the total picture. That's a precarious approach; most good suggestions for improvement originate closer to the actual level of operations (Haberstroh, 1965, p. 1184). Since we expect to achieve an optimum match between operations and need (in the total situation), I suggest that the major group to assist in the design of your model are the "experts," the participants, or the insiders in your particular organization. They are not usually consulted, and that is definitely an emotional (and very irrational) decision.

Defining the Organization

We start with an operational definition of the organization and its purpose, then move along to the general model before concentrating on the specifics of our own bilateral technical function. The components of the model are defined briefly in this chapter but explored further in following chapters, as are the recommended methods for both repetitive and nonrepetitive decision making. Then we will combine the specific situation and you, the technical manager, in order to look at applications of situational theory that are useful in building your personal operating hypotheses.

As I mentioned previously, an operational definition is determined by the operation itself (e.g., the intelligence level of a person is defined by the score that he or she receives on an intelligence test). Since it seems to fit best, we will use an operational definition to define the model, then describe its uses by others, and finally offer prescriptions for you. That description and prescription design process also starts with the relatively independent variables and moves on to those that are supposed to be more dependent. The sequence of the definition and description of the model, therefore, is "people," "organizational structure," and "technology." (See Figure 2-1.)

For our purposes, the structure of organizational models can be defined simply as "people who have fairly stable, regular sets of activities and in-

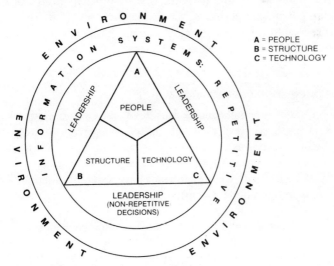

Figure 2-1 Organizational model.

terrelationships" that have expected outcomes (Litterer, 1965, p. 6). Organizations may be said to exist because of the unreliability of the behavior of an individual human being and the consequent need for predictable, stable behaviors in people. Interpersonal cooperation and consequent group behavioral reliability are therefore the central reasons for any organization. Organizations support (and coerce) consistent and well-defined activities. However, organizational purposes often are neither consistent across various groups of individuals nor congruent with the purposes of the people in those interrelationships. Instead, they are more likely to be a complex arrangement of goals, activities, and relationships in which there may be competitions among objectives (Haberstroh, 1965).

But organizations do provide a constancy of existence and purpose greater than that of any individual. Therefore it seems likely that in addition to dealing with the unreliability of the individual, organizations also provide some support for the groups in them which must be greater than the loss caused by intergroup conflicts. Otherwise there would be no reason for the organization to exist. In other words, it's not an internally optimal object but it does result in a greater return to the organizational participants than they could obtain by themselves. The expected outcomes obtained by joining are greater than those obtained by not joining. (This definition is a bit circular and not very exact, but it is still one of the better kinds of definitions that we have in the social sciences.)

Now that we have a definition, the next step is the design of the model. That model should be able to minimize conflict between the needs of the organization and those of the individual. It should be usable as a basis for building the personal hypotheses, which can then be tested. Therefore, we will now proceed to develop a descriptive model of the situation and describe and prescribe some concepts to use as initial hypotheses.

PERFECT AND NOT-SO-PERFECT MODELS

There is a classically simple method of developing a model that was first proposed during the end of the last century (Weber, 1949). This method was to develop a description of an *ideal type* of organizational situation, then measure actual discrepancies from that ideal in order to correct the situation that exists. It is similar to the procedure I have used throughout this book. (This is *really* an operational definition!)

> An ideal type is formed by the one-sided accentuation of one or more points of view and by the synthesis of a great many diffuse, discrete, more or less present and occasionally absent concrete individual phenomena, which are arranged according to those one-sidedly emphasized viewpoints into a unified analytical construct. (pp. 50ff.) It is a matter here of constructing relationships

which our imagination accepts as plausibly motivated and hence as "objectively possible" and which appear "adequate" from the normological standpoint. (p. 92)

The description becomes almost a tautology, but it is easily understandable when you think about it as starting with the best model that you can, defined as well as you can, and then using whatever you have to improve the situation as you find differences between the best model and actuality. It is also possible, of course, to modify that best model if the situation is really unchangeable.

The model of the organizational situation that is proposed as the initial "idea" has three major components—*people, structure, and technology*—and two types of managerial decisions—*repetitive* and *nonrepetitive*. That organizational model resides in a much larger situation called the *environment*, which was covered to some extent in the introductory chapter. In turn, the smaller model of the technical function within the larger organizational situation looks the same and also combines the three major components and two types of decisions. It is like the overall model, just a bit smaller and with slightly different emphases on different components. The components and the decisions are held together with the organizational cement called leadership.

Both the importance of the particular component within the model and the relationships of the components to each other may change, thus temporarily changing the model's shape. The neat, equilateral triangle shown in Figure 2-1 may become acute or obtuse, but the geometry is not that critical. The triangle may even change to another kind of geometric shape if you want to consider other components you think important. The shape is not as important as the idea of having some kind of mental model. Eventually I expect this model or any other model you choose to be changed into another one that may vary, depending upon your design ideas for both the components and their relationships among each other.

This model follows research that suggests how the organizations work and adjust their components to meet environmental changes. That adjustment (Cyert & March, 1963) is proposed to be driven by a humanlike desire to avoid uncertainty through a biased, overall, simpleminded search for a satisfactory short-term solution to organizational problems. Although internal conflict among participants is rife in this organizational model, because of differences in personal theories, that conflict is supposedly minimized through a sequential attention of the total organization to differing goals and the forming of shifting coalitions of participants to share any rewards that the organization gains as problems are solved.

The organization is supposed to adjust itself to avoiding uncertainty as a person does to avoiding pain. I reduce the pain in my head by taking an aspirin and ignoring everything else. The stomach pains are temporarily overlooked until the headache subsides. Presumably, the stomach has formed a temporary coalition with the head not to make trouble till the head pain is reduced. It's just an analogy but these mechanisms are proposed because the organization itself is considered to be personlike in its capacity to learn as a result of experience. There does not appear to be any major conflict between the descriptions in this research and parts of the descriptions of individual behaviors in attempts at self-optimizing. This description of the total organization treats uncertainty as something which should be avoided, like pain to a person. However, we know that real organizations do not make decisions and absorb uncertainty as the theory above suggests, but individuals can and do. Our model, therefore, is only a stage upon which, or the situation in which, the manager acts.

THE SYSTEM AND CHANGES

The model of the technical organization with its three central components of people, structure, and technology is, by definition, a system. That definition of system includes an overall purpose (or purposes) that is different from any sum of the system parts. It also includes a description of the limits between the system itself and its environment. We have discussed how systems are partially similar to human beings in being gestalts and how analysis of them through disassembly or evaluation of separate parts does not provide answers about the overall purpose. It does provide, however, some very general indication of the functions of those various parts that might allow us to deduce the overall purpose. We'll never be sure of our answers, but at least we'll have more information than when we started and a firmer foundation for our consequent speculations about the system's intended and actual overall purposes (they may be different, you know).

Therefore, keeping in mind the limitations of our analysis in order to understand, control, and modify this system, we'll make the useful simplified assumption that each of the three components, their interactions, and the system's environmental limits can be described and prescribed independently of each other. We'll also assume that our model is ideal. With these assumptions, we can start our analysis with a brief definition of the environmental limits of the model—the line at which the organization stops and the environment begins.

No organizational system is totally independent of its environment. Therefore, the limit is always slightly artificial. There are few nonquestionable limits to organizations, just as there are to people. In both cases, there are always additional external factors that could justifiably be included. For example, assuming that a person is a system, if we were to ask that person to define himself or herself, the definition would probably depend upon the person's own peculiar viewpoint. Even outsiders would have difficulty with a definition. The professional anatomist might limit that definition to the physical body lying on an operating table. An economist might extend that definition to the consuming and producing actions of the living person. A theologian would have some ideas about less tangible concepts. Finally, the person might include many elements involving social and political points of view.

The system limits would be in the eye of the beholder. Therefore, in order to have a starting point for our system, we must, understandably enough, have somewhat arbitrary limits. But the artificial definition of the model "limits" is only part of the definition problem. Consider the rest of the model:

- Limits (as discussed above)

- Components

- Interactions of the components with each other and with the environment as we have defined it

The other part is the need to dissect our model for analysis even though we know the dissection process can give us only a limited glimpse of the model's interrelationships. We shall therefore use the phrase "holding all other things fixed" implicitly, if not explicitly, from now on in our definitions and discussions, although that is really an impossibility.

As an example, let us consider the component "structure" in a small R&D company that has the informal organizational practice of issuing purchasing requirements by telephone. We can define that structure as the repetitive human behaviors exhibited in the group and attempt to describe and predict possible changes when the company is purchased by a large conglomerate. Asssuming that this structure will change into a formalized, documented requisition-based, three-quotation process and that a printed purchase order department will be formed when the company is acquired, we can count on equivalent impacts on the other two organizational components of people and technology in the model, even though the acquiring conglomerate seems to change only a few minor internal operating procedures.

However, we shall still implicitly assume "all other things are fixed" as the initial step. We do this as an aid in learning. It simplifies the problems and prepares us all, as learners, to understand these easier ideas before we tackle the real complexities of our organizations and the world. The medical student studies the cadaver, knowing that it is not the human being that lies there, but a part of the human being that must be understood if the student is to become a physician capable of helping live human beings later. No matter how many dissections the medical student who becomes a physician participates in, he or she examines the patient who comes into the medical center as an individual. The individual is a gestalt, more than the sum of the parts. Understand the components of this organizational model so that you can understand the general way technical organizations function. Then you can compare your real, existing organization with this model and eventually redesign that organization to be what you intend it to be.

THE CHANGES AND TIME

Another uncontrollable condition affects our understanding of the model. Changes occur in it as time passes. Some researchers suggest that these changes can be forecast, controlled, and optimized if there is some sort of balance among the various system pressures (including time) that cause change (Kotter, 1978). But for such a balance to exist, there must be an implied state of optimal equilibrium from which change moves the system or model and toward which it should be moved. It's as if there were one best balanced goal to be achieved. This seems to be a very limited view of the situation for the following reasons:

1. *The model we have selected may not be completely representative.* Remember that our model is not the only one possible. We are never really certain that all the relevant variables have been included.

2. *Multiple changes impact every organization, and it is impossible to react to them all.* Management reactions to change may seem to be sequential and almost in a fire-fighting mode (Cyert & March, 1963). If things get out of balance because of multiple changes (e.g., the chief engineer has just resigned, the law on patents has just changed, there is an organizational dispute about new vendors, *and* the data processing division has just suffered a disastrous fire that wiped out all the accounts receivable), which of these changes is to be handled first and how does equilibrium occur if one problem is fixed and another pops up? (When "Murphy" strikes, he rarely strikes in any single place or time, except, of course, if that place and time happen to be the most crucial.) The more apparent questions are what the organizational equilibrium is supposed to be and which of the multiple changes will affect the move toward equilibrium and which are to be ignored.

3. *Equilibrium may not be a desired organizational state.* In your position as technical manager, your major task may be to produce the innovative products and services that continually move the organization into a different relationship with its environment. In effect, your management task could be to maintain a relatively *constant state of disequilibrium:* perhaps balance or equilibrium are not even desirable. You may even

be aiming at a moving target. Additionally, equilibrium for one section of the company is not the same for all the others. Manufacturing managers may be pushing for long production runs to minimize costs while sales managers may be pushing for short runs to meet diverse customers' short-run needs.

4. *Time automatically changes equilibrium.* The model in this book is explicitly defined as relatively static because it is easier to describe and the variables have to be limited in some fashion. That's the way books that are intended to teach are written; but time does affect the organization (and even the contents of books), causing new states of disequilibrium.

Therefore, the concept of change that introduces organizational disequilibrium as a state to be corrected by the system being moved to an optimum state of equilibrium seems to be even less relevant than our less-than-perfect method of trying to understand the total system by analyzing its parts. That equilibrium concept implicitly assumes much more than "holding all other things equal"; it assumes a definable (and thereby fixed) optimum equilibrium for the whole system. There is no such thing. Our model, however, is intended to be analogous to the ones that Weber proposed; it is supposed to be as perfect as it can be. You may have different ideas. We hope so. That is what we are trying to foster.

MODEL COMPONENTS

The People

We start the discussion of the first component, people, with the reasons for people being part of the model at all and how they are tied into it. Then, we'll define people. Of the three components (people, structure, and technology), people is the only one that can exist wholly outside of this model. The other two cannot. Although it may be obvious that the component people is connected to the model through the work they do, the understanding of how that connection is defined has not always been fixed. It has changed over the years from an absolute requirement for hard, grinding physical effort into the creative mental tasks of the modern knowledge worker.

It has an interesting history that seems to have started in high places. For example, "In the sweat of thy brow shalt thou eat bread, till thou return unto the ground; for out of it wast thou taken: for dust thou art, and unto dust shalt thou return" (Genesis 3:19). Now that is grand, and almost overpowering, descriptive and prescriptive theory! It covers a lot of ground, since it ties work to both life and death in a very general working hypothesis.

In perhaps less majestic prose, the following author seems to agree that this working hypothesis connecting work, life, and death exists even today.

> The first time that I ever considered suicide in my life was in that furniture factory as I would stare at the clock and think to myself, "If I have to spend the rest of my life in this hellhole, I would rather end it." Well, of course, I didn't; and as I look back, it was not the worst place I worked. But I was young" (Schrank, 1978, p. 9).

That's quite a change over the years, and even though it would seem that work, as a repetitive behavior pattern, has a consistent, long, and important history, within recent times it seems to have become less important in most of the industrialized Western nations. To be more specific, an inverse relationship almost seems to exist between the amount of physical, "sweat of the brow" work and the degree of industrialization. Work in general and, most important to the readers of this book, technical work have become primarily mental activities. Producing workers are now knowledge workers. (There are those who may say that this kind of work is just as hard, and the sweat of the brow now is psychological, but I've taken a more literal interpretation. Besides, one person's mental sweat of the brow may be another's nectar and ambrosia.) Therefore, even though the work may still be very difficult, that difficulty becomes something to be controlled by the knowledge worker rather than the manager. For our purposes, since this work is controlled primarily by the worker, that worker is the expert who controls its quantity and quality. As a technical manager, you are concerned with assisting in optimizing the efforts of these experts.

Now that we have some rationale for people being in the model, we may move on to define them. The people in the organization are those who contribute (work) in some predictable, cooperative fashion to achieve some organizational purpose and then repetitively share in the results of that contribution. The word "repetitive" is included in order to limit the component to the employees of the organization.

We have defined the connection between the model and the people component as "work" but the definition of *how* that work gets done is a bit more complex. I have suggested that the quantity and quality of work that people do depend on them and, in turn, two major attributes of each person: *abilities* and *motivation*. These are not the only pertinent attributes, but they are assumed to be adequate for our purposes. (See Tiffin & McCormick, 1965, for others.)

Abilities can be defined as a combination of innate human capacities and all the conditioning the person has experienced. Since we are concerned primarily with the technical organization, we shall assume that participants in that function have already demonstrated their abilities through some type of academic achievement and experience. These abilities, therefore, are fairly well-defined for each person, if you accept my terms. Motivation can be defined as an inferred mental selection of a particular behavior from those perceived to be available by the person. This subject deserves more attention. Chapter 3 provides that attention and includes descriptions and prescriptions concerning human motivation.

The Structure

The definition of the next component, organizational structure, starts with the arrangements that are explicitly intended to regulate or control the behaviors of the organization's employees. Formal arrangements typically could include job designs, reporting hierarchies, and directions for departmental goals.

Structure also includes some less formal arrangements that are nevertheless an equal, if not occasionally stronger, regulation of behavior, such as social arrangements (relationships among people based on power, influence, affiliation, trust, mutual needs, culture, etc.). These might stem from the relevant norms and values shared by most employees. Culture is defined as shared ideas that are of long standing and seem to exist throughout the organization. When the ideas are subject to more rapid change and apply to limited groups within the total organization, it's called the informal structure. As you can gather, there are no exact or sharp demarcations between culture and informal structure.

By comparison with either culture or informal structure, the formal structure is comparatively easy to understand and assimilate, especially when it is documented and in accordance with actual operating relationships. However, that is unlikely. Rarely does an up-to-date set of organization charts exist and even when such charts do exist, they are unlikely to reflect accurately and completely what is happening in that company. Moreover, the definitions of the culture and informal social structure are even more difficult to grasp than that of any formal structural design because there are fewer organizational guidelines available. For example, when starting with an attempt to understand the general organizational culture, you may find some broadly understandable commonalities across particular industries. Then there are unique cultural elements for particular companies within those industries.

Consider the cultural similarities in the original development phases of the aircraft manufacturing industry founded by men who were fliers first and managers second. At that time, the culture of those organizations was oriented strongly toward innovative, highly engineered product lines that pushed the state of the art in flying machines. Typical executive officers of those times were Igor Sikorsky and Glenn Curtiss. These men had personally designed and piloted their company's products. The culture for the aircraft manufacturing industry rewarded risk takers and innovative designers.

The culture in the Sikorsky and the Curtiss organizations differed only because of the effects of the markets they served. The Sikorsky organization was modeled to respond to commercial needs and Curtiss to defense needs. But time affects organizational cultures, and they do change. Today, although the Sikorsky Aircraft Corporation is still involved primarily with the development of commercial helicopters (although as a division of a much larger corporation, which tends to limit innovation a bit), Curtiss-Wright Corporation seems to be moving into an investment type of organization and away from its original aircraft culture. In that latter company, the cultural change has been great indeed.

When the organization's culture is congruent with the economic environment and its formal strategies match that culture, it achieves major strengths. For example, Delta Airlines concentrates on customer service and maintaining a constant work force. It fosters teamwork and independent decision making. Employees will even substitute for each other in order to keep baggage moving and planes in the air. This happy situation occurs when the organizational culture (including the informal structure) reflects an overall congruence between the intent of top management (i.e., the formal structure) and the interests of the company's environment.

When there is a separation between them, there are fewer risk takers in management. There is less desire to absorb uncertainty and make decisions. Managers then begin to avoid conflict and instead attempt to please and gain the trust of their own bosses. Managers who have to face these kinds of organizational realities often find that the economic environment doesn't always respond as before, and while the boss's trust is important to keep a job, it is a poor consolation when the whole company goes under because no one is willing to make appropriate and independent decisions. Chapter 4 continues this exploration of structure in a more detailed fashion; we can get effective prescriptions there.

The Technology

One definition of an organization is of a purposeful system that combines materials, knowledge, and methods in order to transform various kinds of inputs into valued outputs. Therefore, the technology of the organization is defined as the processes used for this transformation. Several other definitions of technology assist in classification and consequent understanding. One of these concerns the way that managers make decisions (Thompson, 1967). Since the purpose of our organizational system is just that, this definition could be quite important if our organization's input is problems and the output is solutions. It is probably a good starting point for understanding the actual decision making that I have suggested: the inferred mental process of absorbing uncertainty. Thompson's general typology of technology includes three general, but related, classifications: *intensive, mediating, and long-linked technologies*. Each has its advantages and disadvantages. They offer different ways to employ the overall decision-making technology of converting inputs (problems) into outputs (solutions). They are all used in various kinds of technical operations.

Intensive technology is defined as a variety of techniques intended to change some specific object, with the object of providing feedback for the change. The object itself determines the combination, selection, and order of application of the change method. For example, there is the design and construction of a chemical processing plant where the terrain, the weather, and the interactions of the various labor unions on the site often determine how and if the project will be completed within budgetary constraints.

Intensive technology can be used when the organization has few ideas about how to solve a particular problem. There is maximum, but obviously not complete, uncertainty; and there is very little possibility that algorithms can be developed for a partial solution which might leave the remainder of the problem to be handled under conditions of uncertainty. Intensive technology might be called "therapy" when it is applied to human beings.

In this case, potential solutions can be developed by forming a team of specialists from differing technical areas who can develop a high interdependence with each other. The team members provide expertise in their own areas but are expected to coordinate with and adjust to each other, thereby producing an acceptable solution that no one of them could produce alone. This technique is often used in top management decision making or research and development problem solving.

Within recent years, novel pressures for change coming from unforeseen external environmental elements such as consumer protection agencies and other political bodies and the unavailability of algorithmic types of responses to these new types of changes have resulted in many more or less permanent organizational structures based on intensive technology. Some companies have formed ad hoc product and process technology review committees, for example. These committees use project or matrix teams to work on tasks that are not well-defined, that have high levels of uncertainty. Management feedback as the tasks are accomplished is vital because of the existing need for companies to respond to many external groups besides the usual customer and stockholder groups.

Mediating technology is defined as a variety of techniques that operate extensively but in standardized fashion with multiple clients or customers. It links individuals and groups that wish to become interdependent. For example, computer and accounting departments provide information to and process data for operating departments. All of these operating departments expect to be serviced and treated equally, and data are usually distributed in accordance with predetermined rules that are known to and accepted by the users.

The technology defines the warehousing points for data and the transportation medium for these data among the users. The technology links are known, and the use of this technology does not change the information or other materials being processed by those links. Accounting departments are classically supposed to use this type of technology. If their actions are defined correctly, the technology is similar to that of a pipeline company processing and distributing the materials in the pipeline but not changing them. This technology's hallmarks are standardization in products and services. There is an impersonal professionalism about it that seems to be quite typical in bureaucratic structures, since those structures are intended to provide the same output for the same input on a consistent basis.

Long-linked technology is defined as using serial processes (a process can occur only when the step that immediately preceded it has been completed successfully). In other words, action C can only take place when action B has been completed, and that action depends, in turn, on the completion of action A. This technique is used when the organization thinks it knows how to produce some desired outcome and the total task is broken down into a series of sequential steps. There is a planned flow of work, and the performance of each step depends on the completion of the steps before it. This technology has become very popular within recent times as the emphasis on, for example, long-range planning and project scheduling has increased, and most manufacturing processes do have such an emphasis. Typical examples are automobile mass production and chemical processing in petroleum refineries.

Each of these technologies is useful in different organizational situations or models and the technologies are not mutually exclusive. They only suggest which decision-making technology alternative should be selected first before the manager can begin to deal with uncertainty. Selection of the best technology precedes uncertainty absorption attempts. Of course, as noted before, those attempts begin when:

1. The problem rightly belongs to the particular manager.

2. No algorithm exists.

3. The manager has decided that this problem should not or could not be delegated upward, downward, or even sideways.

These are some general prescriptions for the selection of better decision-making technology. Choosing intensive technology might mean setting up a task force. This task force would probably use some form of nonstructured technique such as brainstorming or Delphi forecasting (Makridakis & Wheelwright, 1978, p. 472). If mediating technology is chosen as most appropriate, the decision might include the development of a special reporting system that feeds data back quickly enough for many other, smaller decisions to be made, thereby removing the requirement for a major and expensive corrective action at the end of an extended time period. An example that fits this type of technology is using inspection data from the incoming inspection department to determine the acceptable quality levels of the products of some new and untried vendor. Finally, long-linked technology is selected when there is a concern with manufacturing processes. It applies even to small quantities of manufactured items such as tooling and special-purpose production equipment made in a fairly well-defined sequence. Moreover, the design of products often follows this technology. In an engineering department, the design will follow something typically called the standard bill of materials. That relatively standardized design sequence defined parallels the methods of assembly of the final product.

Choosing the best technology to fit the particular situation using Thompson's categories is almost as complex as determining the optimal motivational patterns for personnel or developing the best organizational structure for a company. The general prescription in this case would be analogous to that noted in the previous chapter: Select the least complex technology and methods before going on to another. The least complex and most algorithmiclike technology is long-linked technology. The next is mediating technology, and the most complex is intensive technology.

Of course, this sequence assumes that the situation itself has no obvious clues to draw upon. For example, setting up an assembly line (i.e., long-linked technology) in order to solve a problem in marketing (i.e., intensive technology) is not a reasonable choice of technology. Intensive technology is often a choice that is useful across total organizations, is involved in solving unique problems, and is applicable in creating original products. It should be an integral part of your personal theory. Long-linked technology is covered in Chapter 5 (Technology), since that chapter deals with technology concerning production, and Chapter 6 (Information Systems) incorporates information applicable to mediating technology.

REPETITIVE AND NONREPETITIVE DECISIONS

The organizational triangle model of people, structure, and technology is buffered from the external environment by systems of repetitive decisions. These repetitive decisions are the information-transfer processes through which the organizational model and the environment usually interact, in addition to communicating among components. The transfers therefore carry the greatest numbers of interactions among the various parts of the model.

Since those interactions usually begin with some standardized inquiry or input, the information transfer process (or management information system) is supposed to provide some standardized response or output for the organization. If that standardized response is inadequate or even inappropriate, for whatever reason, that inquiry must be directed to some manager for a new response or solution. That always happens when the inquiry is both nonrepetitive and important. (This is fairly familiar ground by now.)

Repetitive Decisions

Most systems that are developed to solve repetitive problems are intended to be closed-loop, rather than open-loop, systems. Closed-loop designs occur when some type of feedback from the output automatically affects or controls the input. In this design, minimal independent intervention is required to see that the action taken either has occurred or is appropriate. Management has set the standard and determined the tolerance from that standard the action may have. The control of the feedback loop that is built into the system corrects the situation.

The classic example of a closed-loop system is the heating thermostat that turns the furnace on when the temperature drops below a predetermined point and (through feedback of the temperature rise affecting the heat-sensing elements of the thermostat) shuts it off when the amount of heat is satisfactory. A variance occurs because of temperature overshoot both when the heat is turned on (because the room is cooling down and the temperature goes through the turn-on point on the way down) and when it is turned off (as the heat builds up and the temperature on the upswing goes through turnoff). The variance in both cases has been accounted for, and the system continually "hunts" for the right answer within predetermined variance limits.

In another kind of system, known as open loop, the results of the comparison against the standard are stored somewhere, and this implied stored standard and other controls affect how the variance is used. For example, when reports showing deviations from budgets are delivered to managers and they take or do not take action on those deviations, an open-loop system is in use. Reports are just indications of the change in the room's temperature as in the thermostat example above. The manager (i.e., the thermostat) is not like the thermostat control in the closed-loop system, since he has the choice of taking some kind of decision or not, or even of letting the whole system fail.

Dynamic or Concurrent Information Systems

When feedback data are being collected while a process is going on and management evaluation is proceeding in accordance with predetermined standards, the controls are dynamic or concurrent. Most closed-loop systems have these controls. An example is the use of statistical quality controls in manufacturing, where measurements of the deviations in process are automatically fed back to correct the particular operation.

These repetitive types of subsystems provide data to every part of the organizational structure that has any contact either with the environment or with another department in that structure. Personnel, accounting, production control, purchasing, and field service are examples of typical departments with many repetitive decisions incorporated in dynamic information systems. Their subsystems are supposed to include rather short time periods between the error or feedback signal and some kind of planned reaction. If the environmental rate of change is increasing, it should mean an equivalent movement toward the design of dynamic, closed-loop systems. One important movement is the proliferation of computers in technical organizations because that kind of equipment is well-suited to the task of providing immediate and dynamic feedback.

Other Considerations

However, all change does not come from outside the organization. Much of it is an internal response to some new demand from the outside, and in some cases it can be even more difficult and intractable. For example, major product failures (i.e., environmental economic input) almost always result in some kind of immediate change within the organization. It may be a change in people ("Who's responsible for this latest fiasco?"), the structure ("Let's increase inspection. Maybe we can pick up the problems here rather than out in the field."), or the technology ("Let's form an ad hoc group to develop solutions to this unique problem."). In this event, the information system cannot be the primary tool for solving these problems since it is intended to handle only problems or changes which have been handled before. It is, however, the first place the problems land.

Changes also occur because of a wide variety of internal positive or negative factors. Typically, an internal positive reason could be the recruitment of a particularly brilliant, intuitive manager who sensitively anticipates problems or reacts to the very small stimuli that he or she recognizes as potentially important. Similarly, negative factors could range from outright incompetence to minor oversights. The repetitive information processes of the model do not apply to these examples.

Adapting Information Processes

Additionally, no information process design can service the needs of all equally (All are not really equal, anyhow; it depends on who was last involved in the process design.) and information processes are often designed to suit the most influential internal and external demands. Consider the rule for computer usage set up at the data processing center that establishes tax reporting ahead of equally vital engineering change notice reporting in the processing queue. In that situation, changes in financial reporting will definitely take precedence over changes in documentation of standard valves for product lists. The financial people are obviously more important to this organization than the technical people. Situations, however, are never static and a rapid drop in product acceptance in the market may result in this organization's engineering people achieving a higher priority than the financial people, at least until the crisis is passed. Therefore, adaptations will always be made for minor information processes, and if these adaptations (or responses to change) are not attended to constantly, the system will invariably begin to suffer major failures after several years. When maintenance and modification of existing information processes have a low priority, trouble in the organization grows.

How to Design Information Processes Badly!

Besides inflexibility and lack of updating, another source of poor design of information processes is an implicit design assumption that it is somehow always possible for positive feedback to result in positive achievements. When profits are reported to be up, there is often additional pressure to increase those profits even more for the coming year. The process is intended to reward more positive achievements and punish negative ones continuously, regardless of the reasons behind these performances.

In a strictly theoretical sense, process stability requires negative, not positive, feedback. Of course, positive feedbacks are not inherently bad. They simply should not be the total basis of any decision making and should not be designed as open-loop processes. However, we can now consider the marvelous thinking processes of some managers when profits begin to go up and provide positive feedback on the company's financial picture. This typical positive feedback could be the one that leads to increased pressure for production with lower budgets and expenditures, such as those for maintenance. When maintenance costs begin to drop and another positive variance occurs between the budgeted number and the actual cost, it is often considered to be a great achievement instead of an indication of impending disaster.

When the catastrophic failure does occur (It always will. Remember, both "Murphy" and Mother Nature rarely take a day off), maintenance is begun again, production is temporarily deemphasized, and the cycle is reversed for a short while. But, unless the failure in the process because of the built-in incorrect emphasis is recognized for what it is, it begins again. Some railroads used to handle their track operations programs this way. Attempting to develop a method of repetitive decision making and designing information processes to assist in making decisions require the inclusion of negative feedback and closed-loop concepts. Easy to say and difficult to do, but we deal with this again in greater detail in Chapter 6.

Why Do We Need People in Information Processes?

Human beings can make decisions in strange and wonderful ways without sufficient data or even without any objectively relevant data at all. (Have you ever wondered what one person saw in another and how that person made those decisions to become married?) Since we rarely have all the data at the right time, there is reason to expect that there will always be a need for managers to fill in the gaps with their mental processes and make decisions. Computers cannot do this, at least not in the same way that human beings can. They can extrapolate, make estimates, etc., but they need definable rules that can be programmed. Either their data must be complete or the decision rules for risk must be programmed before an answer can be provided.

However, there are always many inadequately defined problems that can only be solved by human beings. According to our decision-making rules, those inadequately defined problems are first handled by setting up repetitive systems to screen out the parts that have been solved repeatedly in the past. The remaining parts are solved on a nonrepetitive basis by managers who can and do make decisions under uncertainty, using the appropriate technology.

The organizational model provides for this process because nonrepetitive decisions are handled by people in the parts of the model that are the interstices between the circle of the repetitive systems and the rest of the organization (See Figure 2-1). Examples of this are nonreptitive decisions such as conflict resolution, design trade-offs, forecasting activities, and all the planning processes that precede and support repetitive aspects that typically occur in tasks such as budgeting and forecasting. By definition, nonrepetitive decisions always come before repetitive ones and only people can make nonrepetitive decisions. There have been some description and prescription about the process of nonrepetitive decision in Chapter 1. They are repeated just to illustrate how nonrepetitive decisions fit into the model.

SUMMARY

Using situational theory to analyze some of the data that are available about technical management, we divided this general body of knowledge into two major sections: people and the situation. The technical manager, so far as this analysis is concerned, is one part of the people section, which is the relatively independent variable. The manager's main purpose is to act as a decision maker, categorizing problems according to certain rules that eventually include absorbing uncertainty.

The other section of situational theory is the situation. That variable—the situation—is the one that is relatively dependent in this type of theory. The situation is generally defined in this book as a triangular model that proposes a relationship among three organizational components: people (i.e., the employees), structure (i.e., the organizational design), and technology (i.e., the techniques of changing inputs into outputs), in addition to the repetitive information processes or systems that allow the components to interact with each other and with the environment.

This model has to be defined operationally because its main supports are solidly based (Isn't that a contradiction in terms?) in the social sciences, and these sciences have not progressed to the more objective stages that the natural sciences have reached. Perhaps those more objective stages of the natural sciences are an intermediate step in our learning about our world; they, themselves, are becoming more subjective every day. For example,

> according to quantum mechanics, there is no such thing as objectivity. We cannot eliminate ourselves from the picture. We are a part of nature, and when we study nature there is no way around the fact that nature is studying itself. Physics has become a branch of psychology, or perhaps the other way around. (Zukav, 1979, p. 56)

It should be obvious that when we study our own situations, the art of observation will invariably affect them, just as it does when we study the situations in quantum mechanics. And we can easily understand that, since we, as observers, are part of the situation. However, for the purposes of this chapter, let us assume minimal interaction between the observers and the model being studied. Observing the model's components, we find that those components are intertwined with each other. At this point in our learning program, they must be defined in terms of each other, since there are few independent measures to use. The model is a total model, a gestalt, and the components are quite undefinable outside of the model by themselves. A partial analogy would be to attempt to define a human being in terms of the water and chemicals in the body. That definition helps us to understand the gestalt human being, but it is far from a complete description.

The interaction between the total organizational gestalt and the external environment is through information processes or systems that incorporate the algorithmlike repetitive decisions of the managers who wish to communicate with that environment and with each other. Those processes or systems surround the organization, and they typically include most of the data processing, personnel, financial, and contractual relationships that organizational outsiders first encounter. While the repetitive decisions are like a circle around the triangular model, the nonrepetitive ones fill the interstices between the triangular organizational model and that circle of repetitive decisions. One kind of nonrepetitive decision is that of leadership, which also resides in those interstices. Since nonrepetitive decisions are by definition unique, when they are made they can be automatically placed into the repetitive information process if they reoccur. If they're not handled that way, they are repeatedly handled from scratch, a waste of human effort.

SUGGESTED ANSWERS TO CASE QUESTIONS

1. a. The rate of change of the environment has greatly exceeded the ability of Maximum Motors to respond. The organizational model, including people, structures, and technologies, is becoming unbalanced. I would start the diagnosis by trying to find any commonalities. I would do a frequency distribution of the problems and try to determine:

   • which people they are coming from. Are there particular groups that seem to be in the center of things?

   • Where in the organization they are coming from. Are there particular relationships among groups that seem to be causing conflict?

   • Whether the technology of processing inputs into outputs is effective. Is it necessary to increase the response speed by installing computer systems?

This could be the initial process to select the right kind of thinking technology, intensive, mediating, or long-linked. What problems have you come up with?

   b. It is difficult for any one person who has been in the organization for any period of time to change it alone. A new president who brings in a new supporting staff can sometimes do it quickly, but the on-site manager has a very difficult and lengthy diagnosis, design, and implementation program ahead of him.

2. The agenda would probably look like this:
   a. Problem definition
   b. Recommended solution, including costs and benefits
   c. Preliminary system design and timetable

d. Resources required (i.e., personnel, facilities, funds)

e. Tentative implementation program after the design is approved

3. The problems seem to center in the technical groups managed by these three people. If that is so, they are probably the experts in determining what is happening and are best equipped to evaluate potential solutions. They are the ones to absorb uncertainty in their departments.

4. At this point, intensive technology is being used, since no one seems to have any definitive answers and looking back to past solutions did not help. Perhaps a change is required in the mediating technology to include cost data in computer memories, so that any change is automatically costed when the prints are changed. Finally, it might be justifiable to consider changing the company's long-linked technology and manufacture the product in subassemblies that can be held as inventories and then assembled in small lots very quickly to meet tight time schedules. This solution could reduce the ability to respond to special designs and there might have to be a trade-off. There are, of course, many other possibilities. What would you do?

5. Walter seems to feel that the amount of uncertainty is too high to make a decision at this time, so he has delegated the problem to the experts for their opinions and recommendations. The group's answers may reduce the uncertainty, but eventually a decision will have to be made. At that time, unless it is shown that parts of the problems can be resolved under certainty and risk, a final decision under uncertainty will have to be made.

REFERENCES

Corporate culture. *Business Week,* October, 27, 1980, 148–160.

Cyert, Richard M., & March, J. G. *Behavioral theory of the firm,* Englewood Cliffs, N.J.: Prentice-Hall, 1963.

Drucker, Peter. An informal talk. *Boardroom Reports,* July 1981, *10*(15).

Haberstroh, Chadwick J. Organization design and systems analysis. In James G. March (Ed.), *Handbook of organizations.* Chicago: Rand McNally, 1965, pp. 1171–1211.

Helson, Harry. Why gestalt psychologists succeeded. In Mary Henle, Julian Jaynes, & John J. Sullivan (Eds.), *Historical conceptions of psychology.* New York: Springer, 1973, pp. 74–82.

Kotter, John P. *Organizational dynamics: diagnosis and intervention.* Reading, Mass.: Addison-Wesley, 1978.

Litterer, Joseph A. *The analysis of organizations.* New York: Wiley, 1965.

Makridakis, Spyros, & Wheelwright, Steven C. *Forecasting: methods and applications.* New York: Wiley, 1978.

Massie, Joseph L. Management theory. In James G. March (Ed.), *Handbook of organizations.* Chicago: Rand McNally, 1965, pp. 387–422.

Schrank, Robert. *Ten thousand working days*. Cambridge, Mass.: MIT Press, 1978.

Simon, Herbert A. *Administrative behavior* (2d ed.). New York: Free Press, 1957.

Thompson, James D. *Organizations in action*. New York: McGraw-Hill, 1967.

Tiffin, Joseph, & McCormick, Ernest J. *Industrial psychology* (5th ed.). Englewood Cliffs, NJ.: Prentice-Hall, 1965.

Weber, Max. *The methodology of the social sciences* (Edward A. Shils & Henry A. Finch, trans.). New York: Free Press, 1949.

Zukav, Gary, *The dancing Wu li masters*. New York: William Morrow, 1979.

FURTHER READINGS

Katz, Daniel, & Kahn, Robert L. *The social psychology of organizations*. New York: Wiley, 1966.

Robbins, Stephen B. Reconciling management theory with management practice. *Business Horizons*, February 1977, 38–47.

Taylor, Donald W. Decision-making and problem solving. In James G. March (Ed.), *Handbook of organizations*. Chicago: Rand McNally, 1965, pp. 48–86.

Woodward, Joan. *Industrial organization: theory and practice*. London, England: Oxford University Press, 1965.

PART TWO

THE GENERALIZED MODEL AND ITS COMPONENTS

3
PEOPLE

Case Study

THE CASE OF THE DISINTERESTED DESIGNER

CAST

George Wilcox: Chief engineer

Marge Gustav: Vice president

Millard Myer: Designer

Phil Josephs: Checker

Peter Maxon: Section chief

It was late Wednesday afternoon when Marge Gustav phoned George Wilcox on the shop floor and asked him to stop by in the morning if he had a chance. George wondered what the call was all about and tried to go over all the open items since he had last talked to Marge. His budgets were on schedule, he had completed the appraisals of his section chiefs, and most of his "hot" design projects seemed to be coming along well. What could she want?

When he got to his office the next morning, he found Peter Maxon and Phil Josephs waiting for him.

George: Well, you guys are here bright and early today. What's up?

Peter: George, I've been trying to keep this problem from landing on your desk but I think that it's gotten to be unsolvable and I want to do something drastic to get rid of it. It's that new designer, Millard Myer; he's driving me up a wall. For the past few weeks, ever since he was transferred in here, he has been coming in late anywhere from half an hour to two hours. When he does show up, he sits at his board all day long and half the night too, so I'm told. I know that when I leave here, sometimes after 6:00 P.M., he's still here. His time sheets are never in on time, and the only way they do come in is when I send my secretary down to get them. He seems to know what he is doing, though. Every time I talk to him the answers are right on the button. I've tried to get him to straighten up on his administration, but after a few days he just slips back into those sloppy habits of his.

George: Wasn't hc thc guy you said would be great for those advanced development projects? You remember—the ones that needed the combination of physical chemistry and hydraulic controls?

Peter: Yes, he's the one, but I didn't think that he would be this much trouble.

George: Well, Phil, I guess that you're here about the same thing. What's your position?

Phil: The guy is a gold-plated A No. 1 genius! It took him about half an hour, using the back of an old envelope to write on, to calculate the residual loadings on that model E that was giving us so much trouble. He is a different kind of guy, though, never has lunch with any of the boys; in fact, he never even goes to lunch. Just sits there at his board, or at the little desk behind it, and stares up at the ceiling. Sometimes he works at the computer CRT. Of course, I don't know what he's doing but I guess that he does.

George: Pete, have you talked to him? Told him what you expect? I'm going to ask an obvious question, but have you showed him how to complete those forms? They are a bit complicated, you know. Those accountants don't seem to care how much time it takes, just so long as they get their data.

Peter: Like I said, I've talked myself hoarse. He's very pleasant and always says that he'll get right to it, but after a few days it starts all over again.

George: OK fellows, give me your suggestions in an informal memo because I'm late for a meeting with Marge. Talk to you later.

Ten minutes later: in Marge's office

George: Hi Marge, sorry that I'm a few minutes late, but some of my boys grabbed me when I first came in, and I just got away.

Marge: That's OK, George. What I want to talk to you about is not something that can be fixed right away, so a few minutes delay is no problem.

The board of directors was complaining at the last meeting that we never seem to have any revolutionary products that sweep the market anymore. I guess they still remember the time that we captured a majority of the market for nuclear control valves several years ago with our new TFE models. They seem to see a more gradual approach to our product improvements now. Our competition has been biting into our markets and we have had to reduce our prices to remain competitive. With our overheads, that has a direct effect on the bottom line. Do we have any answer for them . . . anything new and better that's on our drawing boards? If not, do you have any suggestions? They want to see some results within the next year. How about giving me your analysis informally by the end of next week?

George: OK, Marge, be glad to. We're working on some improvements that should help the bottom line as soon as development is completed. I'll have a memo drawn up for you.

Later that day, as he began to lay out the memo, George realized that some of those new products would be coming from Peter's group. He decided to walk down the hall and check things out with Pete. On the way, he passed Millard's cubicle, and on a hunch, he stopped. Millard was deeply engrossed in something on the computer screen and was slowly pecking in some instructions with one finger. Seeing George, he stopped.

Millard: Hi, Mr. Wilcox, how's it going?

George: Fine, Millard, fine. I didn't want to interrupt you because you seemed so involved in what you're doing. (George recognized the symbols as a second-order differential that applied to turbulent flow in valves.)

Millard: Well, I guess that I do get involved. It's usually pretty noisy around here, and when things calm down a bit, I can lose myself in these equations and then they come out right. I think that I almost have the answer to that cavitation problem. If I can get it, we can decrease the wall thickness by about 25 percent and cut the weight way down. That might even get into that new flight development program that I've been reading about in the papers. Wouldn't that be great? Do you think that the company would let me have an interconnect to our computer, so that I could work on it at night at home?

George: Well, I don't know, Millard. Why not take it up with Peter?

On the way back to his office (it was too late to talk to Peter, since he'd gone for the day), George thought over his conversation with Millard. When he got to his office, he found the informal memo from Peter. It recommended that Millard be discharged.

QUESTIONS

1. What do you think the main problem is in the engineering department?
2. Should there be any changes in the recruiting and/or reporting systems?
3. What differences do you see between the culture of the company and that of the engineering department? How do Millard and Peter fit in?
4. Describe the motivation ideas that might be intrinsic here. How would they apply to all the people in this case study? How would extrinsic motivators be described?
5. What should George do about Peter's recommendation?
6. What should George's response to Marge be? What should he say about future developments?

REVIEW: THE MODEL

This chapter is about people, their motivations for working, and what affects that motivation. The component, people, is the most independent and the most important component in the organizational model. It is also the most complex of the three components and probably the most difficult to understand or to change. We start by describing the behavior observed (what people do) and the reasons for that behavior (why we think they do it). Then we move on to how (or whether it is even possible) to change that behavior to meet some organizational need that we have defined.

The description begins with the reasons for the independence and importance of the people component, then goes on to how the component relates to organizational productivity through the motivation to work, and finally deals with how that relationship might be modified. Since one of the classical methods used to modify people's relationships with organizations is training, we will also review briefly some recent research on that subject. The subject of training is also a change mechanism, so it is looked at again in the last chapter from the viewpoint of change.

So much has been written about human motivation that we must have some kind of classification framework to use in order to categorize these materials. Otherwise we might be tracking through a featureless body of general information, with no landmarks as guides. The framework that I use divides all the models of motivation reviewed into two major areas; those that are supposed to apply universally (i.e., those that affect everyone) and those that are supposed to apply situationally (i.e., those that might or might not affect someone, depending upon the contingencies).

Typical examples of the univeral models are theories about people's physical needs, such as food, sex, and safety. The situational models are typically represented by the conspicuous consumption model of Thorstein Veblen; that is, they depend on the contingencies of the social situation. Admittedly, this classification (like so many others in the social sciences) is a bit artificial since an argument could probably be made for placing the same theory in either division and you could undoubtedly improve it by rearranging the models in those divisions. However, the purpose of this framework is merely to assist you in relating the various motivation concepts to each other and to your own framework or theory. When you develop your own framework, you may want to select some of these concepts. If they work for you, so much the better. After classifying and relating these models, we'll develop some very general prescriptions that can be applied in your own organization.

WHY PEOPLE ARE THE MOST INDEPENDENT COMPONENT

As briefly touched on in the prior chapter, *people* are the only component in our simplified organizational model (people, structure, and technology) that exist separately outside of that model. People can and do do things that are independent of their work. The second component, *structure*, might have some slight independence if work-related social groups continue after working hours. For example, it might be argued that the discussions during the engineering department's weekly bowling tournament about the latest design problems meet the definition of structure (i.e., repetitive work-related behaviors), but this would be stretching it a bit. However, stretched or not, we can agree that structure is less independent than people. Finally, the third of the three components, technology (or the methods by which inputs are converted to outputs), is the component with the least independence of the three.

An opposing viewpoint might consider that the organization's decision-making methods, computers, or machine tools can exist outside of it, but under almost all conditions these are merely parts of the technology component that must be blended uniquely to suit the particular organizational situation. The people component is the most independent one in our model and is consequently least susceptible to management-directed change. It is, therefore, the initial component selected for our descriptive and prescriptive process.

This relative independence always results in fewer data being available for our use than with the other two components, structure and technology. When we observe people's on-the-job behavior and then try to understand the underlying thought processes behind that behavior, we are working on partial information. We never really know if unobserved behaviors are more important than, less important than, or even somewhere in between those we can observe. Since we don't have full data, we can only draw conclusions on the basis of the partial behavioral data we do have. People have past working histories and lives off the job, and their effects are carried into the organizational situation. Although we have some general research findings offering guidance on the effects of off-the-job conditions on on-the-job behavior, we are usually quite sure that the guidance is incomplete. The description of this independent component, people, is, therefore, necessarily quite limited. However, regardless of limitations, people have had extensive research done on them, and we will look at some of those research results next.

WHY PEOPLE ARE ALSO THE MOST IMPORTANT COMPONENT

There seem to be two major, interrelated reasons for this most independent component, people, to be the most important one: as managers, they make decisions that guide or integrate the organization and as knowledge workers, they produce the goods and services that support it. The reasons are interrelated because they are often intertwined when people work in an organization. We'll discuss decision making briefly, since we've covered it before, and then move on to productivity.

Decision Making

In Chapter 1, we defined the major task of managers in the technical functions to be decision making and Chapter 2 defined rules to be used in making decisions. Those rules included using an inferred mental process called absorbing uncertainty. That is, if the problem is yours and apparently repetitive, try to solve it by using either algorithms under conditions of certainty or a systematized decision matrix using calculated risk factors. If the problem is new, the decision process includes absorbing uncertainty.

Implicitly assumed in that process was that the decision maker had both the intent and the ability to make those decisions. If that assumption was valid, the process would be followed and decisions would be made every time. That doesn't always happen. We know that decisions are not always made even when they can be, and that could be because part of our assumption failed. For example, one reason many costly, well-intentioned management information systems fail is that even though the managers get the information on time, there is no assurance that they will do anything with it! Surely it is obvious that managers are not always willing to make decisions. Assuming that they are neglects the possibility of lack of human motivation. We will look next at the willingness to make decisions.

We start with a relatively safe and low-level assumption: that *all* behaviors are the outward results of some internal mental processes of the people component and that those internal mental processes are also at the core of decision making. It then becomes important to learn all we can about those behaviors to increase our understanding of the mental processes behind them. We're concerned with how, where, and why they came into existence. More specifically, we want to define, understand, and (if possible) prescribe how to modify the on-the-job behavioral subset positively in order to provide the organization with better management through more effective decision making.

Productivity

The organization is usually expected to produce something of value, and irrespective of whether that value is in goods or services, one measure of how well they are produced is called *productivity*. One definition of productivity is the relationship between input and output. That relationship has often been improved in two very general ways. We can say that productivity increases when the input is fixed and the output increases, or when the input decreases and the output is fixed or increases. In more general terms, when there is a minimal input for a maximized output, we have optimum productivity. We will not be concerned here with either of those two ways to improve, but with a third way: increasing the human mental input to increase the physical output without changing the economic input at all.

For our purposes, let us define the total input as some function of the product of capital invested and human effort (i.e., capital times human effort). If we concentrate momentarily on the physical capital investment part of this equation, we find that in most Western industries during the first part of this century, this capital investment per worker increased almost every year, with a resulting equivalent or greater increase in production output. This output increased because of capital spending on the methods of production (those methods being one kind of technology) and in internal operations (i.e., more effective and flexible structures).

However, investment in another kind of capital (training the people) was not very great. And even when minimal funds were spent on training, it was almost impossible to connect those kinds of capital costs directly with increased productivity. Measuring the changes in output of improved machines and systems is a lot easier than measuring people's output. Therefore, most economic justifications of capital spending were based in faster and better machinery, in purchased materials, and in reduction in the measurable, direct labor content in the products. Since the worker was considered to be an adjunct of the machine, decreasing labor hours was the same as increasing machine productivity.

Within more recent times, there has been a change in capital spending. Modern technical operations are quite different from those of earlier days. The knowledge worker, and not the production machine, now determines the amount of the input for a proposed output. Investment return on capital, therefore, is closely tied to the more or less independent efforts of the technical worker, who is definitely not a machine adjunct.

Reviewing just a bit, productivity was defined as a function of capital and human effort and that human effort became more critical with the increase in the number of knowledge workers in the technical organization.

$$\text{Productivity} = [\text{some function } f_1] \times [(\text{capital}) \times (\text{human effort})]$$

However, even under the best of circumstances, it is not easy to measure productivity improvement from capital investment, because the situation after the investment is never quite the same as predicted. Materials may change, shipments improve, or other interacting things happen to make measurement complex. It becomes even more complicated when we try to measure the effect of the human effort. Human inputs such as commitment and inspiration are surely as important as the estimates of the cost accountant, yet differences in those inputs do exist and do affect productivity. But by using some very loose guidelines, it is possible to come up with a general definition of some measurements. Those measurements are in units that are quite different from those we generally use in scientific or technical areas.

The measurements are in nominal, ordinal, interval and ratio systems. For example, how can we value art? It's possible to define good and bad paintings and even measure that "goodness," using different measurements that are either nominal or ordinal. A nominal measurement is one that differentiates only between two or more categories. "The red checkers are mine and the black are yours." An ordinal measurement is one that defines *more* or *less* than something else without the amount of that *more* or *less* being linear. "I like Remington's paintings of the old west better than I like Warhol's paintings of soup cans, and although I can't give you any objective measure of that liking process, I'll trade you three Warhols for one Remington." That ordinal measurement may be less exact than even the inaccurate measurements of results of capital inputs, but it's the best that we have. For example, the following is a very subjective, ordinal measurement. Nonetheless, it's useful for comparisons of human effort: "Charlie is one of the best hydraulic systems designers in this company. I think that he's able to do twice the job of anybody else."

There are also interval and ratio numbers, but you know all about those.

Interval numbers are linear but have no absolute zero. For example, 60°F is not twice as hot as 30°F, because there is no absolute zero on the Fahrenheit scale. It is just 30°F hotter. Ratio numbers are linear and do have an absolute zero, such as is found on the kelvin scale. Thus, 60 K *is* twice as hot as 30 K, as well as being 30 K hotter. Ordinal measurements are similar to those in decision making under conditions of risk, where the decision maker has to estimate the value of the possible solution and the probability of its occurrence subjectively. Therefore, since we have neither an absolute zero nor any way to measure linearly between two points in our measurement system, we'll obtain the best measurement of productivity using the subjective ordinal measurement system to determine the amount of human effort. As managers, we are concerned with optimizing that human effort, and that is where we go next in our search for understanding.

Motivation Defined

A useful definition of *human effort* includes both the mental and the physical inputs of people and is a function of their abilities and motivation:

$$\text{Human effort} = (\text{some function, } f_2) \times (\text{abilities} \times \text{motivation})$$

Abilities are defined here as the combination of both the inherited characteristics and the training and conditioning of the particular person. Productivity can therefore be defined now through the substitution of the formula for human effort into the prior formula:

$$\text{Productivity} = f_1 \times [\text{capital} \times (f_2 \times \text{abilities} \times \text{motivation})]$$

For example, the concert violinist's abilities include an innate, inherited eye-hand coordination *and* years of corrected practice. The effort the violinist exerts in making music results from using those abilities and from a directed mental process. That directed mental process can only be inferred from observation of the violinist's behavior. The inferred process is called *motivation*—the selection of goal-directed behavior from the repertoire perceived to be available to the person. Obviously, motivation is the key to human effort, which is, in turn, the key to productivity. Productivity depends directly on people. People move the organization. In other words, this rather lengthy development has resulted in an intuitively apparent idea: Human motivation determines the technical organization's productivity and we can measure it only through some kind of ordinal system (i.e., by determining whether it is better or worse than some predetermined standard).

MOTIVATION AND BEHAVIOR

Motivation includes the idea of choice. In our example of what managers do, one choice is either to make or not make a decision. Those choices are vital in determining organizational productivity in the technical organization.

We assumed that those choices can be inferred from our observations of the incomplete set of people's behavior on the job. There are quite a few uncontrollable variables in these behaviors that we can only guess at without even considering if we have been able to interpret our observations correctly. For example, we know that behaviors are affected by time, off-the-job experiences, and the job itself. Some recent research (Kohn & Schooler, 1973) found that four facets of the job have important psychological effects. They were:

1. Organizational characteristics (position in the hierarchy, ownership, bureaucratization)

2. Degree of self-direction and occupational commitment (closeness of supervision, complexity, and work routinization)

3. Amount of job pressure

4. Level of uncertainty

But the data also show that with education and other factors of occupational experience controlled, the degree of self-direction had the strongest influence. We will now do some theorizing of our own. *If* self-direction is similar to independence, and *if* occupational commitment is related to motivation to work, we have some research justification for a preliminary assumption that people seem to be motivated when independent and self-directed. There are a lot of *ifs* but they seem to be reasonable, at least with respect to technical operations in which the processes of production are controlled by the person himself.

The effects of the present job can be added to the effects of the work that occurred earlier in the person's professional career. One experimenter (Kaufman, 1974) investigated the relationship of the work challenge that a varied group of eighty-five engineers had at the beginning of their careers to their consequent work performance, professional competence, and contributions. For those who possessed high technical ability, early challenge reinforced professional contributions and competence later in their careers. This type of research seems to say that it is possible to have long-term effects from early work experiences, at least for technical personnel with ability. These two examples of research findings indicate some of the complexity of the connection between observed behaviors and inferred motivation.

UNDERSTANDING MOTIVATION: IDEAS AND CONCEPTS

The motivation to work, and even work itself, has had a spotty history at best. During the early Greek and Roman eras, work was something to be done by slaves, farmers, or the lower fringes of society. It was not accepted as necessary for free men and was not considered to be the best way to live. Politics and philosophy were important; work was not. As an expected part of life, work had a bad name, and that attitude remains with us to some extent (although not by emphasizing one's interest in politics and philosophy). For example,

> This book, being about work, is, by its very nature, about violence—to the spirit as well as to the body. It is about ulcers, as well as accidents, about shouting matches as well as fistfights, about nervous breakdowns as well as kicking the dog around. It is, above all (or beneath all), about daily humiliations. To survive the day is triumph enough for the walking wounded among the great many of us. (Terkel, 1972, p. xi)

However, these attitudes are not unanimous. They contrast with spontaneous comments I have heard in some technical organizations; e.g., "This place is so much fun, I even might consider paying admission to join it." Differences in attitudes and approaches to work spring from many places. Freedom to control one's own work and early work experiences (see above), social or cultural norms, and other factors are part of that inferred totality called human motivation. Even interpretations of our observations of behavior may not be valid.

> . . . Any reader must be aware of the assumptions undergirding observation. One person's burdensome work may be another's releasing joy, for the human spirit works in strange ways between euphoria and martyrdom. In many respects, of course, the problem of work . . . is really the human task of responding creatively to existence in the cosmos. It is the problem of living. (Fairfield, 1974, p. xiii)

I have noticed in various books about motivation theory occasional implicit prejudices of the researchers when they were describing inferred human motivations connected to difficult or physically demanding jobs. My discomfort was eased by the following quotation.

> There is a peculiar arrogance in those who discourse on the brutalizations of work simply because they cannot imagine themselves performing the job. Certainly workers often feel abstracted out, reduced sometimes to dreary robotic functions. But almost everyone commands endlessly subtle systems of adaptation; people can make work all their own and even cherish it against all academic expectations. (Morrow, 1981)

These quotations throw some light on the potential motivational differences between the head of the design standards department who seems to delight in repetitiously fitting every screw and valve into its proper specification sheet and the advanced product designer who seems to achieve equal delight when he can use novel screws or valves that are not even on those sheets. Now that you have been warned about the possible biases of everyone (including you and me) both for and against work, we are prepared to deal with motivation theory and findings. But, as I said before, what you see in either the theory I will describe or the work behavior you observe may only be what you interpret. And it can be measured only subjectively, with ordinal measurements.

In Table 3-1 we use two broad vertical columns for classifying *universal* and *situational* theories, and then we will classify motivation theory horizontally into rows across those columns. The first, or *cultural*, row includes theories that apply to major themes of life, such as religion and general social culture, for large groups of people. The second, or *working groups*, row applies to relatively stable peer groups and their effect on individuals. This row includes themes such as morale and participation. The third, or *individual*, row applies to the psychological characteristics of the person. It includes many of the inferences that we have discussed before. The vertical column classification, therefore, moves from the general (i.e., culture) through groups into individuals. This classification scheme gives us the ability to relate one motivation theory to others.

For example, considering the two broad vertical columns of universal and situational theories, a theory that states that there is an inherent "need for achievement" (McClelland, 1961) in managers would probably be universally applicable and relatively independent of any situations or contingencies. On the other hand, the suggestion that an unmotivated person at work would respond quickly to the pressures for increased production of his work group (Schachter, 1959) would probably fall into a situation-dependent category. However, before going further, let me repeat that none of the research we shall classify suggests that motivation is *completely* independent of either the person or the situation: After all, we do not live in social vacuums and we all interact to some extent.

This classification is useful because it seems to allow different ideas to be understood more easily. The theories do not always compete; in many cases, they supplement each other.

TABLE 3-1   MOTIVATION THEORIES

	Universal	Situational
Cultural	Religion-culture	Society-class
Work group	Morale	Human relations
Individual	Psychological	Social-psychological

Cultural Motivation

In the first horizontal row of theory called *cultural,* the person's motivation is assumed either to be *universally* applicable or to depend upon some contingency or interaction with the work-related *situation.* It's a matter of scale. If the theories to be explored assume that the motivation to work is universal in that culture, the motivation should be relatively independent of the organizational work situation, since it is an almost constant requirement of that social environment. On the other hand, if the source of motivation in that culture is assumed to be within some identifiable class within a society, motivation will not be universal. Only those who interact with that class in society are influenced by it. In the latter case, motivation is defined as society-class. In both columns, however, there is always some overlap.

UNIVERSAL INFERENCES

The classification of universal motivation (religion-culture) to work as part of a culture is easily illustrated if one looks at the historical record that "millions upon millions of humans have worked at some of the most onerous tasks imaginable to perpetuate their society's values, both tangible and intangible" (Fairfield, 1974, p. xiv). Religion has often resulted in the expenditure of tremendous effort, with the only reward to those involved being promised for some afterlife. According to recent archeological interpretations, the pyramids were built by Egyptian farmers during the annual flood of the Nile River, when farming was impossible, as a way to share in the eternal life of the Pharoah in the Land of the Dead.

In more recent times, the Protestant or work ethic has become part of Western working culture. The concept has existed for a long time, but it was very well described in seventeenth-century Protestantism, which proposed that the route to salvation and escape from eternal damnation in some afterlife was a laborious one in this life involving continuous mental or physical labor. Time spent in leisure and enjoyment was wasted and justified moral condemnation (Weber, 1930). This work ethic was a major motivational push during the westward drive across the North American continent during the last half of the nineteenth century. It became an implicit part of early management theory (Taylor, 1911), and while it may seem to have lost much of its force in relatively recent times, it is still part of the motivational culture of this country.

Another of the broad bases for these universal motivations to work are those inspired by patriotic considerations. Of course, that could be a subset of culture. For example, in a country at war, there are often reports of factory workers who labor almost unceasingly and of members of the armed services who seemingly give their lives willingly because of responses to the seemingly universal motivational forces that they feel. Similar but, of course, less emotional forces can be detected when an entire industrial organization happens to be caught in a temporary economic emergency (though not to the extent of requiring one's life, I hope). Overtime may be volunteered for extended periods of time or other personal sacrifices offered.

In both of these examples—a nation at war and a company in economic trouble—the motivational time span for individuals to participate is much shorter than the length of their working life, which is affected by general cultural and religious forces. Unless there is some additional payoff over the regular rewards that they receive, this motivational pressure quickly subsides. With no such payoff to the individual (either tangible or intangible), demotivation begins within a relatively short time.

SITUATIONAL INFERENCES

General cultural forces may also affect situations. Consider observations of behavior directed at joining a different social class. In this country, the social class might be defined partly by the amount of money that you have, spend, or are purported to have. If there is a social class with expensive habits and the person wants to join it, one obvious path would be to adopt those expensive behaviors too. The relationship between behavior and social-class culture that leads to these expensive habits is beautifully described as follows:

> The quasi-peaceable gentleman of leisure, then, not only consumes of the staff of life beyond the minimum required for sustenance and physical efficiency, but his consumption also undergoes a specialization as regards the quality of the goods consumed. He consumes freely and of the best, in food, drink, narcotics, shelter, services, ornaments, apparel, weapons and accoutrements, amusements, amulets, and idols or divinities. Since the consumption of the more excellent goods is an evidence of wealth, it becomes honorific; and conversely, the failure to consume in due quantity and quality becomes a mark of inferiority and demerit. (Veblen, 1935, pp. 73–74)

It's not too difficult to find this type of behavior. Conspicuous consumption behaviors are alive and well today. And this is only one of many cultural determinants of human behavior.

In the more limited areas of the working organization, company cultures exert pressures on behavior just as those larger cultures do on life outside of work. These are relatively specialized, but they are quite important to organizational participants. For example, there was a fairly large company that had promised to provide almost permanent employment for those participants who cooperated with both the written and unwritten company policies, defined here as the company culture. That promise was never completely and explicitly documented, but it was similar to other company cultures, in that everyone in the company just "knew" it existed. In this case, personnel policies were designed to reward human conformity and employment longevity, using undocumented controls not usually considered to be company concerns.

Everyone in the company for any length of time knew what the policies were. These nondocumented policies directed the expected hours of unpaid overtime work to be contributed by technical professionals, suggested the dress code, and even recommended educational institutions for employees' children. It was a patriarchal or familial type of organizational culture; one that generally promises more than it can ever deliver. (Your family is supposed to take you in when there is no one else, but a company doesn't have to, no matter what familial relationship is implicitly promised.)

As it turned out, that semifamilial culture was a major factor in the company's eventual economic failure. When the market moved away from its major product line, there was no one in the company who would point out to the patriarchal company president that the old products were no longer the best in the trade. Sales and profits dropped alarmingly, and when the president became ill, there was no successor to take charge. Then the company was bought by a large conglomerate and the extensive employee benefit plans (both explicit and implicit) were scrapped quickly in order to "bring this division into line with all our other divisions." New top management was brought in "to shake things up a bit," but the shaking up didn't work well. The ingrained company culture required a patient and lengthy gradual weaning away from the familial culture into a more objective one, but the conglomerate was not structured for patience. The conglomerate's culture was for the here and now, with profits that had to increase every financial quarter. Therefore, it attempted to change the company's culture radically by introducing new policies and planning methods. These were quietly sabotaged by the employees. That new process had violated their cultural expectations, and even though the people realized that the company itself had to change, the change was too fast for them. With more time and training, it might have succeeded. The actions of the conglomerate, of course, were not entirely wrong either, since it had to move quickly because of the loss of the company's markets. Things continued to go downhill, and the company was finally dissolved by the conglomerate after several years of poor performance.

SUMMARY OF CULTURAL MOTIVATIONS

Cultural pressures affect people's motivations in both the larger environments of society and the limited ones of the company. Those pressures have deep roots and must be considered in developing your personal theory about motivation in technical operations. Any hypotheses that include modifying company culture rarely succeed without the cooperation of both the formal (i.e., management) and the informal structures of your knowledge workers. There cannot be rapid violations of existing organizational mores.

However, mores can be changed over an extended period of time. One division general manager applied three years of staff training, repeated functional problem-solving meetings, and personal corrective action to his management group before he was able to move that group from the closed style of management that he had found when he arrived in the division to the more open style of problem confrontation and team cooperation needed to provide new products and innovation. Therefore, a hypothesis requiring modifying major parts of the company culture is probably not within your capability as a technical manager. We discuss it here only as general background material so you'll know what to do when you become the general manager. We now move to a slightly smaller arena and describe some effects of working groups within the company on the person's motivation.

WORKING GROUPS

The effect of the small group on the individual's motivation was carefully reported in the research done during the Hawthorne experiments, which were first started in 1927. The Hawthorne Works was Western Electric's largest manufacturing plant. Western Electric itself was part of the Bell Telephone System. The experiments continued for about five years, and their purpose was to test the effects of physical changes in the worker's environment on the work output. Those changes typically included temperature, humidity, and number of rest breaks during the day. The experimental environment of a small working group was selected in order to obtain better control over the data.

> In a small group it would be possible to keep certain variables roughly constant; experimental conditions would be imposed with less chance of having them disrupted by experimental routines. It would also be easier to observe and record the changes which took place both without and within the individual. And lastly, in a small group there was a possibility of establishing a feeling of mutual confidence between investigators and operators, so that the reactions of the operators would not be distorted by general mistrust. (Roethlisberger & Dickson, 1939, pp. 19–20)

Initially, the output of a small experimental group of five women in a relay assembly room was measured. The experimenters found that changes in the physical environment, rest periods, improved lighting, and so forth produced no directly related change in output. There was, however, a general and persistent tendency of output to rise during all the experiments. This happened irrespective of the increase or decrease in the physical variables. Increased lighting increased production, but so did decreased lighting.

The explanation for this peculiar result was quite novel. It was proposed that the group had achieved a shared work morale (informal support) that was perceived by the women to be reinforced by the management structure (the researchers as the formal organization). They increased their productivity. The informal and the formal organizational structure were mutually reinforcing. The researchers reported that "In human organizations we find a number of individuals working together toward a common end: the collective purpose of the total organization" (Roethlisberger & Dickson, 1939, p. 533).

The research direction was quite obvious. Strengthen the work group, build internal cohesion, and ensure harmony between the plant management and the group goals. However, there was a problem in implementing these ideas. Work groups and plant management goals do not always coincide. This was illustrated by another experiment completed in the Hawthorne studies. In the bank wiring room, the output did not have the same general and persistent tendency to rise during the experiments that had occurred in the relay assembly room. It remained relatively fixed throughout the experiments, even though all the variables that had been tried before were used again. The researchers even reported that whenever one worker in the wiring room experiments exceeded the production standard established by group norms, he was disciplined by the group through physical horseplay and discussions in order to reduce his output. The group's norms (informal organization), which were intended to "protect" them from their perceptions of an oppressive management (formal organization), were controlling output even though it was economically advantageous for the worker to exceed those norms. Payment was based on individual incentives, but the worker was apparently more concerned with group pressures than with personal income.

The Hawthorne experiments suggest that optimum productivity occurs when the pressures exerted by the informal and the formal working groups *coincide*. During the time of these experiments, there was relatively little management concern with the individual, since it was assumed that he or she was modifiable by the group, which in turn was controlled by the formal organization. Management concern for that individual was limited to determining what social pressures could best be utilized to promote group harmony. In other words, individual differences, which were supposed to result in discord, were minimized, since it was assumed that the group controlled the person and management could control the group.

These experiments fostered a human-relations school of management that was supposed to be universally applicable. Personnel departments were established to influence the attitudes of workers' groups. The company was supposed to interview the worker, discover the worker's problems, and then attempt to alleviate them within the existing formal structure. In effect, the company culture was supposed to be integrated into the group's culture. Individual differences were discouraged if they didn't match the norms of the company culture.

Further research, however, attacked the conclusions of these experiments, which supported changing individual and group cultures to match the company's. And reported consistent differences in individual social norms between workers in smaller communities and those in larger, industrialized cities made the task of making the individual conform through the group to company culture very difficult. If workers were really that different, no uniform company policy would work for all. Blood and Hulin (1967) found that the humans-relations concepts of matching group and company goals are more easily accepted by workers in smaller communities than in larger ones. This meant that the findings of the Hawthorne experiments were less universal than originally predicted, since the company now had to account for the individual's cultural background *before* he joined.

The Hawthorne results were not as universally applicable as had originally been claimed, but the company-to-group-to-worker motivation sequence was not entirely eliminated. It was just limited in that it had to make room for the person's antecedents. This is a classic example of original data being reasonable but incomplete. The theory, then, is also incomplete, and there are probably other data yet unexplored that will help to provide more complete answers.

Universal Effects

The Hawthorne studies found that group norms might not coincide with those of management, and suggested that this happened because management had not been able to show the workers that coinciding goals were best for everyone. Management's formal goals were supposed to prevail of course, and it was management's problem to change the motivation or attitudes of the workers to match those goals. Somehow the manager's task was supposed to be to redirect the group's norms to those of management. That didn't always happen. The direction of the goals may have been from the individual to the group instead of the other way around. On the other hand, when group norms are accepted by the individual (when there is high group cohesiveness) and those norms are opposed to the production goals of management, the individual probably is also opposed to those production goals.

Sometimes the employee is even more concerned about meeting group norms than about accepting some individual kind of payment from management as a result of his or her own behaviors. The group norms become an end in themselves and are as highly valued as any kind of compensation. That compensation may take several forms, such as exchanging social benefits with each other (Homans, 1958; Schrank, 1978) and attempting to gain control over one's work (Schachter, 1959). For example,

> I have found that it is terribly important to feel part of a community in the workplace. There is something that the work itself can never provide us with . . . for instance, . . . the work as it was organized did not permit me enough schmoozing time, time to wander around the plant, visit and talk to people in other departments and not be stuck in one spot doing the same thing. We need to see workplaces as communities where people go to carry out tasks for which they are paid . . . what is most neglected . . . is the nature of the human relationships: the rituals such as greetings on arrival, coffee breaks, lunch time, smoke breaks, teasing, in-jokes, and endless talk about almost everything. (Schrank, 1978, p. 78)

These needs (as part of the individual's self-interest) are proposed to be universal and not dependent upon any situation. They are typically more easily satisfied in relatively open technical groups than in the more tightly structured auto assembly groups, but there is a common element that comes only from a group, according to some researchers. That is the social support on the job that comes when a person accepts the group's behavioral norms in return for the social interchanges he or she values. It is suggested that the social support exchange and the need to compensate for lack of control over the work are universal motivational factors associated with groups. While this seems to apply less to our types of knowledge workers, it may be true for them as well. Therefore, we shall review it.

Some research indicates that the worker needs the informal group to help in a defensive reaction against apparent loss of control over the work. When the worker joins others who are in the same predicament, he or she receives some psychological support against feelings of inadequacy. "Misery loves company" is a truism, according to this research, since the group is also miserable (Schachter, 1959). This could explain some of the person's dissatisfaction when a work system excessively simplifies and rationalizes the flow of work to be done. If the worker's efforts are totally defined, he or she is totally controlled by the system, rather than the other way around. It is only when he or she becomes associated with an informal supportive group that the worker senses any capacity to affect the job, since the instructions of the formal management group are only confining.

The Situation

While there are understandably some universal effects of groups on the individual's motivations, they are not all-inclusive. Other effects are not classified as universal because they depend on the situation. And other research seems to show that individual motivation(s) can be modified when the individual in the group perceives a change in the personal situation that affects his or her self-interest. In other words, the person now feels that it is not to his or her benefit to adhere as closely to group behavioral norms. When that happens, there is less universality in those group-person motivational effects.

Two major research areas indicate how the situation can effect a lack of coincidence between the person's motivation or behavioral norms and those of the group. These are participation and psychological fit: *Participation* refers to the opportunity a person has to share in group decision making. That person supposedly helps to change the situation. *Psychological fit* refers to how well the individual's personality matches both the group and the organizational situation.

This research goes beyond the relatively simpler universal ideas where, for example, management has to change the person's attitudes and then those of the group to match the formal organizational goals. It now follows the reverse route, where the person is supposed to be motivated to change his or her own attitudes to attain and keep the group's support in gaining control over his job. In this case, the situation is held constant and depends on the tasks of groups *within* the company. These tasks are rarely the same for each group (the engineering group surely has different tasks and a different situation from those of the purchasing group) and the attitudes or motivation of participants depend on how well they fit into these different situations. Here we are no longer talking about *all* groups of workers, which the Hawthorne experiments tried to do, nor are we concerned about *all* individuals satisfying personal social needs or controlling their jobs. We are concerned now with a particular situation's effect on attitudes and motivation of the individual through the intervening variable of the group.

WHAT IS PARTICIPATION?

Many times when problems arise that seem to come from some sort of human motivation, one of the first suggestions is to get everybody involved or *motivated* through participation. It's proposed as a simple solution to what is usually a very complex problem, and sometimes simple solutions just don't work as expected. The research shows that this technique does not always produce an unequivocably positive answer. It has mixed results.

In perhaps the best-known experiment in participative management (Coch & French, 1948), the researchers used participative management to increase worker acceptance in the introduction of new production techniques in a manufacturing organization. Four groups were involved:

1. A control group, group A, was simply given the new techniques and ordered to comply.

2. An experimental group, group B, elected two members to confer with management and assist in working out the details of the new techniques.

3. Two other groups, groups C and D, participated completely in decisions regarding the change.

The largest increases in productivity were in the participative groups, groups C and D. The control group, group A, stayed below prechange levels, and a high proportion of them quit in the first months of the change. The representation group, group B, was closer to the full participation groups than the control group. The answer was obvious; participation increases productivity.

But the story does not end here. When the experiment was replicated in other situations, the preference for participation that appeared to be common to the American workers in the prior experiment was not repeated by Norwegian workers (French, Israel, & As, 1961), by South American workers (Whyte, 1959), or by German workers (Weiss, 1956). While these latter results may point to cultural differences, it is also possible that the reluctance of workers to accept participation in order to increase productivity is a result of their perception of the situation. It could depend on many factors, such as:

Is this really participation or is it a management manipulation to get me to agree with decisions that have already been made?

Do I have the authority to implement any decisions made during participation or do they have to be approved farther up the line?

Does management really know what it is doing? If so, why is it asking me? Its job is to plan and mine is to do the work. Why should I do its job too?

Will I be blamed if the proposed change doesn't work out as expected?

The real question management rarely sees in its drive for worker participation is, what's in this particular situation for the worker who participates? Many of these activities bring no rewards to the participants except the psychic (or intrinsic) rewards that many researchers (and some managements) believe are very important. It is quite possible that at best the worker does not share this need, and at worst feels that any increases in productivity become a detriment to future employment. Why cooperate in working yourself out of a job, especially if the situation promises no personally valued return?

A researcher describes the reactions of a worker named Toy Wynn as a typically motivated employee after she had participated in suggesting an improvement:

> And what did the originator of an improvement receive for her efforts? Not any money, as one might expect, except through the company's employee profit sharing plan as savings in operation costs affected profit margins, a long term consequence. No, the reward was intangible. Toy Wynn had the stimulation of looking at her job as something that she could improve for her own benefit and for the company's. She had the fun of shepherding a suggestion through team discussions and before the supervisor. And she had the satisfaction of being recognized by the supervisor and her peers for having come up with a good idea. (Wass, 1967)

In all fairness to the company, it is probably true that this achievement could be part of that worker's annual evaluation, but that kind of evaluation is not directly related in time or results to her contribution. A major problem in this kind of program is that there is an implicit assumption of a coincidence of goals between the person and the organization when that organization embarks upon any kind of participation program. The group (or in the case above, her "peers") is assumed to have nothing to do with her attitudes. It is ignored. There is just as little validity in the idea of a coincidence of individual and organizational goals as there is in the idea that there is no direct connection between the worker's and the group's self-interest. Both ideas are invalid. Why should people accept goals when they often lead to consequences that are unrelated to their effort (i.e., higher company profits at the end of the year) or possibly to negative consequences ("We did such a great job of job simplification that we no longer need as many of you!").

I have often wondered about this selection of participation as a first answer to apparent productivity problems, because many managers do not act as if they really believe in participation at all. Haire, Ghiselli, and Porter (1963) reported on a study about participation covering 3600 managers in fourteen countries, including the United States, West Germany, and Japan. They found that in all the countries, managers had a relatively low opinion of the capabilities of the average person, yet when they were questioned about their own attitudes, they stated that they believed that participative, group-centered methods of leadership were more effective than traditional directive methods. How can one have participative leadership methods if the people being led don't have the abilities to participate?

One interpretation could be that those managers accept participation *intellectually* as a future method of managing, but right now, they will not or cannot follow up that intellectual acceptance with appropriate behavior in the particular situation. The participation these managers usually expected could be similar to the universal, human-relations motivations assumed in the Hawthorne experiments, which attempted to manipulate the motivations of the working group (and the worker within it). Possibly participation is used to obtain that harmony between the group and the formal organization, and since the workers do not have very high individual capabilities, the manager must provide the guidance and goals that everyone is supposed to follow. Participation in this case becomes manipulation. This is an extension of the Hawthorne findings.

> Satisfaction for the individual, cohesion within the work group, fusion of the work group to supervision, and collective purpose were believed to be mutually consistent as well as necessary conditions for plant survival. But satisfaction concerns individuals, cohesion is related to groups, and morale to collective plant purposes. These three possible levels of meaning were not distinguished; rather they were merged. (Krupp, 1961, p. 25)

Therefore participation, like beauty, is in the eye of the beholder. When the person considers that participation will lead either to no direct personal reward (however that person measures it), an eventual loss of job, or a substitution of company goals for his or her own, it would be stupidly self-destructive to participate. This is particularly applicable to the people in technical departments. They have the physical freedom to move around the plant, to talk to others, and in general to learn what the past history of the results of participation has been (if it has been tried before in this place). They can easily predict the possible future course of further participation.

On the other hand, there is research that indicates a predisposition of technical workers to accept some positive results from participation. Those data show that "job involvement was more strongly correlated positively with participation in decision-making in the subgroup of the most highly educated participants" (Siegel & Ruh, 1973). What these findings may be saying is that people with higher educations feel a closer coincidence between personal and organizational goals.

That seems to apply to technical groups. Either there is a positive mental alliance of goals that the person assumes when joining the subgroup or the subgroup is modified to fit the goals of the individual because of the person's high value to that subgroup. The reasons are not as important as the result: higher goal coincidence (as a function of higher education) between the subgroup and the individual's goals. This would suggest that attempts to install participative management in technical organization will generally fail unless attention is paid first to achieving some agreement between the common goals of the person and those of the technical group (or of the person's unique situation) before attempting movement toward organizational goals. That usually means:

1. A personally valued reward can be distributed upon the achievement of specified common goals.

2. Achievement of those goals is controlled by the person.

3. Cooperation of or participation in the person's technical subgroup is necessary.

We can conclude that participation to increase people's motivations is not a simple technique. Its effectiveness depends, to a great extent, on how it is treated by the organizational participants. Those participants are all different from each other, but within those differences we have already found one common characteristic that affects their thinking—the level of education. The higher the education level, the closer the alliance with the professional subgroup. But other important criteria distinguish participants (and their managers) from each other within the same company. Those criteria can define the amount of *fit* between the person and the situation in the subgroup.

SOCIAL-PSYCHOLOGICAL FIT

Some very interesting recent research deals with this fit of behaviors of people in different departments or subgroups *within* an organization (Lawrence & Lorsch, 1967) with that subgroup's situation. The researchers defined the organizational economic environment as an independent variable that is seen from the viewpoint of members of a functional group as they look outward. They identified three main functional groups as *marketing, technical,* and *scientific* groups, which generally correspond to the sales, production, and research and development functions of most firms. Among other variables, the researchers investigated the ways of thinking and behavioral patterns of these three functional groups.

> First, we have investigated the differences among managers in different functional jobs *in their orientation toward particular goals*. To what extent are managers in sales units concerned with different objectives (i.e., sales volume) from those of their counterparts in production (i.e., low manufacturing costs)? Second, we have been interested in differences *in the time orientation* of managers in different parts of the organization. Might production executives not be more pressed by immediate problems than design engineers, who deal with longer-range problems? Third, we have been concerned with differences

in the way various managers . . . typically deal with their colleagues, that is, with the *interpersonal relations*. Are managers in one part of the organization more likely to be preoccupied with getting the job done when they deal with others, while those in another unit pay more attention to maintaining relationships with their peers? (Lawrence & Lorsch, 1967, pp. 9–10) (Italics mine.)

They were attempting to define "the difference in cognitive and emotional orientation among managers in different functional departments" (p. 11). They found (as many of us did when we changed jobs and worked for different companies) that for two companies in two different industries, "these two organizations were quite different places in which to work" (p. 155) and (as we could only suspect before) that the managers themselves were also quite different; i.e., "the managers in the two organizations had somewhat different personality needs" (p. 155). They also found that the managers *within* functional groups thought and acted more like each other than like managers in other functional groups *in the same company*. The differences among groups in the same company were called *differentiation*.

The greater the intergroup differences within an organization on the three attributes noted above (goals, time span, and interpersonal relations), the greater the degree of differentiation. Organizations that exist in a fast-changing R&D-oriented environment (i.e., a plastics manufacturer) need high differentiation to be successful. The managers in each of the three departments were very different from each other because of the need to respond to rapid change and innovation. Organizations in a more stable, production-oriented environment (i.e., manufacturing cardboard containers) required low differentiation for success. The managers in the same three kinds of departments resembled each other greatly in the need to respond to little change within their own departments.

The sales department managers, the R&D department managers, and the production department managers of the plastics company differed quite a bit from each other in terms of their attitudes toward goals, time span, and interpersonal relations. While similarly titled managers in the cardboard container manufacturing company varied also, they differed much less among groups than the managers in the faster-changing environment of the plastics company. But in all cases, irrespective of the amount of *inter*group attitude differences, the amount of *intra*group differences seemed to be minimal. Thus, production engineers, whose job functions were definitive, would *all* be concerned mainly about getting the job done and less about maintaining social relationships, no matter what company they were in; sales personnel would all be concerned about maintaining good social relations with customers; and research scientists would fall somewhere between the two with respect to getting the job done and maintaining good relationships.

I have found confirmation of these results while consulting for a diverse cross section of technical and manufacturing firms. This confirmation, of course, was subjective and very informal, since I didn't use the researchers' rigorous measurements and tests. Generally, when I found the same cognitive and behavioral patterns among managers who work in the same functional area, there was a higher probability of group success. In these research experiments, each group of managers (i.e., production, sales, and R&D) differed from the other two, but when the organization was in an economic environment of fast change (such as plastics), the more successful companies had group managers with greater differences (differentiation) *among* the groups and greater similarities *within* groups than companies that were not as successful. Conversely, when the organization was in an environment of slow change (e.g., cardboard containers), the more successful companies had functional groups of managers whose differences among groups were much smaller. They had less differentiation.

To reiterate, although in both examples the three functional groups had differences among each other, the amount of difference among groups was related to the rate of change in the environment; fast change equaled high differentiation and slow change equaled low differentiation. We have two intertwined ideas here:

1. Managers in functional areas really do think alike within their functions. That's not so surprising and is the probable reason that people in the R&D engineering group, for example, prefer to spend their time with each other. They think alike.

2. Managers in one functional area really do not think like managers in another functional area. Continuing the example above, "Those people in the R&D group are not being uncooperative with those of us in production or sales. They just don't think the same way we do!"

These research findings are explored further in the next chapter, where we can describe the organizational (the structure component) tools used to integrate these differentiated groups into one organization. At this point, your theory should be concerned with determining if you detect a similarity of thinking of functional managers within your group, and if the thinking, behavior, and motivation *among* functional groups of managers is differentiated. (See Lawrence and Lorsch for the measurement devices of this differentiation.)

The amount of differentiation is positively related to the rate of environment change *as perceived by participants*. The participants, therefore, fit into their functional groups in terms of their own cognition and behavior. This seems to support the idea that goals are individual- and group-oriented rather than being oriented toward the overall company. That makes sense.

How can goals be constant for the whole organization and still be meaningful in terms of an individual's behavior? Increasing company profits, for example, may mean one thing to the design department and another to the finance department. Goals can now apply only to functions *within* the company and are related to the self first, the group second, and maybe in some general sense the whole company last. But how does this happen, and how does the individual fit the group norms?

SUMMARY OF THE FIT BETWEEN A PERSON AND WORKING GROUPS IN SPECIFIC SITUATIONS

There is a fit between the various functional groups and the person in a particular group. To review briefly, when organizations succeed, an important factor is the managers' (and participants', since they control themselves in technical operations) similarities within groups and the differences among groups with respect to thinking patterns and behavior. Fast situational change (as in plastics and chemicals) requires greater differences; i.e., differentiation. Slower change (such as with food and containers) requires fewer differences, but there always is some amount of intergroup differentiation, and it is defined by the perceptions of the organizational participants.

And since for all of these successful companies, there is always a greater similarity of thinking *within* groups or functional areas, it seems reasonable to believe that there is a greater concern with the goals of their own departments than with those of the whole organization. As the rate of environmental change and the differentiation *among* groups increase, it seems reasonable that the goals of functional groups will also agree less with an overall set of organizational goals.

Applying these findings to human motivation simplistically, in a production-oriented company, the manager of a functional group should be more of a team player who directs his group as part of the total company team. On the other hand, a successful innovation-oriented company has managers who are very similar to others in their departments but consider the company goals as secondary to their own departmental needs. Similarly, companies in environments that change more slowly probably would have more overall team spirit than those in fast-changing environments. Those in fast-changing environments probably would be more entrepreneurial.

Summary of Working Groups

The person's working group is an important link between that person and the overall company. The amount of control that her or his group exerts seems to be linked to many variables, such as group behavioral pressures (Hawthorne experiments) and internal needs (to socialize) (Schrank, 1978;

Schachter, 1959). To optimize this link to increase human productivity, you might include hypotheses about participation that have been modified to show a connection between positive participation and reward to the person (Coch & French, 1948) and fitting in (Lawrence & Lorsch, 1967). Those hypotheses would be helpful but incomplete. They do not include enough information about all the reasons behind the wide variety of behaviors of people at work. Groups are important, but they provide only some of the reasons individuals have a whole range of unique characteristics. The next section takes up some of these other characteristics, dealing with the person aside from his working group.

THE INDIVIDUAL

If we could control for the effects on motivation of the overall organizational culture and the working group, we would be able to describe the person independently of any external influences. Of course, as is the case with so many of our techniques here, that is really not possible. However, it is intellectually useful, because we can then use a simplified classification scheme or model to untangle the many inferences about the internal motivations of people. That's one of the advantages of using models; we hold the rest of a model fixed while we examine the parts.

For example, starting with one of the oldest models of individual motivation, called *hedonism*, we find that it can be traced as far back as the Greek philosophers. Hedonism means that people are supposed to select alternatives that lead toward maximizing their pleasure and minimizing their pain. William James (1890/1950) agreed. He said,

> But as present pleasures are tremendous reinforcers and present pains tremendous inhibitors of whatever action leads to them, so the thoughts of pleasures and pains take rank amongst the thoughts which have most impulsive and inhibitive power (p. 550).

The problem with this view of motivation is that there is nothing to test. The researcher-reporter can never be wrong. Nor can he be right, because any form of behavior can be explained as either pleasurable or painful after the fact and remain impossible to predict in advance. Of course, the pragmatist would say that the particular individual's assessment of pleasure or pain could be based on her or his past behavior, and that past behavior would therefore be a good indication of future behavior. While this might be arguable, we have made no progress, because we might not have an adequate sample of past behavior.

Some obvious problems that we have explored to some extent immediately come to mind. What does the term *adequate* really mean? Will the individual always connect the past with an expected future, and will the connection remain the same? How are pleasure and pain connected with our subject of work and life? Sigmund Freud's general answer to this last question was that effective work was one of the three positive solutions to the human dilemma of life (Fancher, 1973, p. 227). (The other two, by the way, were creative play and enduring love. Not a bad combination.)

These views of work (and the implied motivation to spend time at it) that place it as one of the central thought processes of people was supported by some classic research on the motivations of the American worker to work (Morse & Weiss, 1955). About 80 percent of the respondents reported that they would continue working even if they had enough money to live comfortably. Perhaps even more interesting was a finding that the percentage of people who would continue working was positively related to the amount of training required by their particular occupations. These respondents therefore considered work to be important and they seemed generally to be satisfied with their jobs, especially those that were highly trained. Other research (Katzell, 1979) reported that overall satisfaction levels remained high and basically constant between 1958 and 1976. But is this the whole picture? Perhaps not. Some critics of this research say that the interviewees really were not satisfied but reported satisfaction in order to justify their work.

An interesting and subtle point might be made here:

> When up to 90% of the workers are reported to be satisfied with their work, the behaviorists say that workers do not really know what satisfaction is and they will lead them to a superior kind. That sounds oddly like the proselytizing of a missionary. (Fein, 1974, p. 76).

Considering both sides of the question could bring up some problems. Do the interviewees really know? And are the behaviorist-missionaries right? Research should always be "tested" mentally against your own experiences and situation. Let me give you an illustration. In one company, all the technical managers complained bitterly to me about the administrative restraints on their ability to make decisions; e.g., all purchase requisitions had to be approved by three levels of the hierarchy, even though this delay often resulted in additional costs and late materials. The amount shown on the face of the requisition made no difference, either. However, on the whole, all the managers reported, in confidence, that they were well satisfied with their jobs and the organization, and their behavior on the job seemed to agree with those reports.

Only much later, during a monthly progress review meeting, did a casual remark, which was quickly seconded by most of the meeting attendees, lead me to realize that most of them accepted the administrative problems rather than having to relocate away from the sunshine location of the plant. "Well, this is a lot better than moving to cold weather country." The reported complaints were not as important as the unreported satisfaction with the company's location. Therefore, can we really accept research without some personal interpretation? Some results attempt to describe internal states of mind, and these also often depend on the interpretations of the researcher and/or the subject.

These interpretations (of both researchers and subjects) might be understood better by using the next set of ideas about motivation. These ideas deal with perspectives of personal success at work (called *competence*). White's theory (White, 1959) could be an introduction into this very personal area, which deals with the person as an individual.

Proposed Universal Motivations

White suggests that people want to understand and manipulate their physical and social environment, and they interpret the results of that understanding and manipulation according to their perceptions of their own success. They like to make things happen—in a very broad sense of the word—to create happenings rather than wait passively for them to occur. Both Freud and White agreed upon the importance of creative work to the individual's personality, but there was also a difference between their theories. The difference between White's and Freud's ideas of motivation is that while Freud considered the emerging personality pretty well molded at a relatively young age, White's idea of the person's sense of competence is one of ongoing development. That seems to match most of our experiences working in technical groups, because that sense of competence apparently can even get off to a bad start and then recover to develop strongly as a result of later successes. The sense of competence, however, does not keep on growing forever, because we are physically limited beings. It is a kind of self-fulfilling prophecy. The individual rarely achieves more than he or she expects to, because that's what the individual thinks can be done. (Failure, of course, is also a self-fulfilling prophecy in this theory.)

On the other hand, we know that achievement does not always occur just because the person thinks that it is possible. Physical, psychological, economic, and social limitations are always there. There is only one explicit limitation in this motivational theory; you cannot achieve more than you think you can. This sense of competence can be a major factor in jobs in technical operations, where innovation and creativity are required. Successful innovation begins if the person thinks that it can be done. Then, if the person is right and there *is* success, the sense of competence becomes stronger. In turn, that supports further personal growth and change. (See Gellerman, 1963, pp. 111–114, for further descriptions.)

Turning now to other findings and concepts that are closely allied to ourselves and our people component, it has often been observed that equal rewards and achievements are not accepted equally by all organizational participants. Even if it were possible at a given time to reward everyone equally and have them value the reward equally, its worth to people would change with time and circumstances. Therefore, we will now deal with the changes in people's values.

NEED HIERARCHY

The need hierarchy motivation theory attempts to deal with this (Maslow, 1971, pp. 35–58). It suggests that there is a pyramid or hierarchy into which five classes of human needs are arranged. Those needs at the bottom of the pyramid take precedence over those next farther up and are the focus of the person's attention until they are satisfied. When that happens, the second higher class of needs becomes most important. When these are satisfied, the next class of needs takes over and so on until the top or fifth class of needs is reached. A lower or satisfied need is no longer a motivator, and while more than one class of need may be acting on a person at a given time, the dominant need is the lowest one in the hierarchy. The most powerful and first needs are the elementary biological needs which provide for life itself: food, drink, shelter, and so forth. Where life is primitive, these needs are never completely satisfied and they are the preoccupation of almost every waking hour.

When these very basic needs are satisfied and one can eat, have clothing, and have a roof over one's head, the next class of needs becomes of pressing importance: that of security. As long as the lower class of needs is satisfied, these needs are no longer as powerful motivators as the class now occupying the person's attention, but if the hunt for security begins to clash with reawakened and unsatisfied needs for food or clothing, security again takes second position. The five-step hierarchy is sequential:

1. *Physiological needs:* The basic requirements of self-preservation that can be described by needs for food, water, and shelter.

2. *Security needs:* Both physical and economic, although only the economic requirements concern us here. People want to be assured that they will be able to have an acceptable living standard which will be maintained in the future. Examples of this class of needs are pressures for maintenance of income, jobs, and general employment security.

3. *Social needs:* The need people have to be accepted as part of a social group and to be able to influence other group members.

4. *Esteem needs:* The need to feel important in society is next. Status, importance, prestige, and recognition are the motivators here.

5. *Self-actualization:* This is supposed to be the pinnacle of the need pyramid, where people want to feel that they are trying to reach their full potential and they have jobs they like that are well-suited to their skills and abilities. Job and individual importance, responsibility, advancement, growth opportunity, and challenge are included here. Although Maslow called this last need *self-fulfillment* in earlier works, in his later work (1965) he renamed it *self-actualization*. That seemed to be a more appropriate name for the needs behind people in this class of the hierarchy because these needs are attached to striving for goals rather than obtaining them.

The idea of the need hierarchy is intuitively appealing because it promises all kinds of explanations for previously unexplainable motivational drives. For example, few technical workers in Western cultures are driven by physiological needs. Therefore, security needs could explain the fear and despair of people who are out of work for extended periods of time, notwithstanding the size of their savings or the income that they might receive from unemployment compensation. It could also explain the greater success of unionization attempts during and immediately after a depression.

The social needs of people could explain the desire of workers to move around and socialize with others (Schrank, 1978) and their desire to form cohesive informal work groups or continue social relationships formed during work after working hours. Esteem needs could explain the comments on conspicuous consumption (Veblen, 1935). Finally, self-actualization could explain almost everything, since it is all-inclusive. And that is the problem with this theory; it is so all-encompassing it is hard to use it for specific situations.

For example, the idea that some needs can be satisfied with other needs rising to take their place, with the final self-actualizing need including all the economic and personal goals that one could have, tells management that there is no *single* solution to the problems of motivation. Needs are continually changing, and as soon as one company benefit is distributed or a raise in salary is achieved, that benefit or salary change no longer is a motivator.

Consider the manager attempting to use this theory with his technical personnel. How does he determine at which level of need (other than physiological) his people are? And if he determines that, there is the never-ending management task of evaluation, determination of reward, and delivery of the reward that is tied to some need. Finally, unless the recipient interprets the reward the same way the manager did and agrees that it meets his or her needs, the manager's actions may have no effect. The theory is simplistically appealing and has had some slight support from qualitative research, but the one quantitative test (Hall & Nougaim, 1968) did not support it. The qualitative support for the theory reported in a research summary is as follows: "Workers in higher occupational prestige categories, and those who have more education, place greater emphasis on challenge, autonomy, and other intrinsic rewards" (Mortimer, 1979, p. 5).

However, Hall and Nougaim (1968) tried to test Maslow's theory by evaluating the needs of managers working in a division of a large organization over an extended period of several years. Their hypothesis was that as the managers were promoted, their needs would change in accordance with Maslow's predictions. It didn't happen. The researchers reported that the managers apparently adjusted their occupational values, expectations, and aspirations over time so that they became more congruent with the rewards and satisfactions that they perceived actually existed in their work. The progression was *not* to satisfy a need and then work on another one higher up in the hierarchy as it arises; it *was* to adjust their needs to fit the opportunities available to them. That's quite a difference.

NEEDS FOR ACHIEVEMENT, POWER, AND AFFILIATION

Needs can be interpreted in different hierarchies other than the five noted above. One research effort (McClelland, 1961; 1962) indicates that most people have three related needs that affect the way they work. According to this concept, most people have one of these needs as a relatively constant and dominant force in their motivational and behavioral patterns. Although these needs may sound as if they came from a horoscope, they did not. There is substantial research that supports those descriptions (Litwin & Stringer, 1968, pp. 14–24).

1. *Need for achievement:* These people typically make good executives, particularly in challenging or difficult industries. They enjoy activity and like work in which they can take personal responsibility for finding solutions to problems. The idea of winning by chance does not produce the same satisfaction as winning by their own personal efforts. They prefer moderate achievement goals and calculated risks. They also want concrete feedback on how well they are doing.

2. *Need for power:* These people will usually attempt to influence others directly, and they are seen by others as forceful, outspoken, and demanding. There is little concern for warm, affiliative relationships and they tend toward having strong authoritarian values.

3. *Need for affiliation:* Since warm, friendly, companionate relationships are primary goals, these people often take supervisory jobs in which maintaining good relationships is more important than making deci-

sions. This need does not seem to be important for many managerial positions but it is a fair assumption that building good working relationships with both superiors and subordinates is required to obtain other, broader kinds of personal satisfaction.

The research does not propose that any individual has only a single one of these needs; only that one of them is usually dominant. It is quite possible for a person to have all three needs. An example is the case of one organization's president who had a very high need for power combined with a moderate need for achievement. In that company's dynamic economic environment, he created a thriving and successful business. In another firm, the president had an equally high need for power but not as much need for achievement. That president dominated every meeting, personally made every decision, and finally became the limiting factor in the growth of the business because of the physical impossibility of handling all the details of a growing business the same way he had when it was much smaller.

This need theory differs from Maslow's in that it has methods to measure the presence of the three needs. This method uses a derivation of the Thematic Apperception Test (TAT) developed by Murray (1938). The person being measured is shown a series of pictures, usually of people in ambiguous social and work situations, and is asked to make up a story for each picture in the series. The stories are either written down by the subject or recorded by the experimenter and then analyzed for evidence of the different kinds of imagery associated with the various motives. This test is projective because the person "projects" into the story her or his own thoughts, feelings, and attitudes. The stories are a sample of the kinds of things the person thinks about and include her or his dominant needs. Numerical scores of the imagery noted are matched to various scales, and the nominal amounts of the three needs (i.e., achievement, power, and affiliation) can be determined by a qualified experimenter.

Summary: Universal Motivation Factors

Need theories are universal descriptions of people's motivations. They include many findings as to why people work, but that is their main problem as well as their attraction. Too general a description provides insufficient specific guidance for the next step of applying that description to your own situation. That step requires more tools than these theories can offer at this time.

The Situation

Most ideas about motivation imply a relationship between the person and some situation, since no one ever really works completely alone. And the ideas we cover now treat that relationship between the person and the situation explicitly, but without referring to specific groups or cultures. Those ideas are the ones suggested by equity concepts (Adams, 1963), two-factor theory (Herzberg et al., 1959), infantile treatment by the organization (Argyris, 1957), and best person-leader patterns (Likert, 1959). All these theories seem to apply best to the relationship between the individual and his or her situation and all have been expanded by other, more recent researchers, but the theoretical cores remain unchanged.

Equity theory was selected because it deals with a one-on-one valuation of the exchange of services for reward in a job. Two-factor theory was selected because of the seeming applicability to technical personnel, since many of the subjects were engineers. Infantile treatment is also concerned with exchange between the person and the organization, but that exchange is considered to be an immature one. The knowledge worker is supposedly treated almost as a child, which results in conflict as an endemic part of the organization. Finally, best person-leader pattern theory was selected because of the vital exchange between the person and the organization through the behaviors of his supervisors or leaders.

EQUITY THEORY

Equity theory generally is based on relative rewards and deprivations. The employees provide the *inputs*, which typically include particulars such as education, experience, training, seniority, skill, creativity, sexual differences, ethnic backgrounds, social status, intelligence, and the efforts expended on the job. The employer provides the *outcomes* or rewards to the employees for those inputs. These typically might include pay, fringe benefits, intrinsic rewards, seniority, and status symbols.

An employee perceives that inputs related to outcomes are inequitable "whenever his perceived job inputs and/or outcomes stand psychologically in an obverse relation to what he perceives are the inputs and/or outcomes of others" (Adams, 1963, p. 424).

If the employee's inputs exceed those of another, but his or her outcomes are the same or less, inequity exists for that employee. Similarly, if his or her inputs are less than another's, with the same or greater outcomes, inequity also exists. There is a type of mathematical formula:

$$\text{Inequity} = \frac{\text{person's inputs} - \text{person's outcomes}}{\text{another's inputs} - \text{another's outcomes}}$$

This perceived inequity involves tension, and the persons are assumed to be motivated to reduce this tension through psychological and/or actual physical changes in the inputs, the outcomes, and/or the selection of the comparison "other."

Although the theory doesn't provide any general guidelines between the amount of tension resulting from overreward and that resulting from underreward, there have been findings that fit an intuitive answer. The overreward tension may be reduced more easily by the person's increasing the subjective value that he or she places on his or her input. It's easier to come to the conclusion that you're worth more than that you're worth less. This theory provides an explanation of the way in which people develop standards against which they judge the fairness and equity of the rewards received on the job.

When unfairness and/or inequity occurs, the increased internal tension is reduced in two general ways: Change either the cognitive framework or the actual physical conditions. Changing the cognitive framework means rethinking the value of the input and/or the output, or even changing the mental relationship to the "other," who is used as a standard. One way that it can be done is described in another idea involving cognitive dissonance (Festinger, 1957). According to this theory, if a person feels that he or she is being treated inequitably because, for example, that person's salary is less than that of an associate who seems to be doing the same work, that feeling is often reduced through these mental alternatives.

One alternative is to find something about the job outputs, other than money, to which the person can attach a higher value. For example, members of the armed services who are paid less than equivalent managers in commercial organizations may regard themselves as defenders of freedom and thereby attribute additional social value to their work. That changes their cognitive framework and may decrease feelings of inequity. Another alternative, according to the theory, could be to place a higher value on the work of those equivalent industrial managers, thereby concluding that, for example, a member of the armed services is not worth as much as those managers. And, finally, a third alternative could be to pick a different working group or associate as a comparison. One could also use any combination of these alternatives. While those changes would definitely modify the cognitive framework, taking physical action to change the situation could also change feelings of dissonance. Such action might take the form of decreasing the quality or the quantity of work done, asking for a raise, or, in some cases, changing jobs either within or outside the company.

An interesting and rather surprising prediction of equity theory is that people do not seem to attempt to maximize their outcomes (such as money or status) but instead try to obtain only an equitable or fair amount. With less than the "fair amount," there is a feeling that an injustice has been done and with too much, potential guilt about receiving it.

This theory might be particularly applicable in describing motivations of individuals working in situations where everyone is doing essentially the same type of work. These conditions will occur less in the future in technical organizations but presently they might exist in functions such as administration or in engineering standards, where it is easier for one person to compare herself or himself against another. However, if the technical group is involved in creative design or innovation, this motivation theory is unlikely to be a central one. It is much more difficult to compare one's inputs and outputs against another's in that situation (although the human mind is capable of many unusual things), besides which many of the people that are drawn to this kind of work (creative and innovative) seem to be less concerned about comparing themselves to others than about satisfying some inner drive. Therefore, it seems reasonable that equity theory might apply best to people in production-oriented tasks. For creative or innovative poeple, some other theory of motivation might be more suitable; possibly the two-factor theory.

TWO-FACTOR THEORY

This research was based on an investigation into the reported causes of job satisfaction and dissatisfaction of *engineers* and accountants. The researchers (Herzberg et al., 1959) asked these two groups to tell them about the times when they felt very good and very bad about their jobs. The reports about the "good" times frequently concerned the content of the job. Achievement, responsibility, recognition, and the work itself were most frequently mentioned. Conversely, the "bad" times were concerned with the job context or the situation, not the job itself. Frequently mentioned in this latter category were supervision, company policy, working conditions, and salary.

These findings seem to mean that the job itself tends to produce satisfaction (and motivation). However, unfavorable job contexts such as poor supervision and bad working conditions tend to produce dissatisfaction, although good supervision and good working conditions do not produce satisfaction. These findings could be confusing until we realize that they indicate a nonlinear effect of both job context and job content on job satisfaction. In other words, improving the job context or surroundings will increase the person's job satisfaction only to the point at which that person becomes almost neutral about the job. At that point, the job itself must produce any positive feelings of job satisfaction and related motivation.

There were several problems based in the assumptions in the data-gathering process behind this theory. For example, there was the assumption that the people who were questioned had the ability (and the motivation?) to report the conditions of their jobs that satisfied and those that did not satisfy them accurately. They might not have reported objectively (if that were even possible), since it also seems reasonable to suggest that the people questioned might attribute successes and accomplishments to their own efforts (i.e., as a result of their own motivation on the job or the job itself) and dissatisfaction or inadequacies not to themselves or their work but to obstacles represented by company policy or supervision (i.e., the working conditions or the job context). These assumptions were the subject of some criticism in the psychological press.

In spite of this, the theory does seem to be applicable to some extent to the part of the technical group concerned with creativity and innovation. It intuitively fits better with the description of the creative physical scientist who is very involved with her or his work and values the freedom to create more highly than the physical surrounding of that work. It is also a step forward in terms of measurement since we can determine by questionnaire what is needed to satisfy the requirements (or hygiene factors, according to two-factor theory) of the job context before setting up interesting jobs that have the intrinsic motivating factors in them (or those factors described in this theory).

This theory assumes that people are mature, since they are supposed to be able to report their needs and independently respond to both the job context and the job content. It also could connect with Maslow's ideas about a hierarchy of needs (see Figure 3-1). If the lower needs of physiology and safety (as defined by Maslow) are satisfied, they will no longer motivate. If the person is assumed (in the framework of the two-factor theory) to have those needs satisfied (part of the hygiene needs for most technical personnel), the next motivating factors should be social. Most technical personnel have the freedom to socialize on the job if they wish to do so. Therefore, two-factor theory could have been tapping either esteem or self-actualization needs, which is reasonable, since these needs would be related to the job content and not the situation or job context. Eventually, we have come to the conclusion that it is the job itself and not the context that is central for creative and innovative people.

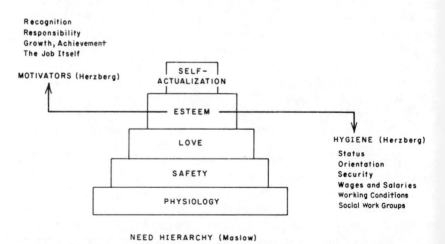

Figure 3-1   Maslow's and Herzberg's theories: A possible combination.

In contrast to two-factor theory, Argyris (1957) proposes a framework that suggests that the person is treated and responds as if he or she were a child in most organizations. In this theory, the job context and not the job is central.

MATURITY AND SUPERVISION

Argyris proposes that most traditional industrial organizations treat people as if they were irresponsible children and not the mature adults that they really are. For example, he suggests that seven changes normally occur as an individual matures:

1. . . . he moves from a state of passivity as a child to one of increased activity as an adult.

2. . . . gradually outgrows his total dependence upon others and develops a capacity to shift for himself.

3. . . . has a limited repertoire of behaviors, but as he grows up, he can respond to a situation in different ways.

4. . . . does maintain his interest for a very long period of time.

5. . . . time perspective is very short (but increases with maturity).

6. . . . develops from being everybody else's subordinate to being an equal or even a superior.

7. . . . does not have much of a "self" to have attitudes about, as a child [but apparently will when he or she becomes an adult]. (Argyris, 1957, p. 73)

He suggests that these changes are completely ignored by organizations and people are treated as children and consequently are coerced to be dependent, subordinate, and submissive.

> There are three major sets of variables which cause the dependence, subordination, etc. The formal organization-structure is the first variable. . . . Directive leadership is the second, and managerial controls (budgets, incentive systems, quality control, motion and time studies) is the third. The degree of dependence, subordination, etc., that these three variables cause tends to increase as one goes down the chain of command and as the organization takes on the characteristics of mass production. (Argyris, 1959, p. 119)

According to this theory, the person's reaction is appropriately immature, involving informal activities such as apathy, goldbricking, and/or rate-setting. However, if the organization changes its treatment of people, the following "predispositions" (Argyris' word) or mature feelings and attitudes of persons should appear:

1. The need for togetherness in relation to other employees.

2. Viewing wages as guaranteeing a fair standard of living and a secure job.

3. Noninvolvement about upward mobility in the company.

4. Control over one's immediate work environment, including the need to be left alone.

If these predispositions are encouraged by the organization there will be "the combination of the informal employee culture and the psychological work contract" (Argyris, 1959, p. 148), the employee will accept change, and (implicitly) productivity and organizational growth will improve. This connects to the Hawthorne experiments, in which human-relations theorists attempted to show that the best organization was one in which the person, the working group, and the total organization had the same goal: increased productivity. Of course, in that case, the formal organization defined what productivity was and in this one, it's supposedly the individual. This theory seems to be more applicable to technical operations and could be placed in groups in which either production or creativity is stressed.

In my opinion, there are still many examples of attempted organizational treatment of technical personnel as immature. But I believe that as time goes on they will diminish, because the demographics are against it. With most of our production now dependent upon the output of the "human capital" of our industry, treating this "human capital" as if it were childlike would only cause our productivity to drop. That has not really happened to any great extent yet in human services. The pockets of resistance in which an immature type of organizational climate exists can only diminish in the future because the technical people in those organizations will not be producing. Those organizations will not be able to exist in an environment with a growing need for creativity and innovation. How can anyone *demand* creativity?

There is a major difference between the viewpoint of those researchers who assume the inner motivations of people at work and those who ask the people themselves to report their motivations. Two-factor theory asks people and seems to assume that the respondent's internal mental state depends on her or his work. On the other hand, other theories of Argyris and Likert (whose research we will evaluate next) state that everyone has inner motivations or basic requirements for independence and opportunity, and these requirements are not being satisfied in the work situation at all, but are related to it.

That position is partially supported by research that investigates the centrality of work to the person's life interests. We have already noted the results of Morse and Weiss, who found that most people would continue to work even if they had no economic motives, and recent demographic data show that there are indications that work itself is becoming more important even when there is a diminished economic need. In all occupational categories and especially in technical areas, people seek ". . . both intrinsic and extrinsic satisfactions from their jobs" (Mortimer, 1979, p. 16). Intrinsic means things like self-satisfaction and achievement. Extrinsic means pay, position, etc. This seems to give more support for the two-factor theory and less for the inner motivation theories, which are not as job-dependent.

But intrinsic and extrinsic satisfactions are defined differently, depending upon the person and the theory selected. In Argyris, we noted that the descriptions of the sources of motivation include the effects of management and supervision. That is different from the ideas in the two-factor theory, which place supervision in the "hygiene" category, as a demotivator rather than a motivator. But the definition of what is intrinsic (besides the obvious definition of being within the person's mind) and what is extrinsic (what he or she physically receives from work) could determine which theory applies best. Definition and measurement seem to be the major sticking points in all these theories and obviously the complexity of the human being has not been completely described so far. However, supervision does play some role in all the theories covered. That is the concern of Likert's person-situation interactive theory. As in other theories, there are some interesting definitions of sources of motivation that he claims everyone has in the work situation.

LINKING PINS AND SUPERVISION

Likert (1959) is concerned with reinforcing major motivational forces that he says everyone at work has. These forces include economic motives, ego motives (such as status, recognition, approval, and acceptance), security motives, and a desire for new experiences (i.e., creativity or curiosity). His position is that "an individual's reaction to any situation is always a function, not of the absolute character of the interaction, but of his perception of it." It is how the person sees things that counts, not objective reality (if that can ever be defined).

Consequently, "an individual will always interpret an interaction between himself and the organization in terms of his own background and culture, his experience and his expectations" (Likert, 1959, p. 191). If the person is to satisfy his internal motivational forces, the organizational situation should be viewed as supportive and one which contributes to the individual's sense of personal worth. Part of that support is assumed to be a satisfactory level of economic reward, but so is the decision process by which the levels of economic reward are established, according to Likert.

In other words, people have intrinsic needs; but the satisfaction of those needs is related to the person's perception of organizational support (the interaction between person and situation). And that support is shown primarily by enough income and influence in the process of deciding what that income should be. That influence occurs mainly through the person's direct supervisor. This concept is the basis for a "bottom-up" directed management structure.

The supervisor is a "linking pin" (Likert, 1959, p. 203) between the person and the next higher level of management in the organization. Therefore, a person is supposed to be able to influence his or her supervisor, who in turn influences *his* or *her* boss, and so on. Similarly, the supervisor's boss is the linking pin for the supervisor up into the next highest level. When this situation applies, the major inhibiting factors against productivity, such as punitive budgets, become unnecessary. Productivity increases, because the linking pins ensure that the great majority of the people have been able to influence the relevant goals of the entire organization. This tends to produce greater cooperation with the organization in attempting to achieve these goals. Gellerman sums it up:

> The real business of management is to assure the profitable use of its assets; therefore productivity is the goal, and control is merely one of the possible means to achieve it. But the effect of control concepts is too often an inhibition, rather than a stimulation of productivity. . . . Sustained high productivity eluded the man who was preoccupied with it and fell lightly into the lap of the man who was chiefly concerned with creating an atmosphere. It was almost as though the most sophisticated way to serve management was to ignore it and serve the employees themselves instead. . . . At the root of this paradox is a simple idea (so simple as to seem almost self-evident) which actually has some rather revolutionary implications. That is the responsibility for production is inherently the province of workers, not supervisors. The proper function of supervisors is to provide information, materials, and organization that workers need to do their jobs and otherwise stay out of their way (Gellerman, 1963, p. 45–46).

Summary and Caveats: The Individual

There seems to be some common problems with most of the descriptions that we have been reviewing:

1. *Standard internal motivations:* Whether you assume that all people have the same basic internal motives either by themselves or interacting with the work situation (Argyris, Freud, Likert, White) or a succession of motives (Maslow, Herzberg), there really are no tools to determine what those motives are or how to measure them.

2. *Changing of motivations:* The relationships of various motivations within people (McClelland) or how they are changed with circumstances (Adams) is also subject to question. Needs for achievement, power, and affiliation have, of course, repeatedly been found in people and the Thematic Apperception Test that is used is perhaps one of the best tests in terms of validity and reliability that psychologists have. The test results, however, are subject to fairly casual and transient kinds of unintentional manipulations by experimenters, most of the validation studies were done on college students, and there is some uncertainty as to whether they are as valid for females as for males (Sechrest, 1968, pp. 529–628).

3. *Reporting:* People do not perceive situations the same way, nor can they be completely objective in their reporting of feelings, interpretations, reasons, and so on. The subject is, in effect, the object.

4. *Research descriptions:* As I have pointed out before, researchers are just as human as the subjects on whom they report. They (and I) have implicit (and sometimes explicit) positions from which data are recorded and analyzed, theories formed, and hypotheses tested. These positions sometimes include a moralistic tone that should be considered.

> Humanists' goals and behaviorists' objectives appear similar. Both accept Maslow's self-actualization concepts as the preferred route to self-fulfillment. But by what divine right does one group assume that its values are superior to others and should be accepted as normal? Both the selection of goals and attitudes toward work are uniquely personal. The judges of human values have no moral right to press their normative concepts on others as preferable. (Fein, 1974, p. 72)

However, in summarizing the data (and in all fairness to the researchers noted above), there do seem to be repetitive and important themes that appear in most of the theories on motivation. My interpretation of these themes would be:

1. Work is important to people, and that importance generally seems to be related to the amount of training, personal investment, and organizational position a particular person has.

2. Economic security and satisfactory compensation levels are prerequisites for most people before they can begin to devote their full efforts to the job.

3. There is an important motivational interaction between the person and the situation, and that interaction is interpreted differently according to the perceptions of the person concerned. The perception, in turn,

is modified by work experiences rather than inherited characteristics of the person.

4. People function best in groups in which there is a similarity in attitude among the members of the group. However, different groups within the same organization do think differently from each other and usually do not have the same overall goals as the organization does.

5. There is no one single way to describe motivation. Organizations can move toward getting more effort from technical people by developing structures that are more open. However, there are important considerations of the needs of the cultural, social, economic, and technological environments that affect the amount of openness.

6. Managers can attempt to gain motivation from employees through the manipulation of mutually agreed upon contributions and rewards associated with those contributions.

All the descriptions that we have looked at so far seem to indicate that motivational concepts that are wholly based either on situational or personality variables are not likely to be of too much help to us in building our own motivational theories. They attempt to explain a complex problem with an oversimplified description. The complexities of the subjects being studied—human beings—tend to minimize the use of simple theories. They rarely can explain complex phenomena. They can be used, however, as building blocks for more complex theories and as guidelines for the construction of specific hypotheses to be used in specific situations.

Then, of course, we may use more than one theory to account for differences in individuals and in situations. When we do that, we are no longer even dealing with the relatively simple person-situation interaction, as in situational descriptions; we are moving past description and using theoretical building blocks to predict and possibly to change and control. We are dealing here with a theory subset, and that is called contingency theory. The difference between the two theory types may be understood as follows: Situational theory describes and contingency theory prescribes.

Situational theories usually are meant to show what happens as the result of the person-situation interaction. Contingency theories usually suggest: If this situation arises, this is what to do. These presentations are not that clearly stated, since situational theories usually have some implicit contingencies. And contingency theories, of course, are always based in situational variables. There is another, smaller subset of contingency theory, called expectancy theory, which begins to provide some special tools for personal theory building. Using all the other theories that we have discussed as bases for analysis of the person(s) and the situations, we now come to the personal theory and the development of the motivational hypotheses to test in your own organizational situation.

EXPECTANCY THEORY: PRESCRIPTION FOR MOTIVATION

People (personnel, managers, etc.) in technical organizations are supposed to make decisions. There is always an element of risk in decision making and whenever an "individual chooses between alternatives which involve uncertain outcomes, it seems clear that his behavior is affected not only by his preferences among these outcomes but also by the degree to which he believes these outcomes to be probable" (Vroom, 1964, p. 17). In very general terms, the decisions and the motivations behind them depend upon some function of the person's subjective *total value* of the probable results of that decision multiplied by the *total expectancy* that the results will actually occur. The subjective total value is the sum of values of that decision. The total expectancy of occurrence is the sum of the potential achievement of those values. In the following equation, M = motivation, A = sum of values, and B = sum of expectancy of achievement.

$$M = f \times (A \times B)$$

Using a simplified example, if the alternatives for a decision to complete or not to complete a machine design today are limited to whether the designer will work overtime this weekend or let it go until Monday, the selection of the alternative chosen *could be* as follows:

1. Summing all these subjective values (to the designer) for working overtime:
   a. Overtime pay (positive)
   b. Compliment from the boss (positive)
   c. Increased status and recognition in the drafting department (positive)
   d. Loss of free time (negative)
   e. Complaints from the family (negative)
   f. Others, both negative and positive

   As in all motivation theories, implicit assumptions abound here. That it is possible for people to define and add both negatively and positively valued outcomes to determine a sum of values seems like a reasonable assumption. People do make value judgments all the time, and although the cognitive mechanism by which they do this is not really known, it does happen. In this theory, when the mental sum of the values is positive, motivation is increased; when negative, motivation is decreased.

2. Summing all the potential expectancy of achievement (again to the designe

   a. The design will be completed (positive).
   b. The design will not be completed (negative).
   c. The design will be partially completed (either way).
   d. Others, both positive and negative.

In addition to being able to add expectancies of achievement cognitively, there is another characteristic that applies only to expectancy. The range of expectancy can vary, in my opinion, only from $+1$ to zero to -1. In other words, a $+1$ means that if the decision is made, the person absolutely expects the achievement to occur. A zero means he or she expects that nothing will happen. That occurs when you multiply any value by zero. Multiplying a positive value by a positive expectancy should result in some positive number that determines the person's motivation. On the other hand, nothing will occur when multiplying by zero, and demotivation should occur when the expectancy is less than zero.

Since the person's values and expectancies are based on that person and on that person's perceptions of the situation, the idea of "expectancy" as a part of motivation neatly includes the whole range of potential causes. The historical context of an individual's expectancy is quite important, since it affects how he or she now values future outcomes.

As an example, the management of a listed industrial manufacturing company attempted to develop a motivational plan based on this theory. It used a modified type of management by objective to establish measurable goals and the suitable rewards for achieving those goals. But it discovered that very few of the technical personnel were willing to estimate the expectancy of achievement (of the contingent reward for reaching the predetermined goal) to be greater than zero, because in the past, reaching a goal was not rewarded. Changes had always occurred between the setting and the reaching of the goal, and the variances between the two had become excessive. Because of this there were few rewards, and the people felt that management never either maintained its goals or paid off on its promises. Their perceptions were very different from those of top management.

The changes in direction of objectives and goals during the time period between setting objectives and measuring progress toward them have to be considered if the final variances are to be separated into those due to the changes in the objectives by top management and those due to the employee concerned. The emphasis must be on today's happenings and the reasons behind them. Implementing these expectancy concepts takes time. In my experience, it takes about three years of consistent effort before they are accepted. The administrative management follow-up involves restructuring goals and objectives as required, but also *paying for partial achievements* if the restructuring caused by management greatly modifies the original objectives. Of course, if the restructuring is caused by the employee, there may not be any payment at all, since the objectives were not met. I say, "may not" because sometimes restructuring is required because of unforeseen technical problems and not personal motivational failures.

THE DESIGN

Expectancy

The closer the direct connection between some organizationally valued goal and the person's expectancy that she or he can achieve it, the higher the probability that she or he will subjectively assign to it some value between zero (it will never happen) and $+1$ (a sure thing). The goal must be both definable and measurable, so that interpretation of the degree of goal achievement is minimized. Optimally, the goal should also be highly valued by the person. "A better attitude toward the company," for example, is not exactly a measurable goal unless there is an agreed upon attitude-measuring instrument, but completing a specific design (including documentation and lists, as defined) within a specific time period given specific resources is measurable.

The probability score assigned in the mind of the person doing the work is, by definition, subjective and almost impossible to determine objectively, but it can be done and documented by the person who uses subjectively valued ordinal measurements. But regardless of how it is done, you can be sure that the employee mentally calculates the probability of achievement and the next part of the equation, the payoff.

Values

The other part of the equation, the values, is just as difficult to handle as expectancies. Those values are limited by the rewards available to and deliverable by the organization, since industrial organizations have legal, economic, ethical, and other limitations. Some rewards may be impossible to deliver. No matter how much of a motivational value the person places on the painting of Mona Lisa by da Vinci, it's not likely that the painting will be a reward in any motivational scheme. Conversely, the person might regard the limitations positively, because he or she recognizes that some things cannot come from the job and some can, and it is always helpful to get a clear understanding of which is which. (On the other hand, considering it strictly from a theoretical viewpoint, if the problems and risks in removing the painting from the Louvre were less than its value to the person, a strictly logical approach would be to attempt to remove it—an example of a completely valid but almost certainly unachievable goal.)

Therefore, the motivational system design includes negotiating a measurable, mutually agreed upon, and valued goal, supplying the resources necessary to achieve it, and establishing an organizationally acceptable reward for achievement. In simpler terms:

- Set up goals.

- Determine the rewards, and measure achievement.

- Deliver!

The last item is difficult if you, as the manager, do not have control over such things as salary changes, office assignments, time off, or any of the other rewards that could be valued and are within the ability of the organization to provide. It is even possible that public approval of a job well done could be a valued reward, and that is certainly within your ability to provide, as a manager. However, even if you do have control over extrinsic rewards, the next pitfall to avoid is the obvious one of goal change during the achievement process. If that is inevitable, a partial reward must be delivered and a new goal established.

THE CAFETERIA APPROACH

In the cafeteria approach, the person is involved in both the goal definition and the methodology for achieving that goal. (This is based on a management-by-objective concept.) The goals are a joint determination between the person and the technical manager, but since a person's values may change between the time that the goals are set and when they are achieved, the rewards do not have to be a joint determination at this time. Within the rewards that the organization can deliver, it should be possible for the person to be able to set his own values by selecting the reward that is most appropriate when payoff time comes. This happens to some degree when someone gets a raise in salary because of goal achievement, because that person can spend the raise any way he or she likes. The employee is selecting the eventual personal value (e.g., the new car) through spending the additional funds to get that value. Commercial establishments have successfully used trading stamps in this kind of a motivation system. In order to achieve their own sales goals, they tied customers' purchases or increased sales to some universal medium of value or trading stamps that customers could use any way that liked.

But money, although very important, isn't supposed to be everything.

Assuming that the basic requisites of satisfactory compensation and security are achieved, some of the theories suggest that people may want things other than money. If those things are of equivalent value to each other and can be provided by the organization, it seems reasonable to design a plan that would provide them. Therefore, if the person can pick either the rewards or the medium of value that she or he considers appropriate for reaching a predetermined goal, at least that part of the motivation formula is satisfied. Examples of equivalent types of rewards might be getting additional vacation time, attending technical conferences, getting a bigger office, or receiving a reserved parking space. The medium of exchange could be any kind of company scrip and the selection could be from the organizational "catalog" of rewards. (I say again, however, that this does not eliminate the need for satisfactory compensation and job security.)

Getting Started

According to Perham (1978), several major companies have begun to implement this *cafeteria* approach to rewarding people, but so far they have limited it to selection of fringe benefits. People select the benefits they want from a catalog of available alternatives of medical, disability, vacation, savings, and pension plans. They use "company dollars" to buy them and may change the mix they choose every year. Those company dollars are equivalent to a medium of exchange and are distributed in accordance with salary level, tenure, and other variables of employment. The employees "buy" different packages of benefits as they wish, with only a minor stipulation by the companies that certain minimums in insurance and personal protection must be maintained. As a person's circumstances change over the years so can the benefits she or he selects.

This concept should be equally applicable to goal setting and achievement; and would be particularly applicable in a technical organization, where effort and productivity are difficult, if not impossible, to measure objectively. In that situation, the end results could be predefined and the general methodology chosen to achieve them. A program of implementation might involve the steps of goal setting, value setting, measurement, and payoff:

Step 1: Setting the Goals

The problems are no different from those in any goal-setting program. They include, using McGregor's (1972) format:

1. Arriving at a definition that is mutually agreed upon

2. Determining the measurement of achievement and allowing for variances both above and below goal achievement

3. Setting the methodology, including resources available and procedures to be followed

Step 2: Setting the Values

1. *Determining the values available in the organization.* Those values can be almost anything that is within the management prerogatives of the manager concerned. (This is not a straightforward task. I remember one manager who confidentially told me that he had no control over salaries, positions, hiring and firing, desk location, or other rewards. These rewards were all controlled by company policy. Therefore, there was no possibility of his getting a better output from his people because of their own input. Of course, that was a self-fulfilling prophecy if equity theory could be applied here. In that situation, since no person's rewards could be affected directly by that person's own performance, most people's performance had sunk to some minimum level. Any performance above that level depended solely on the person's internal programs and any personal influence exerted by the manager himself.)

2. *Defining how the value shall be connected to the goal achievement.* For example, for every hundred valves assembled during the next three months in your assembly department, there shall be a maximum of three functional failures at final test, using test procedure 8364-B. This goal shall be worth fifty points. Each failure over three shall lead to the deduction of twenty points, and each failure less than three, to a gain of twenty points, e.g., no failure = 160 points.

Step 3: Payoff

Define the value of the points received toward the potential rewards; e.g., at that person's grade level, fifty points may be worth an extra week's vacation, a raise of a predetermined amount, or some other reward. Of course, when all that can be given is the salary increase, the "cafeteria" approach begins to resemble closely a management-by-objective program that is completed honestly and paid from salary increases.

Advantages and Disadvantages

Some of the relative advantages and disadvantages associated with this approach are:

ADVANTAGES

1. *Definition of needs.* There is no necessity to define the status of the person's present needs or historical antecedents. Only those defined *now* by the person in making choices of values are relevant.

2. *Definition of goals.* Only those goals that can be defined and measured can be used.

3. *Other considerations.* These ideas utilize the other motivational concepts as raw material for goal and value setting. A cafeteria approach helps the person to achieve personal as well as organizational goals. It helps the person to control the job.

DISADVANTAGES

1. *Openness and trust.* This approach requires an environment quite difficult to achieve in many industrial organizations. Machinelike organizations that are concerned primarily with high production and very tight product quality controls might not respond well to creative kinds of organizationally open activities. Conversely, a typical advanced research and development group could not operate without it.

2. *Value statement and goal setting.* Although this is intended to be a one-on-one process, the group and cultural environments often provide an all-too-effective limitation on it. A culture that is tied to company policy and not supportive about viewing people differently will not allow this approach.

3. *Administration.* This system is more complex and costlier than adopting a uniform or an anniversary-based performance review system.

SUMMARY

I started this chapter with the intent of assisting you to develop your own theory of motivation and possible hypotheses to test it. To do that, I used a descriptive, then a prescriptive, approach covering ideas in the literature and in other managers' experiences that could be used as building blocks for your theory. We recognized that each organization and situation is unique, so the applicable theory probably will be as unique as the situation to which it is applied. The building blocks were put into *universal* and *situational* interactive columns in a descriptive framework (see Table 3-1). That framework had rows labeled cultural, work group, and individual concepts. Classification by authors who espouse these ideas is shown in Table 3-2.

TABLE 3-2   MOTIVATION THEORIES CLASSIFIED BY AUTHOR

	Universal	Situational
Cultural	Weber	Veblen
Work Group	Roethlisberger and Dickson, Homans	Coch and French, Lawrence and Lorsch
Individual	McClelland, Maslow	Adams, Herzberg, Argyris

My general prescriptive suggestions are based on the concepts discussed and my general conclusion is that no one can ever be motivated, since it is literally impossible for any manager to define for any other person the variables that will effect positive change for that person. The usual notion in most of industry is that somehow managers must motivate their employees, but it actually works the other way around. Motivation comes from within the person; it is an urge to accomplish something, to do something that will reach some goal. Managers, therefore, can only manipulate rewards; they cannot change internal motivation directly. In effect, we can see the output of motivation, but not the motivation itself. And the rewards a manager can offer are relatively limited: money, position, a few others. It depends on how much the person values those rewards. However, I do feel that these suggestions are valid:

1. Any method for changing the behavior of the person requires consideration of the interaction between the person and the situation. Universal theories are just too general to be of immediate use.

2. By attaching a reward the recipient values to mutually agreed upon, measurable goals, behavior can be modified.

A contingency cafeteria approach seems to resolve the problems of internal motivations and their change over time. It promises to be a self-fulfilling prophecy if limitations on achievements and rewards are well-defined in the context of the situation. The responsibilities for implementation are clear: The person achieves, the organization delivers. Both sets of implementation tasks require definition, and while that definition is difficult, it is not impossible. For example, can there really be an equivalent definition of the value of an extra week's vacation and, say, a larger office or a private parking space? The first stage in the solution of any problem is definition. I hope that we have at least defined our terms through the general prescriptive approach. Implementation depends on the motivational hypothesis selected, and that is up to you.

The repetitive interactions among people define the next most important component of our organizational model, *Structure*. We will, in the next chapter, follow the same descriptive, then prescriptive, method with the same intent: to build your theory and testable hypotheses.

SUGGESTED ANSWERS TO CASE QUESTIONS

1. The engineering department has not adjusted to the changing economic environment. Both the company and the engineering group have stopped being innovative. The engineers seem to be quite able to support a production-oriented environment, but are not "different" enough from production to begin to produce new products.

2. Is it necessary for everyone to be treated alike? What does that mean? I have been in organizations where administrators were required to fill in forms, even to the extent of sitting down with designers once a week to find out what they were doing. As far as recruiting is concerned, my feelings are that creative people are the best evaluators of other creative people, and I would suggest that Millard be one of those selected to interview potential new employees. Reporting systems should fit the situation. If time sheets are inappropriate, perhaps weekly progress review meetings with resulting minutes providing the reporting would be sufficient. Systems should suit the people. When the people have to suit the system, there is little creativity.

3. Peter seems to fit the existing culture of the engineering department, which in turn fits the company culture very well. Millard fits neither. It is ordinarily difficult to change culture quickly but the present situation George faces calls for drastic action, such as forming a separate and special new-products group, physically away from Peter and the existing group.

4. Intrinsic motivators could be a quiet place to work and no requirement for filling in time sheets. Extrinsic motivators could be adequate salaries and an interconnect to the computer from the home. Can you predict what would happen with the people in this case study?

5. George should discuss his own evaluations of the situation and compare them with Peter's. If I were George, I would suggest that Peter come back with other alternatives to solve the problem of Millard.

6. While it might appear that Millard is on to something, I would recommend that you follow very conservative accounting concepts when responding to upper management requests; i.e., report a loss as soon as it occurs, but don't report your profits until you're sure of them. I would respond by telling Marge about several potential possibilities for improvement and what would be required to achieve them; for example, different recruiting and training practices, changing the company policies on computer interconnects, implementing a different "cafeteria" motivation program. He can then subjectively assign risk and value factors to each of the recommendations for future developments.

Those are my suggestions. There is another question you might want to work out for yourself. How would you develop a motivation program for these people? Select one person and work out a hypothesis. Could you use these ideas in your company? Why?

REFERENCES

Adams, J. S. Toward an understanding of inequity. *Journal of Abnormal Social Psychology*, 1963, 67, 422–436.

Argyris, Chris. *Personality and organization*. New York: Harper & Row, 1957.

Argyris, Chris. Human behavior in organizations. In Mason Haire (Ed.), *Modern Organization Theory*. New York: Wiley, 1959, pp. 115–154.

Blood, M. R., Hulin, C. R. Alienation, environmental characteristics, and worker responses. *Journal of Applied Psychology*, 1967, *51*, 284–290.

Coch, L., & French, J. R. P., Jr. Overcoming resistance to change. *Human Relations*, 1948, *1*, 512–532.

Fairfield, Roy P. *Humanizing the workplace*. Buffalo, N.Y.: Prometheus Books, 1974.

Fancher, Raymond E. *Psychoanalytic psychology—the development of Freud's thought*. New York: W. W. Norton, 1973.

Fein, Mitchell. The myth of job enrichment. In Roy P. Fairfield (Ed.), *Humanizing the workplace*. Buffalo, N.Y.: Prometheus Books, 1974, pp. 71–78.

Festinger, L. A. *A theory of cognitive dissonance*. Evanston, Ill.: Row Peterson, 1957.

French, J. R. P., Jr., Israel, J., & As, D. An experiment in participation in a Norwegian factory. *Human Relations*, 1961, *13*, 3–19.

Gellerman, Saul. *Motivation and productivity*. New York: American Management Association, © 1963.

Haire, M., Ghiselli, E. E., and Porter, L. W. Cultural patterns in the role of the manager. *Industrial Relations*, 1963, *2*, pp. 95–117.

Hall, Douglas T., & Nougaim, Khalil. An examination of Maslow's need hierarchy in an organizational setting. *Organizational Behavior and Human Performance*, February 1968, 12–35.

Herzberg, F., Mausner, B., and Snyderman, B. *The motivation to work* (2d ed.). New York: Wiley, 1959.

Homans, G. C. Social behavior as exchange. *American Journal of Sociology*. 1958, *63*, 597–606.

James, William. *The principles of psychology* (Vol. 2). New York: Dover, 1950. (Originally published, 1890.)

Katzell, Raymond. Changing attitudes towards work. In Clark Kerr & Jerome M. Rosow (Eds.), *Work in America: the decade ahead*. New York: Van Nostrand Reinhold, 1979.

Kohn, Melvin L., and Schooler, Carmie. Occupational experience and psychological functioning: an assessment of reciprocal effects. *American Sociological Review*, February 1973, *38*, 97–118.

Krupp, Sherman. *Pattern in organization analysis*. New York: Holt, Rinehart & Winston, 1961.

Lawrence, Paul R., Lorsch, Jay W. *Organization and environment*. Boston: Harvard Univ. Press, 1967.

Likert, Rensis. A motivation approach to a modified theory of organization and management. In Mason Haire (Ed.), *Modern organization theory*. New York: Wiley, 1959, pp. 184–217.

Litwin, George H., & Stringer, Robert A. *Motivation and organizational climate*. Boston: Harvard Univ. Press, 1968.

McClelland, D. C. *The achieving society*. Princeton, N.J.: Van Nostrand, 1961.

McClelland, D. C. Business drive and national achievement. *Harvard Business Review*, July–August 1962, *40*, 99–112.

McGregor, Douglas. An uneasy look at performance appraisal. *Harvard Business Review*, September–October 1972, *50*(5), 133–139.

Maslow, Abraham. *Eupsichian management: a journal*. Homewood, Ill.: Irwin Dorsey, 1965.

Maslow, Abraham. *Motivation and personality* (2d ed.). New York: Harper & Row, 1971.

Morrow, Lance. What is the point of working? *Time*, May 11, 1981, pp. 93–94.

Morse, Nancy C., & Weiss, Robert S. The function and meaning of work and the job. *American Sociological Review*, April 1955, *20*, 191–198.

Murray, A. H. *Explorations in personality*. New York: Oxford Univ. Press, 1938.

Mortimer, Joylin T. *Changing attitudes towards work*. New York: Work in America, 1979.

Perham, John. New life for flexible compensation. *Dun's Review*, September 1978.

Roethlisberger, Fritz J., & Dickson, William J. *Management and the worker*. Boston: Harvard Univ. Press, 1939. Excerpts reprinted by permission.

Schachter, Stanley. *The psychology of affiliation*. Stanford, Calif.: Stanford Univ. Press, 1959.

Schrank, Robert. *Ten thousand working days*. Cambridge, Mass.: MIT Press, 1978.

Sechrest, Lee A. Testing, measuring and assessing people. In Edgar F. Borgatta & William W. Lambert (Eds.), *Handbook of personality theory and research*. Chicago: Rand McNally, 1968, pp. 529–628.

Siegel, Alan L., & Ruh, Robert A. Job involvement, participation in decision-making, personal background and job behavior. *Organizational Behavior and Human Performance*, April 1973, *9*, 318–327.

Taylor, Frederick W. *Principles of scientific management*. New York: Harper & Brothers, 1911.

Terkel, Studs. *Working: people talk about what they do all day and how they feel about what they do*. New York: Reprinted from an edition published by Pantheon Books, A division of Random House Inc., 1972.

Veblen, Thorstein. *Theory of the leisure class*. New York: Viking, 1935.

Vroom, Victor. *Work and motivation*. New York: Wiley, 1964.

Wass, D. L. Teams of Texans learn to save millions. Reprinted with permission from the November 1967 issue of *Training, the Magazine of Human Resource Development*, Copyright 1967, Lakewood Publications, Minneapolis, Minn., 612-333-0471. All rights reserved.

Weber, Max. *The Protestant ethic and the spirit of capitalism* (T. Parsons, trans.). New York: Scribner, 1930.

Weiss, R. S. A structure-function approach to organization. *Journal of Social Issues,* 1956, *12,* 61–67.

White, Robert F. Motivation reconsidered: the concept of confidence. *Psychological Review,* 1959, *66,* 297–333.

Whyte, W. F. *Man and organization.* Homewood, Ill.: Irwin, 1959.

FURTHER READINGS

Bell, Daniel. *The cultural contradictions of capitalism.* New York: Basic Books, 1976.

Berg, Ivar. Worker discontent, humanistic management and repetitious history. In Roy P. Fairfield (Ed.), *Humanizing the workplace.* Buffalo, N.Y.: Prometheus Books, 1974, pp. 7–15.

Fleischman, E. A., Harris, E. F., & Burtt, Harold. *Leadership and supervision in industry.* Columbus, Ohio: Ohio State Univ. Press, 1955.

Haire, M. E., Ghiselli, E., & Porter, L. W. Psychological research in pay: an overview. *Industrial Relations,* 1961, *13,* 3–19.

Kaufman, H. G. Relationship of early work challenge to job performance, professional contributions and competence of engineers. *Journal of Applied Psychology,* June 1974, *59,* 377–379.

Leavitt, Harold J. Applied organizational change in industry: structural, technological and humanistic approaches. In James G. March (Ed.), *Handbook of organizations.* Chicago: Rand McNally, 1965.

Levinson, Harry. *Executive.* Cambridge: Mass.: Harvard Univ. Press, 1981.

Robbins, Stephen B. Reconciling management theory with management practice. *Business Horizons,* February 1977, 38–47.

Scanlon, Burt, & Keys, J. Bernard. *Management and organizational behavior.* New York: Wiley, 1979.

Skinner, Wickham. Big hat, no cattle. *Harvard Business Review,* September–October 1981, 106–114.

Strauss, George, & Sayles, Leonard R. *The human problems of management.* Englewood Cliffs, N.J.: Prentice-Hall, 1972.

Widick, B. J. *Auto work and its discontents.* Baltimore, Md.: Johns Hopkins Univ. Press, 1976.

Yankelovich, D. The meaning of work. In J. M. Rosow (Ed.), *The worker and the job.* Englewood Cliffs, N.J.: Prentice-Hall, 1974, pp. 19–47.

Yankelovich, D. The new psychological contracts at work. *Psychology Today,* May 1978.

4
STRUCTURE

Case Study
The Case of the Inadequate Turbine

The scene is the Amalgamated Machine Company. The company produces a broad variety of capital equipment. Power turbines are just one product line and each turbine development project has its own project manager.

CAST

Milt Cowan: General manager, turbine division

Eric Redder: Project manager, "Blue" turbine

Maryann Lane: Field support manager

Alan Bonson: Corrosion engineering specialist

Clyde Miller: Chief turbine engineer (Alan's boss)

SCENE: MILT'S OFFICE

Eric Redder was nervous. There was something he couldn't quite understand in the tone of voice that Milt used that morning when he phoned Eric and asked him to step into his office about 10:00 A.M. When Eric arrived, he found Maryann Lane and Milt going over some official-looking papers on Milt's desk.

Milt: Oh hi, Eric, come in. You know Maryann here?

Eric: Sure. We have worked together on several projects. (Maryann nodded at him.)

Milt: Well, how's the "Blue" turbine coming along?

Eric: Just fine. We just finished life tests on the pumps and we'll be ready to order major components for the shell and casing by the end of the week.

Milt: Well, that's great. You're getting this project going, just as we had expected, and I'm sure that this little problem that's come up won't be much of a bother. However, Maryann insisted that we talk about it a bit. The "Blue" turbine is mighty important to this division's future because of its light weight and dependability. Those are major assets when we sell it to less developed nations for use in energy recovery activities. But you know all about that. Now, about this little problem. Maybe I'd better ask Maryann to explain it.

Maryann: Well, in field support we are very concerned about training local third-world operators who don't have the basic skills that we would normally expect to find in more developed industrial areas. These operators have to learn how to maintain the turbines in the field and therefore we have to consider some aspects of turbine maintenance that just don't come up often with our other customers in more in-dustrialized nations. For example, we have just learned that the turbine model before yours, Eric—you remember the model Q—has a bad reputation in the field because it rusts in semitropical environments.

Eric: Well, how does that affect the "Blue" turbine? Our project charter definitely states that the turbine is to be operated in an air-conditioned, environmentally controlled power station only.

Maryann: We can't always control how our customers use our products, and even though the operating instructions are clear, if the turbine begins to corrode, our reputation suffers—even if it's the customer's fault. We don't want our reputation to suffer, so how can we prevent this type of problem from happening to the "Blue" turbine?

Eric   Now look here, I've been working with this project for over a year now. We've already had our first design review and we're pretty well along in procurement. I don't know if anything can get done now about this type of problem but really, why should we in project management get concerned? I always thought that the education of the customers and their operators was the responsibility of field operations. We build them, write the manuals, and then turn them over to your group. Why bring me any more problems?

Milt: Of course, you're right, Eric, but could you think about this problem a bit? Maybe this is one of those things that we can resolve here in the plant rather than out at some desolate power station.

Eric: Well, I suppose that I could have our design evaluated for corrosion resistance, but I'm not sure what the effect will be—especially now that we're ready to order major components. They might all have to be changed to stainless steel and the delay and cost could be appreciable.

Milt: OK. Take a look into it. Coordinate with Maryann here. And, oh yes, by the way, top corporate headquarters has been watching this project since it could be so important to us; and I'm sure that they wouldn't like it if costs were to increase very much.

Seeing that the meeting was over, Eric and Maryann left Milt's office. Eric was very unhappy. Maryann was sympathetic but offered no immediate advice. She did, however, give him some standards on field corrosion and offered to help in any way that she could later. Eric went back to his office and put in a phone call to Alan Bonson.

Eric: Hi Alan, how are you doing? I'm OK, but say, we have a minor problem that just came up and I'd like to get your opinion on it. Can you meet me here in about half an hour? Fine, see you then.

Later, Alan came into Eric's office, his usual twenty minutes late. After exchanging greetings, Eric covered the problems with Alan and asked for an analysis of potential corrosion problems if the turbines were operated under the new standards that Maryann has passed on.

Alan: Look, I'd like to help, but I've got four other project managers to satisfy. Your budget for corrosion testing is about used up, and I'm leaving on a two-week vacation tomorrow, so you'll have to hold everything until I get back. Sorry, but you should have thought of this before.

He then got up and left Eric's office. Eric was really angry. He sat for a while cooling off, then got up and walked across the building to the office of Alan's boss, Clyde Miller. Luckily, Clyde was in and there was no one else with him at the moment. That was unusual, since Clyde was handling about seven major turbine projects in addition to the "Blue" turbine.

Eric: Hi. Got a minute to talk?

Clyde: Come in. Why not? You're a project manager. I'm supposed to report to you and the rest of you guys who have only their own projects to consider. I'm trying to keep you guys happy in addition to administering and training the technical prima donnas who really turn out the bright, creative products that we need to keep us in business. (Noticing that Eric was not responding to his joking around, Clyde stopped; then continued in a more serious tone.)

Clyde: What's wrong? It looks like somebody has been raining on your parade.

Eric explained what had happened at the meeting that morning with Milt Cowan and the subsequent meeting with Alan (who reported to Clyde). Clyde offered to look into the situation and report back by that afternoon. Clyde then called Maryann and Alan and spoke to both of them on the phone. Later that afternoon, he stopped in to see Eric.

Clyde: Hi there. I may not have all the answers but I have a few suggestions about that corrosion problem.

Eric: Well, what do I do now? And remember, suicide is not a viable management alternative because it sort of keeps you from getting any promotions in the company.

Clyde: Very funny. You have several problems. Maryann's specifications would require a complete redesign of the housing. It also needs new operating manuals and extensive accelerated corrosion testing of the pumps that we just finished testing and accepting. Alan says that he's sick and tired of doing all your work in a hurry, but that's no real problem, since he says that about everybody. I did find out, however, that he hasn't made any definite plans for his vacation. There's always another problem, though, and here it is: According to company policy, there's no way that we can reschedule him without his approval. There might be a way, though. He's always wanted to visit our plant in San Diego because he has a sister there. Of course, I'm not telling you how to spend project money but we do have some vendors in that area. What do you want to do?

QUESTIONS

1. Are there any procedures Eric should follow to resolve the corrosion problem?
2. How should he handle Milt's request to minimize costs and delays?
3. What should be done about Alan? Are there any procedures to follow? Who is responsible?
4. What should Maryann's role be in this project, now?
5. If you were Milt, how would you have handled this problem with the "Blue" turbine?

REVIEW

It's obvious that no organization exists without a *people* component. In the previous chapter, we were mainly concerned with the behavior of the individual at work. This emphasis on the individual is relatively modern. Only in recent times has there been any appreciable management attention paid to people as individuals and not as economic resources for the organization to use: They were no different from other raw materials.

Prior to the beginning of this century, most management tasks were concentrated on the optimum acquisition of inanimate resources or facilities and minimal effort was put forth to develop the human organizational structures intended to transform those resources into revenue. The organizational structure, defined as the repetitive interactions among people in the organization, was a secondary consideration. Manufacturing plants, cash, tools, materials, and processes were primary and highly valued commodities that deserved close attention from management. Growing markets, relatively inexperienced (and therefore interchangeable) labor, and an extremely limited number of organizational designs made any extensive attempts to develop specialized organizational structures unnecessary and even wasteful.

There were reasons for this. In addition to the stable social culture of the times, there was a much slower rate of environmental and market change. Markets *grew*, but they did not change much. A product that served a market need had a reasonably long life. Obsolescence was not a major factor then. If the organizational structure could meet current demands with slow market and environmental change and the people were relatively interchangeable, the time spent on improving human interactions would be wasted, since those interactions contributed so little to overall productivity.

And the organization design of that time mirrored the existing society. That design was primarily a single proprietorship organized like a patriarchy. While there were some differences due to the organizational size, even the larger corporations were generally organized internally as if they were single proprietorships. The manager or the foreman of each group was really the boss; he hired, directed the work, fixed the wages, and fired if necessary.

There were some differences between industries but most companies in the same industry were alike. Therefore, organization design choices were obviously not very important. The important decisions were about machines. If production equipment efficiency were improved through redesign, that would be important! The people could easily be replaced with others, and they often were. It was a rational or logical point of view and the organization's structure was designed rather quickly by top management fiat. Concern for individual human motivation was not required, since productivity was built into the machinery. Human beings were intended to supply and service the production equipment as it transformed raw materials and power into finished products and, eventually, organizational revenue.

The logic and efficiency of the machine were understood easily, and consequently it was the model for attempts to develop equally logical and efficient organizations. Management first designed the machines and then the organizations; and these designs produced. But environments and markets changed at a more rapid rate in this century and industry has gradually moved from the production of goods to the production of services. People have displaced the machine as the organization's capital resource and have become the human capital that drives most industrial organizations. These people are no longer interchangeable, since the process of developing and producing the innovative products and services that support the organization's growth is a creative one that is unique, not interchangeable. (Human interchangeability, of course, never really existed.)

This is supported by the notion that creativity is not even distributed normally in the technical population, when measured by the output of research personnel (Shockley, 1957). Each one of the people is unique and since the people (and their creativity) have displaced the machine as a major source of organizational growth (and consequent revenue), the attention paid to improving their effectiveness must increase. This would lead to management attempts to develop the best structure to foster this effectiveness, since that structure is partly responsible for the economic success of the organization. Therefore, *structure* is the second component of the organizational model, described briefly in Chapter 2. We will cover it here in more detail.

This chapter begins with a description of various structural designs and how they are applied to technical organizations. As before, description will be followed by prescription: that general prescription modified by you to fit your own situation. The description used here first covers structures that were expected to be universally applicable to all types of organizations (i.e., the recommendations for the "one best way" to manage), then those structures that are situationally based and designed according to existing contingencies (i.e., under these conditions or situations, these are the suggestions for design). Contingency theory (as a subset of situational theory) will also cover structures that are expected to be modified as a function of the organization's development process. For example, small companies are structured one way and, as they grow, they become organized differently.

STRUCTURE: INTRODUCTION AND DEFINITIONS

Formal Design Elements

Even with the relatively recent emphasis on *structure*, the formal organizational designs are not extremely varied, since there are so few elements with which to build a structure. For example, the structural design may be either a tall or a flat configuration, i.e., there may be many levels of responsibility between the chief executive officer and the person on the shop floor (tall) or very few levels (flat). The structure may be market or production directed. That is, it may be intended to serve different markets or different products. It may be centralized or decentralized, hierarchical or diffused. It may have limited or extended control spans. It may have line or staff and functional or project orientations. There are several other possibilities, but in the aggregate there is an unimposingly short list of design elements available.

Informal Organizational Design

However, these few design elements do provide some variety of operating structural designs, and in recent times that variety can occur *even within the same company*. We will cover the reasons for this later (see Lawrence & Lorsch, 1967). But, continuing a bit further, we often find that the intended formal designs of management vary significantly in their operations from the way that they are supposed to operate. These differences between formal intentions and actual operations are due to the continually changing, dependent elements that define the informal organization—the organization designed by the participants. In effect, we have a limited number of basic formal design elements that interact with many other, less well defined, informal elements to form the total organizational structure—the repetitive interactions among people. An analogy could be made that the formal organization is the skeleton (about the same for all people, with minor variations) upon which the flesh and blood (or the differences among people) of the informal organization is built. Neither the skeleton nor the body survives without the other, and changing one changes the whole organism.

The Totality

The total organizational structure is then a network of patterned behaviors (Katz & Kahn, 1966, p. 51) with management-intended (i.e., formal) and nonintended (i.e., informal) elements. A note of caution here: Sometimes documentation is considered to be the mark of the "approved" formal organization. That is not always so. There are organizations that have never had an organization chart, just as there are others that update their organization charts and job descriptions regularly. However, in both of these examples, organizational participants quickly learn who are the chiefs and who the Indians.

Therefore, although documentation is one, and perhaps in most cases a major, indication of the existence of a formal organizational structure design, it is not the only one. However, it is rarely if ever used to define the intricate web of informal relationships that must often be learned on site by the newly recruited organizational participant. Interestingly enough, the design of the informal organizational structure, or its style, does not occur by chance. While there are many contributing factors, according to one author it seems to depend primarily on the past leader of the company "who shaped and reformed the organization during the initial periods of growth" (Meir, 1967,

p. 471). It could also be due to present needs, the thinking of present participants, and/or other less obvious criteria that only those initiated into the particular organization would be aware of. The institutional style is a major part of the total organizational culture, and it includes both the formal and informal designs.

In order to have a starting point for our descriptions, we will assume that the formal structural design is the determining factor of the total organizational structure, with the informal structure a secondary design intertwined with it. Therefore, we'll start our description-prescription process with formal design ideas and concepts first, following with informal design ideas. Another reason for this sequence is that since there are only a limited number of design elements to consider in formal designs, it is easier to start with less complex concepts. Informal designs, by comparison, are almost unlimited, since they can be affected not only by all the elements of formal design but also by typical variables such as organizational history, cognition of participants, and the almost daily changes that every organization experiences. However, I repeat that this informal design (i.e. the organizational "flesh and blood") is very well thought out, since it is intended to solve problems that the formal design does not (or cannot) handle. Since these informal designs really do accomplish their purpose in most cases, they are remarkably resistant to change by management.

Analyzing the Totality: Formal versus Informal Elements

One definition of the formal structure is the institutionalizing of decision making. In terms of repetitive interactions, the formal structure is the procedures, communication patterns, and suggested behaviors that are expected to be used in an existing situation, that have been faced before, are expected to occur again, and are considered to be relatively important to the continued operation of the organization. It's the management behavioral norms that are defined by the organization's top management. The formal organization stresses predictability, efficiency, and the coordination of the efforts of many people. There is little provision for individual differences, spontaneity, social needs, or the unique and changeable patterns of creativity that are so vital to the success of a technical organization. It directs and constrains behavior. Conversely, the informal structure does support those unique relationships that are a function of the individuals who happen to be in the organization at that time. (But if those individuals are self-constrained or lack spontaneity, the informal organization will obviously also be similarly constrained.) When the organization succeeds in achieving its goals, a necessary but incomplete support for that success is a major congruence between those goals of the formal and those of the informal structure. These goals are averaged, of course, since they reflect the varying goals of the organization's internal groups. While this congruence is rarely the sole factor supporting success—there are such other factors as markets, resources, and politics outside the organizational design,—I believe it is one of the more important ones. Congruence suggests that the goals of the repetitive behaviors that are required formally and then are provided informally by organizational participants are about the same. With increased divergence between these two sets of goals, there is sure to be less success than possible, because internal conflicts and friction due to this lack of goal congruence decrease the cumulative effect of positive contributions made by decision-making organizational participants.

For example, decreased congruence may happen as the result of a change in management personnel. That affects the informal organization. Although the memo from top management announcing the appointment of a new chief engineer had little impact on the formal structure of the engineering department, the informal structure involving the communications channels, methods of working, and administration have surely changed with that appointment. The former chief didn't really care about getting time sheets in on time just so long as projects came in under budget and met the design specifications. The new one is a stickler for project completion too but also wants the time sheets every day, no matter what. Apparently the formal design has not been altered, but the informal one surely has, and in this example, conflict is sure to increase (at least in the near future, until things simmer down).

Other changes in the organizational environment can revise an informal design almost as fast as changes in management personnel. Changes in the economic environment such as what happened to the petroleum industry can do it. Increases in crude prices, for example, pressured many petroleum companies to restructure their technical organizations away from an emphasis on refining and manufacturing toward one that emphasized exploration. Of course, there was probably less difficulty in the informal design of the exploration group in accepting its new and expanded responsibilities than there was in the refineries, since the exploration group increased and the refineries decreased in importance. And yet, as far as the formal organizational structure was concerned, there were relatively few design changes.

Conflict and Variety: Formal versus Informal

When there is little congruence between the informal and the formal organization, most research indicates that the informal organization will probably control. However, "probably" does not mean completely, and the formal organization is supposed to have the power to alter or disband the informal design through replacement or changing of personnel and/or the situation. On the other hand, there is the well-known example of control of the formal by the informal organization that occurred in the Hawthorne experiments, discussed in the last chapter (Roethlisberger and Dickson, 1939). The workers' informal structure was able to determine the rate of production in the bank wiring room regardless of the inducements offered by management. That organizational structure was based on the social (or informal) needs of the working group, which were opposed to the needs of the formal structure. That's typical in many industrial organizations when the goals of the formal and the informal organization don't agree. In the Hawthorne experiments, the informal organization became a defense against the production-oriented goals and the "machinelike" design of the formal structure.

With the increasing value of human capital and the more flexible structures of today, there probably should be closer goal congruence. The increasing value of the highly trained, relatively independent technical person should result in more flexible, human-oriented structures and increased congruence. That can never be complete, however, because there are differences in the speed of change between the informal and the formal organization. The formal structure's changes are based on some change in the organizational economic, political, and/or social environments. The informal structure changes seem to depend primarily on the internal social environment, or in other words, upon who is in the organization at the time. Generally, people's ideas and expectations at work change more quickly. In summary, the formal and the informal structures change at different rates of speed and for different reasons.

If the goal here is to understand the alternatives available in describing and then prescribing organizational structures that include both formal and informal designs, the first step should be some type of definition or classification process. By analyzing applicable formal structural designs, determining where and how they fit, selecting the elements from the list available, and developing a model, we have begun an iterative organizational design process of the formal structure. The next steps are to define the organizational criteria that should support an informal structure closely matching the formal model and to attempt to support a design with those criteria. Since it is almost impossible to actually design the informal structure because individuals, not organizations, control it, the best that we can do here is to try to use these criteria to affect it. This should improve the total organizational design.

After the designs are completed, we would lay out a testable hypothesis that this design is supposed to satisfy and gather operating data on how it works. When the data come in and the variance between the hypothesis predictions and the data exceeds that which is acceptable *to you as the designer-manager*, go back to the formal design to modify it. This is an ongoing process because of changes, such as those in technology, participants, or markets. It also involves continual testing of the informal organization to determine how close these goals are to those that you have included in the formal design, and so on back to the formal, then informal iteration of testing. The process of organizational design never ends. It can start anywhere in the cycle of design, develop hypotheses, test, and redesign. However, this process can end if your two interacting structures (formal and informal) no longer match the overall environmental needs as well as those of your competitors that do. If this happens, your organization will decline and eventually die (Chandler, 1962); this really stops the process.

GENERAL CLASSIFICATIONS OF FORMAL STRUCTURES

We start the classification process with two major categories: those that are proposed to be universal (i.e., a universally usable model) and those that depend upon the particular situation and contingencies. The universal models are typical of the classical management theory developed during the end of the last and the early part of this century. They reflected the general approach to technical problem solving of that era: suggesting there was always some definite answer out there if the manager were only able to find it. Newtonian physics, with its emphasis on certainty, was a typical example of this deterministic thinking. Universal models suggest that there is one best way to organize (Scott, 1975).

The situational-contingency models are relatively new and suggest that there are many ways instead of just one best way, and that the optimal design depends on the particular contingencies or situation. There is a situation and a time; an "if, then" kind of design (Osborn, Hunt, & Jauch, 1980). Contingency models could almost be analogous to the models of modern physics, since they now include uncertainty rather than Newtonian certainty. The contingency models reflect many of the concepts about the primary importance of people to the organization. A familiar word of caution though: Some of these concepts include assumptions that are inherently normative and therefore depend on the particular researcher's point of view. A normative point of view is not always testable, and therefore may provide results that are not as replicable as those resulting from an experiment in modern physics. In other words, normative theories suggest that there are approachable universal ideals like participation, openness, and conflict confrontation, while contingency theories suggest that people's behaviors depend upon the environment and the participants' characteristics (Beer, 1980).

Some structures, such as those in process industries like refineries and chemical processors, cannot stand very much openness, participation, or confrontation of conflict. Therefore, the normative viewpoints suggested by supporters of those research theories do not always apply. However, they can be a starting point, since there may be no other data ("I think that's the way it should be since there are no relevant findings that I can use"). But since we do have some research findings to start with, we begin our design by describing those structural building blocks suggested by various researchers that you might decide to use to build your unique organization (but remember the source of the blocks). The formal model that you design may typically include blocks such as technology, organizational history, markets served, background of organizational participants, legal restraints, and/or product life cycles.

Strategy Defines Structure (The Most Important Block if You Want to Succeed)

If there were only one factor that affected the organizational structure it would be the organizational strategy, and that strategy is always initially determined by the market served. When the management of the organization understands the needs of its markets and then develops a strategy to meet those needs, the structure selected as part of that strategy can be a major tool to convey that strategy to organizational participants. The structures are the documented decisions intended to guide similar decisions in the future.

For example, there was the deliberate strategy of Alfred P. Sloan, the president of General Motors in the 1920s, that transformed General Motors from "an agglomeration of many business units, largely automotive, into a single, coordinated enterprise" (Chandler, 1962, p. 130). He set up uniform accounting procedures for all the divisions but supported decentralization to serve separate segments of the automotive market. The central offices were intended to define lines of organization and maintain communications *between* divisions, but each division had its own authority to operate autonomously to service its markets and, in some cases, even compete for markets with other divisions.

This structural design quickly impacted the market positively and the competition of the centralized, more rigid design of the industry leader of that time, the Ford Motor Company, in a negative fashion. General Motors quickly passed Ford, and became the industry leader within a relatively short time after the new structure was implemented. The corporate officers at General Motors made policy and coordinated the operations of the various divisions within the corporation. The separate divisions were encouraged to develop their own unique organizational structures to meet their perceptions of the needs of the marketplace. The structure of the Cadillac division varied from that of the Chevrolet division, even though both were within the General Motors framework.

However, no matter how important the "market strategy" building block is, it is still only one foundation building block. There are others that also determine the rest of the foundation and the structure built upon it.

UNIVERSAL MODELS: IS THERE A BEST WAY TO ORGANIZE?

The universal models of organization design are those intended to be applicable anywhere. They can range in size from the broad, large-scale, rational, logical, and mechanistic recommendations of bureaucracy (Weber, 1957) and classical theory (Taylor, 1911), through the middle-scale human relations theory (Roethlisberger & Dickson, 1939), into the very specific individualistic open organization theories (Likert, 1961). The common ideas in these models are that there is generally a best way to organize, and the closer the structure is to this ideal of the best way, the more effective that structure will be.

Of course, there are differences in defining what that effectiveness is and how it is to be measured. Therefore, minor differences are allowed within the overall scheme of the universal models, but those overall models are still supposedly universally applicable. That might account for the contemporary existence of organizational structures that are designed in accordance with one or more of these universal models and yet are slightly different from each other. It poses an interesting problem for the organizational designer who wishes to select the most effective structure, since these differences were born in the assumptions behind them and the historical environment in which they were developed. We will start with one very important model in this universal group: bureaucracy.

Bureaucracy—The Professional's Model

There were two major reasons for the birth of bureaucracy: the growth of larger organizations during the end of the nineteenth century and the gradual separation of management from ownership. Something had to be developed that would operate without the continual guidance of owner-entrepreneurs. Nepotism and familial ties had usually been able to guide managers into following the structure developed by the owner-manager. When the owner and the manager were different people, this no longer applied.

A need quickly developed for well-trained and objective managers who could develop markets and service them, provide goods and services at a profit or, as in the case of governmental agencies, at a minimum cost, and be capable of doing so without checking with the nonexistent owner. The development of the professional manager and an ideal structure, quite different from that of the extended family, were the solutions put forth by Max Weber (1957). He described the duties of the managers and that ideal organizational structure in which they were to operate. He provided the theory behind the objectivity and logic that are the hallmark of the professional. The design components have been summarized as contrasts between the professional and familial organizations in Table 4-1.

TABLE 4-1   CONTRASTS BETWEEN ORGANIZATIONS

Professional	Familial
1. Rules, policies, procedures	1. Personal direction by charismatic leader
2. Equal treatment based on performance	2. Rewards based on kinship and/or emotion
3. Division of labor and a stress on expertise	3. Multiple job assignments, depending on need
4. Roles in a hierarchy	4. Roles based on personality
5. Career commitment	5. Temporary association
6. Separation of ownership and control	6. Consolidation of ownership and control

Source: Osborn, Hunt, and Jauch, 1980, p. 277.

The organization staff was expected to carry out the decisions of top management by means of the following:

- A system of rules
- Impersonality of interpersonal relationships
- A well-defined hierarchy of authority
- A division of specialized labor
- A system of procedures that implemented the rules
- Employment and promotion based on technical competence (Shannon, 1980)

Order and discipline were implicit. Information was passed upward and decisions were passed downward. Authority and responsibility increased as one climbed the organizational pyramid and were clearly defined for each position in that pyramid. Conflict and disagreement were dysfunctional and were obviously results of poor organizational design. In Weber's attempt to describe the ideal designs, he included requirements that vertical specialization (i.e., covering a vertical section through the organizational pyramid) was to be matched with more control, and horizontal specialization across various parts of the pyramid was to be matched with more coordination.

Although his work was translated from French only about the middle of this century (1949), Henri Fayol was another early writer of the late nineteenth and early twentieth century on management and organization. His ideas incorporated the professionalism of bureaucracy into classical management. He was also concerned with developing the best administrative structures for any organizational design. He suggested "ideal" design principles that depended on and reinforced the familiar pyramidal structure that is part of many formal organization charts of today. These principles were:

- Division of labor
- Authority and responsibility
- Discipline
- Unity of Direction
- Subordination of individual interest
- Adequate remuneration of personnel
- Centralization
- Scalar authority chain
- Order
- Equity
- Stability of tenure of personnel
- Initiative
- Esprit de corps

With these principles in place, he predicted that an optimal organizational structure would always result. This design approach toward an ideal, the search for the "best way" to develop *the* organizational structure, was typical of these early attempts at definition. Then it was thought that most problems could be solved through a logical, scientific approach. Markets were expanding and companies were becoming larger. The pressing need to coordinate the efforts of larger groups of employees was becoming a major management task, since the alternative was no coordination or, worse yet, coordination by the employees themselves. The expectation was that if either of these alternatives came to pass, production would drop to the lowest level acceptable to an inherently disinterested employee. Managers of that day felt that output would invariably remain low unless there was strong, central direction with an organizational structure to implement that direction.

Science was making great discoveries and industry benefited. No one seemed to doubt that the positive benefits of science's objective methodology could equally be applied to organization design. There just had to be a scientific best way to develop and operate the organization. As a matter of history, many of the companies that were organized in this machinelike model did grow more rapidly than those that were not. The machine model was logical and predictable, and it worked. However, it provided little room for deviation in markets, in technologies, or (especially) in the behaviors of organizational participants. One of the central figures in this push to efficiency using logical and scientific methodologies was Frederick Winslow Taylor.

Classical Theory: The Production Model

Taylor and his contemporaries developed their ideas during the end of the nineteenth century and the beginning of this one (Fayol, 1949; Taylor, 1911). They produced organizational designs that are still being used today. Their models tried to duplicate the efficiency of the physical processes of the manufacturing plant. The efficiency, logic, and objectivity of bureaucracy suited them well. For example, job descriptions were similar to the manufacturing operation sheets provided by industrial engineers. There were detailed prescriptions for managers to follow as if they were like the machines on the manufacturing floor. Cooperation and efficiency were supposed to result automatically from science and rationality—inefficiency, from poor management and inefficient workers. According to Taylor, waste in industry "arose from lack of expert engineering knowledge, failure of management to erect suitable standards, and the withholding of effort by the workers" (Krupp, 1961, p. 16).

Time study measured physical activity, wages were related directly to measurable standards of productivity, and each job and organizational function was supposed to be explicitly defined. The essence of the scientific, mechanical organization was a result of integrating the work to be done by the organization with effective studies of plant layout, materials flows, and work-place design (Taylor, 1911, pp. 36–37). The management task was to determine the *one best way* to organize the work.

All organizational levels in this mechanical, production model had their job descriptions, covering criteria for workers and managers alike. Management's tasks were defined as decision making, goal setting, and control. These tasks were centralized, with direction flowing down from the top through the structure and information about production flowing up. The formal organizational position itself contained the authority and the power. With unity of command (i.e., reporting to only one boss) and unity of direction (i.e., orders downward and information upward), the structure could function efficiently, with inefficiencies or disruptions of the model easily detected and eliminated.

In all fairness to Taylor and his associates, they were quite aware that this organizational structure was a very difficult thing to achieve. They placed most of the responsibility for an organization's failure to achieve the optimum with the managers, not the workers. They were eminently practical people who knew how to solve problems. In their own words,

> The type of management which regards the exact definition of every job and every function, in its relation to other jobs and functions, as of first importance, may sometimes appear excessively formalistic, but in its results it is justified by all practical experience. It is in fact a necessary condition of true efficiency in all forms of collective and organized human effort. (Mooney & Reilly, 1931)

According to this model, the formal directive authoritarian structure leads to high performance, and this structure is universally applicable with respect to all organizations. The traditional roles of the master-servant relationship implicit in the society of those times were embodied into the employer-employee relationship. This was truly grand theory in the best traditions of science, and it matched the social and cultural environment of the day.

Efficiency in organizations was supposedly increased when the wasted motion in poorly defined structural relationships was eliminated. The organization chart was expected to determine the authority and the responsibility of all participants. When I studied this many years ago, it seemed very logical and possible. It all fitted together so well. "In effect, power (authority) is exercised downward through a company's organization and becomes weaker (covers less area) as it travels down the line. . . . Theoretically, authority and accountability are coupled together in their initial travels downward" (Silverman, 1967, pp. 7–8).

While the logic was intellectually inviting at that time, my opinions about their overall applicability have changed. These concepts may apply to some of our organizational designs, but my experience in many technical organizations showed that they are *not* universally applicable. In many technical organizations, authority is dispersed with little concern for the formal organization chart (if indeed one even exists!) and it is not necessarily associated in any direct way with the accountability for results. There are subtle reasons for this that are not always apparent. For example, external uncontrollable events or even mismanagement may cause negative results. Distinguishing the causes between them is difficult, requiring expensive information systems, if it can be done at all.

But even then, several of Taylor's contemporaries understood that the classical model was not as adequate as it could be (Gulick, 1937; Follett, 1942). For example, Gulick injected the variable element of the specific function into the otherwise uniform and coldly efficient machine model. Accordingly, the structure cannot be uniform in all cases but depends on the functions or tasks to be accomplished.

> Students of administration have long sought a single principle of effective departmentalization just as alchemists have sought the philosopher's stone.

> But they have sought in vain. There is apparently no one most effective system of departmentalization . . . organizations must conform to the functions performed (Gulick, 1937, p. 41).

Of course, this doesn't help the organization structure designer too much unless there is a recommendation of how these specific functions affect the design of the structure.

And while Follett shared the general idea of that time that it was possible to define general principles to guide the organizational design, she recognized that the situation was often the major determining factor. When the situation changed, the authority relationships might have to change and become more dispersed and less unified. She and Gulick disputed the assumptions of classical theory that authority was unitary, homogeneous, and explicitly defined by the organizational structure. If the situation defines the structure, universal designs cannot work, since situations are not "universal." They may resemble each other or be typical of some variable such as technology, but the people in them who are part of the situation are not universal; they are unique and that makes the structure unique.

However, in classical theory, both the workers *and the managers* were interchangeable. Human beings and the social groups were to be adapted to the plant structure, and this adaptation was defined by the organization chart. That formal organizational chart defined the relationships coordinating and directing the pieces of the machine to perform most effectively. But, the research by Roethlisberger and Dickson (1939) that was originally based on this theory resulted in findings that were almost in direct conflict with these major parts of classical theory. We have covered some of that research; now we review the structural design concepts.

Human-Relations Theory: The Social-Group Model

Some of the results of the Hawthorne experiments were discussed in Chapter 3, when we covered motivation. Other, equally important results affected organizational structure. We'll review those aspects now. As noted in Chapter 3, the studies at the Hawthorne Works of the Western Electric Company were completed approximately between 1927 and 1933. The original purpose was to test the effect on worker output of single changes in the physical environment. The typical variables were conditions of work, fatigue, and monotony. By varying the lighting and rest breaks, for example, researchers expected the workers (who, they thought, would respond as if they were interchangeable production machines) in the machine model of the organization to respond through changes in output. Output would increase as conditions improved and decrease as they worsened. This was a linear and very classical structural theory. It reflected the structure of the Hawthorne plant itself, which employed about 40,000 people, and most of the other large industries of the time.

The results were interesting. Changes in the physical environment (i.e., rest periods or improved lighting) did not result in related changes in production output. In the first set of experiments, production output gradually increased, regardless of the experimenters' positive or negative manipulations of the environment. The researchers interpreted this as the workers responding to the attention paid to them by the researchers. This was a small modification of the idea that workers were like machines but not much of a modification, since they all were supposed to be reacting in much the same way. The workers, as an informal organization, had responded with more production even though the formal organization (the researchers) had, for example, made it more or less difficult to produce by decreasing or increasing the light where they worked.

> In place of a controlled experiment (there was the) . . . notion of a social situation which needed to be described and understood as a system of interdependent elements. This situation included not only the external events but the meanings which individuals assigned to them: their attitudes toward them and their preoccupations about them. (Roethlisberger & Dickson, 1939, pp. 183–184)

This was a change from the idea that workers (and, by inference, managers) are machinelike parts of the organization. Machines are not affected by the meanings of their work. The machinelike model seemed to be less universal than the classical theorists proposed. However, recently the results of these experiments have been reinterpreted. In the Hawthorne experiments, where production slowly increased over the course of one experiment almost independently of the positive *or* negative changes in the experimental conditions, the original interpretation was that the workers were responding positively to the attention being paid to their ideas and to themselves as individuals by the experimenters.

In another experiment during the Hawthorne series, a diametrically opposed result occurred. Production remained constant under the same varying experimental conditions that were imposed during the first experiment. In fact, the experimenters noted that whenever one of the workers attempted to exceed the group-imposed production levels, one or more of the other workers warned that person to get back in line and decrease his output. The interpretation, in this case, was that the informal organization had frustrated the goals of the formal (i.e., the experimenter's) organization. How could this conflicting result be correlated with the initial result of gradually increasing production?

These seemingly divergent sets of production results support a conclusion that is at direct variance with classical theory. In both cases, the informal organization, which isn't supposed to exist if classical structures are working well, actually controlled the results. The informal organization was really in control. When the perceptions of the informal group indicated closer goal congruence with the formal organization (e.g., "You're paying attention to my ideas. Therefore, your interests must be similar to mine."), production increased. When they indicated goal divergence (e.g., "You're just trying to make me increase production so that the work will be finished and I can be laid off."), production was fixed at a level that was just sufficient to satisfy the experimenters' expectations and continue the experiment.(These experiments were conducted in the middle of the depression and jobs were hard to get.) The informal organization's control over the behavior of individual workers depended on the mutual dependence of the workers upon each other or, in other words, the group cohesiveness. When the workers felt that it was in their own self-interest (however they defined that state) to produce, they did. When they felt it was not, they fixed the level of production at the minimally acceptable level to stretch out the job.

Although the formal organization included all the logical elements of control and harmony of classical theory, the social elements of work or the informal organization reflected the more emotional elements in people. Since those emotional elements really controlled production, they had to be considered by organizational theorists and managers. They were variables that had to be dealt with. Differences between goals of the formal and the informal organizations had occasionally proven to be destructive to increased production. The informal organization controlled output without approvals of the formally constituted management structure. This lack of predictability and control was to be avoided. The formal organization was to be followed, and the informal organization (which management recognized could not be ignored) modified to fit it. Cooperation and logic had to prevail if the best organizational way to manage was to be developed.

> The formal organization of an industrial plant has two purposes; it addresses itself to the economic purposes of the total enterprise; it concerns itself also with the securing of cooperative effort. The formal organization includes all the explicitly stated systems of control introduced by the company in order to achieve the economic purposes of the total enterprise and the effective contribution of the members of the organization to their ends. (Roethlisberger & Dickson, 1939, p. 558)

When the informal structure had goals that conflicted with those of the formal organization, such as the work group rather than the group's manager controlling output, the formal organization was expected to investigate the conflict and through communications and training secure a closer cooperation with stated organizational goals. Interviewing workers, exhibiting management solicitude, and modifying working conditions (as long as there was no interference with production) were the new tools to secure optimum output.

This was the time of the development of the modern personnel department and the widespread use of attitude surveys. Those aspects of human behavior that contributed to conflict and limitations on plant production were to be changed. This was a mainstay of human-relations theory. "Changing plant culture was a one-way stream. The pathology of organization was the disruptive or irrational changes that did not initiate with management— behavior that deviated from the norm of plant structure and business goals" (Krupp, 1961, p. 30).

Human-relations theory, therefore, could be considered as a continuation of the classical "machine model" organizational structure but there were significant differences. It had a broader, more pervasive approach, since it dealt with the thinking and emotions of people. Although it was also a universal model, it provided for greater variability in behaviors and increased the allowance for human differences at work—just so long as production was maintained or increased. It softened the relative authoritarianism of classical theory, improving cooperation and even, in some cases, increasing productivity. But management, not the worker, originated it.

Classical Theory versus Human-Relations Theory and Technology

Human-relations theory, with its attempts at manipulation of the informal organizational structure, is actually a more flexible and limited, midrange theory than the proposed unlimited range of classical management. While similar to it in recommending a universal approach (i.e., improving output by manipulating worker behaviors to match formal organizational needs), it does not lay down the rigid and exact specifics classical theory does. The methodology and the implementation are more flexible, depending to a great deal upon the emotional content.

Moreover, classical theory deals with larger organizational structures, such as the total organization, while human-relations theory deals with the structure of smaller groups within that total organization. The theories' resemblance to each other comes from their common origin: the scientific ideas of the machine model. In both, nonpredictable human behaviors are considered to be dysfunctional, but the cure is different. Classical theory either ignores it or treats it as conflict that must be erased, and human-relations theory acknowledges that it has a right to exist but attempts to modify it to fit the formal organizational goals.

Classical and human-relations theories were expected to cover all organizational possibilities, but different organizations attempted to implement these seemingly universal approaches with varying degrees of success. In some cases, human-relations ideas meant that people's views were listened to by management. This didn't always result in economic success; some companies that used it even failed. Conversely, other companies that didn't seem to use either of these theories succeeded, apparently in spite of the obviously lowered quality of the organizational structural design. As we've said before, when success and failure use similar designs, it seems reasonable to assume that other, interacting variables have not been accounted for.

The work of Joan Woodward (1965), covered in more detail in Chapter 5, showed that one possibility for the differences was *technology*. This was an intervening variable that was poorly understood and seemed to be partially responsible for organizational success. The technology definition she used was the limited definition of the manufacturing processes used by the firm. According to Woodward's findings, the organizational structure suggested by classical and human-relations theory, with fixed, machinelike, formal relationships, is only one alternative in the range of structures to be used. If the firm is in mass or large-scale production, classical theory, with the familiar pyramidal structure, is the optimal choice. (Sometimes it does work.) When small lots or individual units are produced (such as in tool-and-die shops) or production is continuous (as in process production in refineries or chemical plants), classical structures are not appropriate. (Sometimes it does not work.) In small lots, the skill of the people determines productivity, while in process production, the design of the plant does. (More pieces of the puzzle are falling into place!) We now have order and direction (i.e., rigid classical theory), manipulating workers to fit into organizational needs (i.e., less rigid human relations theory) and different structures due to manufacturing technology (i.e., least rigid technology theory, based on the abilities of people to produce goods and services). But this was only a partial explanation of contingencies. Other researchers had developed situational factors that suggested new and different structures based on environmental change and on the development and aging of the organization itself. These approaches are studied next, since they are more adaptable for us in our development of personal theory. They provide a further development of personal management theory in a contingency rather than a universalistic framework: "If A occurs, do B, rather than "This is the best way to optimize."

SITUATIONAL-CONTINGENCY THEORY: THE UNCERTAINTY OF IT ALL

Woodward's work describes the situation in terms of the technology of production. Lawrence and Lorsch's work (1967) describes the situation in terms of products (development and rate of change), economic uncertainty, and the attitudes of organizational participants. These are more limited or less universal and therefore more applicable design parameters for us than the grand prescriptions of classical theory or the conflict resolution methods of human-relations theory. Woodward's work is explored in detail in Chapter 5, so we will move on to a further exploration of Lawrence and Lorsch next. (By the way, discussing a research finding in several areas is indicative of the cross-fertilization of the modern multidisciplinary techniques.)

Lawrence and Lorsch attempted to find out why the division of tasks and responsibilities among organizational departments did not always increase productivity, as predicted by classical theory and the ideas of bureaucracy. In fact, there was even a conflict with human-relations theory here, because sometimes misunderstandings and conflict increased even when the informal structure apparently supported the formal one. There were even departments in the organization that obviously agreed with the goals of the formal structure but exhibited behaviors that conflicted with other groups within the same formal overall organization. This would be a conflict among goals of different informal organizations within the overall organization.

The researchers felt that insufficient attention had been paid to the relationship between the structure (i.e., of groups) and the environment (i.e., the situation). They recognized that organizational participants, the managers in different jobs, had not only different informal organizations but also different personal orientations toward particular goals. The research, therefore, parted from classical theory, since that theory had no provisions for differences in managers. It was also different from human-relations theory in that it accepted differences between groups of managers that could not be (and should not be) adjusted through counseling or organizational modifications. The researchers found that cognitive differences both between managers in different groups and in the design of the organizational structure were important if an optimal match among formal, informal, and environmental factors was to be achieved. The informal organization, of course, is not uniform across the total organization but changes according to the needs of smaller groups within it.

The relatively uncontrolled variable, the environment, was defined from the perspectives of the organizational participants as they looked outward. The researchers defined three main subenvironments (see Figure 4-1): the market, the technical-economic environment, and the scientific environment. The market was concerned with customers, sales, and service; the technical-economic environment with production; and the scientific with research and development or scientific functions.

Managers in three different industries—plastics, food, and container manufacturing—were questioned about how they viewed these three subenvironments and what was important to them in each of the subenvironments. As predicted, the orientations for each industry were different and the orientations of departmental managers *within companies* in the same industry were even different from each other. "Sales personel . . . indicated a primary concern with customer problems, competitive activities, and other events

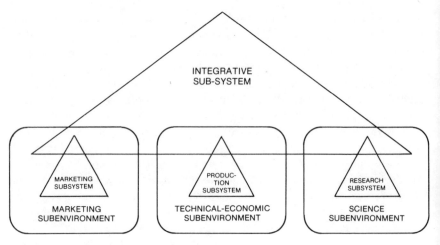

Figure 4-1   Differentiation and integration.

in the marketplace. Manufacturing personnel were all primarily interested in cost reduction, process efficiency, and similar matters" (Lawrence & Lorsch, 1967, p. 37).

There were definite differences in cognitive processes. The cognitive differences were defined as *differentiation:* ". . . the differences in cognitive and emotional orientation among managers in different functional departments" (Lawrence & Lorsch, 1967, p. 11).

This differentiation was measured by various research questionnaires. The variables for individuals were:

1. *The time orientation* of the managers. Do production executives have a shorter-range viewpoint than design engineers, who are concerned with longer time spans?

2. *Interpersonal orientation,* or how people relate to each other. Are there different task and relationship patterns?

3. *Orientation toward goals* that are a function of the particular department. To what extent are sales managers concerned with different objectives (sales volume) compared with production executives (units produced)?

Another variable for groups is of direct concern to this chapter, *formality of structure.* I.e., are there different formal supervisory levels in different departments of the organization? This last variable was measured by the researchers themselves and was not part of the questions they asked of the organizational participants. Resolving interdepartmental conflicts and coordinating the various organizational structures was accomplished by a process called *integration:* ". . . the quality of the state of collaboration that exists among departments that are required to achieve unity of effort by the demands of the environment" (Lawrence & Lorsch, 1967, p. 11).*

Classical theory provided for integration through enforcement of the rules and procedures that governed the behavior of organization members. With a stable and predictable situation, these mechanisms are often effective. With rapid change in the environment and/or in the people in various organizational positions, the effectiveness of these rules and procedures declines. We need other methods to predict and control behavior. One very popular method in classical theory was the use of forecasts, plans, and budgets to predict and control. Plans can be modified and forecasts changed as the situation changes without the need for equivalent changes in the "perfect" organizational structure. The modification process is intended to be one that involves all participants in an adjustment of their goals (i.e., the human-relations theory part) to fit the goals of the formal organization. This process requires more communications within and between departments and greater cooperation among participants.

However, this research by Lawrence and Lorsch showed that managers in different departments really did have *different* thinking processes. These processes were not modifiable by the formal organization to any great extent; the managers were not trying to be difficult, they really were different. With these different viewpoints (differentiation), there is a greater need for interdepartmental conflict resolution (integration). Stability in the environment, with little change in the structure, supports the idea that rigidly defined rules and procedures for all work better, but environmental instability and fast change require the use of more flexible structures and continual coordination of very different kinds of people.

One conclusion of this work is that there is *no one best way* to organize, since there may be one functional structure that is appropriately closed and bureaucratic (e.g., quality control, standards design, computer operations) and one that may be open and relatively unregulated (e.g., quality assurance, advanced design, computer systems development); and yet both types may be within the same company. When this happens, organizational conflict is neither good nor bad but an expected part of the structural design. Uncertainty about the future and in decision making is a major consideration that determines how bureaucratic (low uncertainty) or freewheeling (high uncertainty) the structural design should be.

*From Paul R. Lawrence and Jay W. Lorsch, *Organization and Environment: Managing Differentiation and Integration*, Boston, Division of Research, Harvard University, Graduate School of Business Administration, 1967.

The more differentiated an organization, the more difficult it is to resolve conflicting points of view and achieve effective collaboration [p. 108]. . . . organizations will tend to elaborate and subdivide units that cope with the more problematic or uncertain sectors of their environments [p. 100]. . . . in addition . . . the organization must fit not only the demands of the environment, but also meet the needs of its members [p. 55]. (Lawrence & Lorsch, 1967)

I have taken certain liberties with the sequence of the authors' ideas, but I hope that I have not tampered too much with their intent. Surely these are changes from the relatively stable and fixed formal organizational structures of classical theory. This organization has parts that are differentiated from each other, and the amount of this difference is related to how the participants perceive their environment. An unusual thought to any traditionalist: The people in the organization are the sources of internal differences, and the greater the internal differences among groups, the easier it is for the organization to adapt to rapid change and, if necessary in that situation, to grow and prosper.

This also leads to a familiar and obvious conclusion. Since each organization's rate of change might be different, each organization will certainly be different from its neighbor, not only because of this differing rate of change but also because of the different perceptions of the participants in the groups within the organization.

Therefore, designing an organizational structure for your group that is better than the one you have becomes a very interesting problem. It requires selecting the right interpretation of the rate of change of the environment and a formal structure that matches the differentiation or differences among groups in the informal structure. For example, fast change means wide distribution of power and of the decision-making process. Slow change means consolidation of power and of decision making. The social and psychological variables seem to be almost as important as the main purposes of the technical department.

While this may be quite a bit more complex than the prescriptions of prior theories, that's reasonable, too. More complex situations sometimes require more complex explanations. A simpler answer to a complex problem may be desired, but will it work as well? As we go further with our prescriptions, we must consider the amounts of differentiation among departments and the consequent amount of integration that will be needed to coordinate these different thinking groups. Since those design variables depend on the amount of perceived change in the organization's environment by the participants who will resemble others in their particular group but will be different from other groups, they legitimize the commonsense idea that everyone isn't supposed to be like everyone else or be treated like everyone else. It might even be possible to have a company policy that says "There will be no overall company policies except those required by contract or law. Everyone is be treated differently, as individuals, or at worse, as members of their own particular group within the organization." In the final analysis, creative people who are very different from each other do not seem to respond well to uniform policies, even if those policies are egalitarian in intent.

These design concepts have been supported to some extent by other research findings. For example, there are those involving the attitudes of managers (Morse, 1970) which show that some managers "like" to be in well-defined structures and some don't. Those that "like" it seem to fit better into production-oriented organizations; those that don't, into less restricted groups. There are other prescriptions that research can offer to us.

1. The relationship between functional departments is the most important variable and the various departments' interdependence determines effectiveness (Galbraith, 1970).

2. The "linking pin" concept (Likert, 1961), in which each organizational function has members in it who are simultaneously members of both upper and lower groups in the structure, is most important for greater cooperation and clearer communications. This might be analogous to the "integration" of Lawrence and Lorsch.

3. The maturity of participants (Argyris, 1967) is the major determinant of the effectiveness of the organizational structure, and allowing workers to develop their own methods and procedures improves this structure. This is a bit away from the differentiation-integration scheme but it does support the idea that cognitive differences among participants (such as maturity) is important.

4. There is also the concept that the distribution of power within the organizational structure is the prime consideration (Kotter, 1978). Power in a slow-changing organization is not distributed widely and is usually held at the top.

The Independent Variable—Time

In addition to the human variables of organizational participants that affect the structure design, there is the one very important variable over which we have no control: time. How does time affect the structure?

Of course, organizations are affected by time just as participants are. Organizations grow, shrink, change their borders, become centralized or decentralized, and definitely are changed over the passage of time. One of the reasons for this change is the incorporation of modifications to the structure that resulted from repetitive management decisions. A nonrepetitive decision that originally resulted in a unique solution has been repeated on several occasions, and the structure is now changed to institutionalize that decision. (For example, why decide on a method to place a purchase order for the tenth time if you can solve the problem once by writing a purchasing procedure that the new *purchasing department* will be able to follow without your intervention?) The method of purchasing is no longer a type of nonrepetitive decision. That unique decision has been institutionalized and the structure changed with the development of a purchasing department. The organization has changed over time. However, sometimes when nonrepetitive contingencies are incorporated, one possible result is excess capacity over that needed for the major existing organizational goals (Thompson, 1967, p. 47). Changing the structure may not give you a perfect fit between present expanded needs and the new organizational design. Sometimes overcapacity results.

For example, it is possible, now that you have written the purchasing procedure, that what was once a part-time assignment for you as a technical manager has grown into a full-time assignment for a purchasing manager, a clerk, and an expediter. Very few things are free—especially organization designs. Even though all you needed was two-and-a-half people, people come in whole units, and occasionally overcapacity results. In this example, it's best to change the organization slowly and keep it a bit lean.

Developmental Theory: Another Contingency that Deals with Time

Time does other things to organizations besides institutionalizing nonrepetitive decisions for groups. One approach (Filley & House, 1969) that deals with the whole organization proposed a model of growth and change that could fit many of the overall needs of technical organizations. However, this model applies only to firms that begin their growth with a single-product base. Since that is a typical beginning phase for many technical companies, this model could be very appropriate; it provides a general yardstick to measure where your organization is against where it should be. The measurements, of course, are, at best, ordinal. The single-product model describes three general stages of growth: the traditional (or craft) firm of stage 1, the dynamic firm of stage 2, and the rational administration of stage 3. See Figure 4-2.

The majority of small technical businesses seem to fit into stage 1. The structure is simple and direct; it consists of the owner-manager and the individuals who report to her or him. There is little or no documentation of the formal organization because the personal views of the owner are well-known to everyone in the firm. The future is judged to be like the past, so

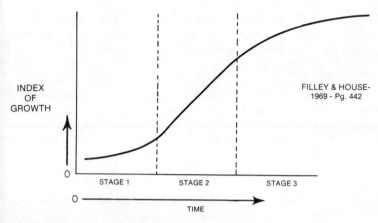

INDEX
OF
GROWTH

FILLEY & HOUSE-
1969 - Pg. 442

STAGE 1   STAGE 2   STAGE 3

TIME

Figure 4-2   Growth over time.

there is little need for planning, and since the organization is relatively stable and unchanging, there is no justification for using planning to modify the organizational structure. Conflict is minimal and communications are good. Any functional activity that will not contribute to the bottom line is not developed. With no planning or marketing staff, the organization is very flat.

Stage 2 occurs when innovation brings growth and change and the owner has to cope with the much greater uncertainty these things bring. The organizational stability of stage 1 gives way to greater managerial differentiation, with an increase in internal conflict as departments (and the departmental managers responsible for them) rise and fall in importance. Growth goals are forward-looking since the future will not be a repetition of the past. The technical staff either comes under the direct supervision of the organizational chief or else is given a free hand so long as the conflict among departments is acceptable and new products and services are being created. Planning is used to integrate and coordinate the more diverse departmental activities. The organizational structure is not as flat as it was in stage 1, since staff functions such as finances, marketing, and planning have been added.

At stage 3, the firm is much larger and more complex. The owner and the entrepreneur have been replaced by trained executives who "plan, organize, direct, and control" (whatever that means to them). The organization becomes the familiar triangular or pyramidal shape. Action is taken only after careful analysis and prediction of probable consequences. Rationality and formal organizational structures are guides for solving typical structural problems such as the "correct" span of control and amount of responsibility delineated for each position. There are many committees and consequently a great number of committee meetings. The structure is minimally adaptive and not very innovative. Technical "experience" is valuable and "innovation" is very carefully considered (if not distrusted) before being implemented. The "not invented here" syndrome appears very forcefully to obstruct the adoption of any ideas not developed internally. At this stage:

> The firm becomes a collective institution rather than a one-man show. Its institutional objectives becomes separated from the needs of its participants and organizational efforts are directed to the changing needs of a defined market, rather than exploiting the short-term advantages of the initial innovation. Its mechanism is adaptive. (Filley & House, 1969, p. 449)

Organizations have reached stage 3 in total size without attaining the more or less collectively rigid mental set that is described. They have done this by dividing their companies into many smaller ones and allowing complete decentralization, an excellent form of differentiation. The corporate headquarters serves only as a resource supplier when the smaller divisions cannot provide sufficient funds, personnel, and/or facilities to sustain their own growth.

Contingency and Uncertainty

Many of the technical organization structures that we discussed could fit into this model. Taking the formal structure as one element, stage 1 could be classical theory, with the owner-manager acting as the organizational patriarch. Stage 2 shows the high differentiation and integration patterns of the contingency theory of Lawrence and Lorsch, and stage 3, the manipulative structures of human-relations theory, with its emphasis on internal harmony. The others that were mentioned are probably equally applicable. The point is not to determine the exact theory that applies to your situation, since no theory will exactly fit, but to be sensitive to the variables and how they relate to you in order to build the best structure that fits your situation as you then perceive it. It's not a one-time job, since all of these things are moving.

We now move into other, more direct prescriptions that apply to the unique needs of the technical organization. These prescriptions assume that the mission of the technical organization in your company is fairly well known to you. If it is to be a source of new product development, manufacturing support, evaluations of competition, or anything else, the purposes have been defined and you know what they are. If you don't know, that would be the best place to start. But assuming that you do know, the next section prescribes several structures that can be modified by you, the designer, to fit these purposes. These are more specific prescriptions.

THE TECHNICAL DEPARTMENT: ORGANIZATIONAL PRESCRIPTIONS

Since the environmental situation is more of an independent variable than the formal organizational structure, the logical design steps would be to define (1) the situation as it presently exists, (2) the appropriate organizational design that you believe will best fit it, (3) the existing design, and, finally (4), some methods for moving from the existing design to the desired one. All these steps require definition and ordinal measurement. You, as the designer, set up the values and the scores.

The design that works best should have less difference between "what is" and "what should be" than other designs. It also requires a close match between the designed formal structure and the informal climate, since our organizational participants are "knowledge workers" who control the major parts of the work quality and quantity. Therefore, the differential motivation plan noted in Chapter 3 or something equivalent is required as part of the structural design. Even so, all this does not promise perfection; we are dealing with perceptions in all cases and we've reviewed some of the problems with those. So while it may be possible that the organization built will get the job done or satisfy the needs of the people, achieving either goal may mean compromising the other (Morse, 1970, p. 85).

The message seems clear enough. We will try to develop a formal organizational structure that matches the needs of the environment first and the informal structural situation second. The environment is always the prime variable. The closer the match, the more effective the structure will be in assisting managers to get their jobs done, since the environment will provide economic rewards, and the informal structure, personal ones. But there will always be compromises that the formal structure must make, depending upon the importance of the economic and personal goals. The closer the goal congruence between formal and informal structures the more effective is the intended formal organization structure that includes them both.

Making Design Goals Operational

However, even beginning the design and attempting to define the limited goals of the technical department (which may be different from overall company goals) is very difficult, since the definitions are always operational ones. (If you recall, an operational definition is one that refers to the thing being measured; e.g., intelligence is measured by the score on the intelligence test, or a successful organizational structure provides more economic and social rewards to organizational participants than an unsuccessful one does.) Perhaps one way to start your design would be to use the general theory behind differentiated and integrated structures and determine whether your organization is in fast or slow change (see Lawrence & Lorsch, 1967, for details). Once you have that definition, follow it by developing an appropriate differentiated or integrated structure as a testable hypothesis. That hypothesis could even be a discussion basis for a preliminary tryout (or test) of your organizational structural design. This is a reasonably logical process: definition of variables and relationships, developing a hypothesis, and trying it out or testing it. This whole process is, of course, based on some prior defined goal(s) that the structure is to help achieve.

As noted before, I believe that most technical managers know the operational goals or definitions or the "what" of their groups; i.e., "developing the best products in the gas transmission industry as measured by industry acceptances of these designs in the new pipelines." That's more limited and somewhat different from the company goal, "to make the best return for our stockholders and employees." But even that limited goal has not been made operational enough to be used as a basis for organizational design.

The answers often result from proposing a lot of questions that begin with "how." *How* will it be done? How will we know that it has been done when it is supposed to be finished? How will progress be measured? And so forth. These are all typical operational questions to be answered if the "what" goal is ever going to be satisfied. Except for the size of the questions, there is no difference in answering these questions for the structure than in answering questions for an individual in a personal motivational system (see Chapter 3). When goals are made operational (for example, develop a line of valves for the gas transmission industry to meet the new functional specification of the latest convention standard, weigh half as much as our present designs, and present no cost increases, and do it by the end of this year), we can begin thinking about the types of repetitive behaviors or structures needed to reach them, or about the differentiation-integration thinking or the technology of manufacture or the time variables or whatever the design hypothesis uses. After goals definition we can deal with defining the various design hypotheses that are easily available. Our further discussion here is limited to the technical operations, not those of the company as a whole.

FUNCTIONS AND PROJECTS

Definition

The term *technical organization* is intended here to include both the scientific and engineering and the production and manufacturing departments of the company. Most technical organizations, such as in engineering and manufacturing, are organized either on a functional or a project basis. Functional structures are continuing structures with group goals that are defined by overall company goals. These functional structures are desirable in stable situations, are survival-oriented, and are intended to be "immortal"; that is, they continue to exist relatively independent of the tasks assigned to them.

These tasks are often not very long-lived. They arrive and are generally solved within the time limits of the company's financial reporting periods.

The engineering department as an organization, for example, goes on as long as the company exists, and though divisions within it may change with time, those changes are usually quite gradual. Even though the goals of a particular engineering department may not be the same as overall company goals (i.e., make reliable products versus expand our share of the widget market) they are fairly constant over time. Someone may decide, "Well, we ought to strengthen our field support of the new stainless steel widgets because they seem to be having some installation problems that will require continuing education of our customers." That particular structural change will probably continue as a more or less permanent one as long as those stainless steel widgets are in use. Functional structures emphasize cooperation, harmony, and logic. They are less expensive to operate than project structures because they are stable and the people can be trained and procedures used can be applied (in a decision matrix) without a great amount of structural change needed over the passage of time.

Conversely, project structures are usually single-purpose designs that have a definite time limit. They are desirable in rapidly changing situations, have limited, specific goals, and, when they achieve those goals, they are dissolved. Conflict is endemic around projects because people act differently in different projects and procedures are rarely applied consistently for more than one project. Modification of past decisions and behavior or the creation of new decision matrixes are more or less required each time. Conflict is also caused by occasional confusion if the projects cut across well-established functional boundaries. The term *project* used here includes task forces, matrix groups, and other limited objective structures.

These two structures, *functional* and *project,* are not an either-or design in many technical organizations. Functional organizations may have project structures within them and project organizations may have functional structures. In the former case, the projects are fairly limited; in the latter case, the projects are probably large and will last for several years, thereby giving the project structure a quasi-functional status. Therefore, these organizational models are part of a spectrum or range of possible design models that can be fitted to the particular situation you perceive.

The Functional Model

Now that we have classified the two major structures of functions and projects, we can define the designs available to each of them: In functions, the least complex formal organization design is the triangular- or pyramidal-shaped model. It's a familiar model that was based in classical management theory and can be either discipline- or product-oriented. This design is effective since it fits the requirements of a relatively unchanging situation. Uncertainty in both the external environment and internally in the organization is relatively quite low. Classical theory, the stage 1 development company, and the Lawrence and Lorsch production-oriented container manufacturing company with low differentiation and high integration are all examples of this very cost-effective design. It fits bureaucratic descriptions very closely.

It is perhaps the oldest design available, possibly having been described in Exodus (18:25) when Moses, on the advice of his father-in-law (could he have been one of the first management consultants?), selected one able leader to rule tens of people, then from that group another leader to rule the tens which now became hundreds, and finally from the group of leaders of hundreds, other leaders to rule the new hundreds, which were thousands. With such leaders being responsible for the limited span of control of ten people, it was possible to communicate to 10,000 people in just four steps. (As an interesting intellectual exercise, consider how many levels of organizational structure your company now has and how many it would need with this span of control of ten people. Do you really think that all those managers are needed?)

A pyramid permits rapid communication and it becomes more effective as the shape becomes flatter (longer span of control), since there are fewer levels through which to communicate. This structure has tremendous potential to use power as a command instrument when it works as it is intended to work, since it is relatively easy to obtain feedback, determine which level of the organization has not responded, and take corrective action. In many

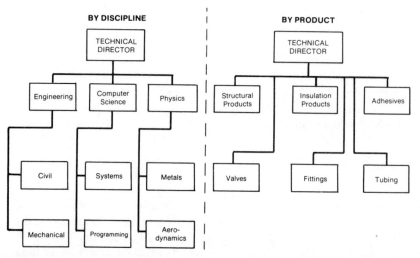

Figure 4-3   Design: Discipline versus product.

cases, there are two major divisions of this design: organizing by discipline (i.e., electrical engineering, tooling design, mechanical design, etc.) or by products and services (i.e., container production, stainless steel valves, refining, etc.). Figure 4-3 illustrates these two divisions.

ORGANIZING BY DISCIPLINE WITHIN A FUNCTIONAL STRUCTURE

Advantages: Grouping of personnel and facilities around a particular discipline provides the organization with expertise that can be applied horizontally across the total structure. For example, anything that the company knows about corona loss in wires can probably be found in the electrical power transmission group of the electrical division of the engineering department. When all personnel with particular skills are grouped, it is possible to hire experts, train technical specialists to achieve the expertise needed, or even to minimize the number of people needed to supply that expertise, since it will be available to the entire organization. Skills can be shared or pooled as required.

There is a clear promotion path for individuals upward that is dependent upon the personal acquisition of professional competency within a vertically defined specialty. Problem solving and decision making are quite straightforward if the problem is classified to be within a particular technical specialty. It is likely that the latest state of the art will be applied to problems, since all the specialists within a discipline are located together. Errors and/ or breakthroughs are easily communicated. Positive and negative experiences are directly available to all interested parties.

It is also possible to maximize the use of expensive facilities and equipment and to justify their procurement. As an example, it is easier to justify the installation of a large computer-assisted design facility economically if many people will be using it continually and the experts are right there either to operate it or to assist the user in operating it.

This structure has another very important advantage: The informal organization tends to be very similar to that of the formal structure. Being trained in a particular discipline for years generally develops behaviors in individuals that are recognizable and accepted by their peers. Joining a structure in which those behaviors are encouraged and specialized technical expertise is respected further supports social behaviors that match the organization's culture. In this case, the culture tends to match the formal structure and formal authority (the power assigned to a particular position) tends to match influence (the power that accrues to a particular person). Social groupings (one aspect of culture) parallel the formal design, since specialists tend to develop recognizable behavioral patterns that are accepted by their group. This explains in part why specialists socialize with each other in after-work activities. The lettering on the tee shirts of the interdepartment bowling or golf competitions often describe a functional discipline such as the "Power Boys" (i.e., the electrical transmission group).

Disadvantages: When a functional discipline is emphasized, the integration or exchange of ideas across the organization can suffer. Cooperation is difficult to achieve, since vertical relationships are emphasized over horizontal ones. Achieving integration (Lawrence & Lorsch, 1967) becomes more time-consuming and more expensive. In some cases, it can even become the major responsibility of the department manager. The development of vertical promotion paths rarely provides the management ability that managers who coordinate the activities of others must have. Occasionally the organization may promote the best technician into a managerial job where technical expertise is secondary to the requirements for skills in dealing with people. Additionally, the common professional and social bonds among technical specialists can develop extraordinary resistance to any change of the status quo. In some cases, self-perpetuation and improvement of narrow expertise support the mental walls that are typical of the *"not invented here"* (N.I.H.) syndrome, which acts as a strong deterrent to any creativity that is not generated internally. The state of the art can be applied but it had better come from within the group and not by any "outsider."

ORGANIZING BY PRODUCT WITHIN A FUNCTIONAL ORGANIZATION

Advantages: The organization structural model that is based on product orientation groups all personnel and facilities according to a particular product, service, or market. This coordinates organizational expertise in much the same way as the discipline-based model did, but while the discipline model applies specialized knowledge across all products, markets, and processes, this product-based model concentrates in depth on a limited number of products, markets, and processes. For example, the product orientation of the power transmission tower group could include the skills needed in structures, metallurgy, electrical and civil engineering, and fabrication.

Therefore, solutions to problems that are concerned with different aspects of a *particular product line* are concentrated. Expensive equipment and facilities still are justifiable but they are harder to justify because the equipment and facilities now apply to more generalized areas, such as industries, and their use might therefore not be as important as in a technical specialty. A high-frequency voltage tester that was, for example, needed by a discipline concerned only with high-frequency electrical transmission (e.g., the specialist in corona loss in the physics department) might only be marginally justifiable for a whole product line that services the electrical transmission industry, since corona loss might be just one of many problems and not even at the top of the list.

In a product structure, the state of the art is less likely to be as current as in a discipline-oriented structure, since all the specialists in that art are not always grouped together. While both positive and negative experience is still directly available to interested parties in the rest of the company, they are transmitted only as they apply to an industry or product line. There probably is a higher level of technical uncertainty in product-oriented structures than in discipline-oriented structures, since changes in products and services often occur much faster than changes in basic disciplines.

Although the informal organization still parallels the formal organization, it is not quite as congruent in its goals as in the discipline-oriented structure. One broadens one's goals in moving from the specialized discipline orientation to the broader product orientation. Although the product orientation may include fewer disciplines than the discipline orientation, which includes all products, there seems to be a wider range of work and problems in product organizations. Another difference is that discipline-oriented groups seem to have higher differentiation; they're more like each other within groups but more different from other groups. They therefore require more integration among various technical groups. Product-oriented groups, however, are less like each other within groups. Their training could have been different. For example, an electrical engineer and a mechanical engineer could both be working on the new hydraulic control product line and probably have less differentiation, thereby requiring less integration.

Promotion paths in product-oriented groups are broader than those in discipline-oriented groups. They not only go up the organization, but might also go into different groups of the same organization. The cost of the technical operations, however, has decreased because very highly specialized (very costly because probably less than fully utilized) technical talents are rarely needed. Product-oriented structures can limit participants' viewpoints as easily as discipline-oriented structures can but increased product or industry knowledge is a powerful justification for the individual's promotion into higher levels of management, since the scope of the problems increases. It is one of those "necessary, but insufficient by itself" criteria that often appear in textbooks. A knowledge of the market environment is a virtual necessity if the individual is to succeed as a manager and product-oriented structures encourage that.

Cooperation and coordination across product or industry lines within a company are probably easier to achieve, since differentiation is lower. The "not invented here" syndrome could be almost as applicable here as in a discipline-oriented organization, but now the new idea is more difficult to kill because products cover broader areas than disciplines and they are therefore not as susceptible to the expert who says that it won't work.

Disadvantages: The major disadvantage of using this design is a loss of specialized and professional skills. Most people tend to lose the ability to apply learning quickly unless they use it continuously. When was the last time you used the partial differentials that you had to learn in the classes in advanced calculus? How much trouble would it be to dig out your old texts and relearn them? (Do you even know where they are?)

Summary: When and How to Design that Functional Structure

The alternatives for functional structures that were mentioned above are useful to you when you have generally defined the situation as you would like it to be and as it is now. They are helpful in that first stage of selecting the structural pieces with which to change the "is" to the "should be." When and how do we get that change started? We start with the "when."

> . . . the organization has two choices. It can adapt continuously to the environment at the expense of internal consistency—that is, steadily redesign its structure to maintain external fit. Or it can maintain internal consistency at the expense of a gradually worsening fit with its environment, at least until the fit becomes so bad that it must undergo sudden structural redesign to achieve a new internally consistent configuration. In other words, the choice is between evolution and revolution, between perpetual mild adaptation, which favors external fit over time, and infrequent major realignment, which favors internal consistency over time. (Mintzberg, 1981, p. 115)

According to this writer, these two choices are obvious. I agree and I believe that "perpetual mild adaptation" is the better one. I suggest that when there is a perception of an increasing mismatch between the situation and the goals of the organizational structure, there will be a negative economic impact on the organization. Since that mismatch is slower, less noticeable and more insidious in the technical group within the overall organization, it seems to me to be easier to assume that it always exists; therefore managers should follow an organizational redesign process that is no different from any other ongoing corrective process, such as budgeting, recruiting, or training. The "when," therefore, is at regularly scheduled intervals.

The "how" is not that obvious, but once you begin to use the methods suggested next, you will find that it is useful in the ongoing processes noted above. We begin by considering how to measure uncertainty. Since we know that it cannot be measured objectively (if we could measure it, it would no longer be uncertainty), we must develop a way that moves this concept of "uncertainty" into the subjectively more definable *risk* concept. Then we can use this risk to evaluate various potential scenarios as you develop different organization designs or hypotheses. This becomes the major part of the "how."

Uncertainty can always be defined as a *nominal* number; i.e., it is different from something else: A red checker is different from a black checker. Risk can be defined as an *ordinal* number, since it is not only different from other risks, but it is also defined as being larger or smaller than other risks. The proposed structure differs from another one and has a subjectively determined higher or lower potential for success. Finally, certainty can be defined as an *interval* number, since there could be a linear relationship between various numbers on any measurement scale measuring various alternatives, but there is no absolute zero, of course, so there are no *ratio* measurements.

As an interesting point, most of our physical measurements are interval based, such as degrees Fahrenheit used to measure temperature. Since there is no absolute zero in interval scales, the numbers in them are not multiplicative; e.g., 60 degrees Fahrenheit is not twice as hot as 30 degrees Fahrenheit. It is just 30 degrees hotter. If we could find an absolute zero for our measurements, we would be able to use ratio measurements, in which there is an absolute zero and the numbers are multiplicative. The kelvin temperature scale is an example of this. We have no such numbering system available in management. We will, however, be able to use ordinal numbers qualitatively in developing a starting point for our organizational design, i.e., this proposed design is going to be better than the one we had and I can guess that its successes will increase our technical output by about 25 percent.

As an extension of my suggestion to assign risk factors to a potential scenario subjectively, you might consider each of the elements of the structure, such as job specialization and behavior formalization, or whatever you wish to use as a measurement, and subjectively "score" them. Then you as the designer can come up with a total "score" for the structure that *exists now*. The assignment of the subjective risk factors can be done by using a Likert-type scale (Likert, 1961) for each element, as shown in the following table:

Score	Description
1	Very low
2	Low
3	Average
4	Above average
5	Very high

When you have assigned all the scores for *what is*, you can do another score for *what should be*. The difference between the scores is an indication of the amount of structural redesign needed to fit the structure to the situation as defined. The differences between the scores on each element are indications of specific corrective actions that should be taken to modify the structural design. All these numbers are subjectively assigned, it should be understood.

Just as an example, let's assume that we have subjectively decided to use these typical elements to measure our organizational structure. Using that subjective scale of one to five points, we arrive at the results shown in Table 4-2.

The results might indicate that there should be a structural redesign toward a project orientation (notice the positive score on faster response to market and the negative score not conforming to department policies), using people who are highly qualified in a particular discipline such as high voltage lines (new product output) and are willing to act independently of department policies. The methodology suggested above is not the only way, nor has this way been tested extensively or empirically. Some of the problems to consider in using this method are that you might not be able to attach numerical values to elements, they may not be additive, or the values may not be equivalent. (Is a 2 for element 1 above equal to a 2 for element 4, or is a 4 at least two units better than a 2?)

However, whether the methodology has been tested or not is really not as important as developing some kind of useful tools that result in a method for organizational design. What is important is that somehow you absorb uncertainty and move measurements into risk, *documenting your thought processes*, as you propose your hypotheses for a new design. The documenting process helps you to learn. This approach potentially includes all aspects of the organizational structure.

There is another approach (Miller, 1981) that is directed at the redesign of control systems (not total technical structures, as we have been doing) to fit changed goals. Although that approach is concerned mainly with the manufacturing sections of the technical organization, the analytical procedure used seems to be equally applicable to the design of systems for the formal organization of any technical function. In effect, this approach is a classical one since it tries to fit the internal functional systems to the needs of the manufacturing group for information. It compares various system alternatives

TABLE 4-2   SUBJECTIVE MEASUREMENT OF ORGANIZATIONAL STRUCTURE

Elements	What is	What should be	Differences
1. Fast response to market	2	4	2
2. High voltage line expertise	5	5	0
3. New product output	1	3	2
4. Conform to dept. policies	3	2	(1)
5. Etc.			
Total	11	14	3

with some predetermined manufacturing goals and selects the best fit. But systems are only a part of the formal structure. The task of the systems designer includes, for example, the communications systems, training operations, and possibly the costing systems. But since systems and structures are interactive, those systems eventually are modified by and become part of the structural design. As an example, the organizational model that responds to short production runs with many different products might require "manual" systems, maintained by human beings who perform a broad set of management tasks under high uncertainty. When those tasks are performed, the structure is partially defined since that is one definition of structure and that would definitely affect the organizational structure.

Other methods of organization design are more qualitative than those noted before. The answers to the questions below could be the design criteria for a changed structure, for example:

Means-end schema: What business are we really in?
Authority structures: Who bosses whom on what matters?
Job-slot pyramid: What is the extent of individual responsibility?
Communication-network: Who talks to whom, when, and about what?
Group-linkages: What teams coordinate, and how?
Decision rules structures: When internal conflict threatens, what gives?
Program inventory: What tasks are we proficient in?
Spatial organization: Where do we operate from?
Style: What special character shall we assume?
Socio-political environment: What kinds of public service are worthwhile for the organization? (Meir, 1967, p. 477)

It is then possible to use a similar Likert-type score here for "what should be" and "what is" and again work from that position.

Another method of coming up with an organizational design involves charting the various responsibilities of managers. This charting process enables managers from the same or different organizational levels to participate in identifying their roles as well as the roles of others in making particular decisions. The process assists the organizational designer by clarifying roles and relationships, both formal and informal, as perceived by the managers themselves. The way it's done is to have the respective managers complete a form (see Table 4-3) that lists all their decisions in a column on the left side of the page. Against each decision, in a horizontal row, they score whether their participation is required (score = 3), for approval (score = 2), for information (score = 1), or not needed (score = 0) (Galbraith, 1970). Management loads are distributed according to the numbers of 1s, 2s, and 3s. According to our chart, Sam seems to be the major decision approver with the most "ones," Mike may be able to take on more work, and Charles could be the important decision maker. This is very subjective, as are all the other measurement schemes.

**TABLE 4-3 SUBJECTIVE MEASUREMENT
OF RESPONSIBILITIES**

	Charles	Mike	Sam
Budget/cost decisions	1	0	1
Recruiting decisions	2	1	2
Technical decisions	3	0	1
Totals	6	1	4

We have reviewed several paths to use in structural design and the one to choose may not be that clear. However, some type of selection must be made, because if you don't (or can't because of other duties) accept the idea of periodically scheduled redesign, the "when" to design will quickly select a time of its own. That will be when the overall organization suffers economic losses as the fit between it and the situation deteriorates with time, and your mismatch contributes to those losses. Sooner or later, something will give. If the redesign direction comes from your own evaluations, the task will probably be much smaller than if it comes from the external economic environment. In any event, that environment cannot be ignored for long; eventually, something must be done. If you wish to start slowly at your own group level, here are some potential questions to be used in determining the need for organizational redesign before it begins.

1. What is the most important contribution of your group?

2. What is the most important problem?

3. If your group were eliminated, what would the result be?

4. How can your group easily be improved?

5. What recommendations for improvement would you be comfortable in suggesting today? (Galbraith, 1970)

Another way to start a redesign could be to start with your major variable: people. Table 4-4 shows some of the assumptions about people in several very generalized structural designs. If you see a fit, you may have a starting point. People are at the basis of any organizational design, and these assumptions should be understood before attempting to use any of the designs shown in Tables 4-2, 4-4, or any other designs. Scoring of the elements selected can be done for your organization as suggested before, by scoring all the statements for "what we have," then doing it again for "what we want." The arithmetic differences indicate where to redesign.

The idea that there is no "one best way" to manage has therefore been expanded to include no "one best way" to develop the organizational structure. Moreover, designing an organizational structure that is applicable to

TABLE 4-4 ASSUMPTIONS ABOUT PEOPLE IN VARIOUS DESIGNS

Design	Element	People
Cottage industry	Paternal, familial	Children, need guidance
Classical theory	Hierarchical power, limited span of control, owner is chief	Have only economic needs
Human relations	Social concepts, conflict resolved through managerial solicitude	Satisfy social needs on the job, cooperate with group norms
Technology	Manufacturing process controls the design (see Chapter 5)	Select their own jobs, using their own needs for guidance or freedom
Bureaucracy	Discipline, professional, documentation and procedures	Highly trained, experts in their jobs
Project-matrix	Authority follows responsibility, constant conflict and restructuring	Need for achievement, self-actualization

the technical department presents a greater problem for the designer than perhaps any other department of the company, since the repetitive decision making (that the formal technical organization is supposed to handle) is often split into two orientations: functional and project-matrix. We do, however, have some guidelines.

Determining whether to select the functional or project-matrix structure depends first upon the time constraints and then upon the cost factors. A functional structure is better when a continuing or never-ending series of relatively smaller tasks, such as product improvement, standards development, drawing controls, vendor support and evaluation, or technical training, must be completed. It's best described by the highly integrated, thoroughly controlled, mechanistic model of an organization involved in mass production. This is the least costly structure to operate and to control, since it is a top-down directive structure that does not (and cannot) consider major differences of the participants. It assumes that their personal welfare coincides with that of the organization. Interestingly, effective participants in those organizations usually agree with that (i.e., the company man). It tends to prevent dysfunctional change and emphasize harmony and cooperation.

On the other hand, project-matrix structures are specific, nonrepetitive, and short-term. They have major and fairly well defined tasks to complete or accomplishments to attain. These are more costly to operate than functional structures, since they tend to be bottom-up structures, incorporating differences among participants. They gain the advantages of creativity but also attain the disadvantages of potentially increased organizational confusion. The structure itself may vary from loosely organized task forces through various modifications, finally becoming a separately housed project concerned with extremely large, costly, and relatively lengthy tasks; but there is always a time limit on the life of the project structure. One definition is ". . . an organization designed to accomplish a specific achievement, created from within a functioning parent company and dissolved upon completion of that achievement" (Silverman, 1967, p. 1).

The Project Model

Time, as the major independent variable of the situation, is the primary consideration in the choosing or the design of a project structure. The sequence of the design is (1) the amount of time or funds that the project will need (independent variable) (2) related through the organizational structure (intervening variable) (3) to cost and performance (dependent variable).

A project structure, by definition, always includes higher uncertainty and greater complexity than a functional structure and often results in more or less institutionalized conflict. This could be an asset (if innovation is desired) or a liability (if predictability, coordination, and lowest cost are desired). Project structures support both individual flexibility and balanced, open decision making, but at a higher cost than the more orderly functional structures. These structures include an unusual characteristic: everyone in them reports to at least two supervisors, and this happens even if the technical organization has no formal structure for it. When there is no formal structure, project managers tend to exert more personal influence on project-assigned personnel, while functional managers have the only formal control. The effect is still the same: two bosses, even if the formal structure doesn't show it.

A typical project or matrix organizational structure (the terms *project* and *matrix* are used interchangeably here) could look like the one in Figure 4-4. The project managers for each project report to the chief project manager (and implicitly to the client or customer). When formally organized this way, each engineer reports both to his or her functional manager and to a project manager. Each finance representative and each quality control person also reports to a functional manager and a project manager. This structure provides for a typical project organization in this particular company to consist of a project manager, an engineer, an accountant, and a quality engineer.

Each person in a matrix has at least two sets of responsibilities and these sets of responsibilities can often clash with each other. Managers head up and attempt to balance the dual chains of command (both functional and project). This dual reporting structure provides the organization with the capability for meeting the varying needs of changing technical requirements or goals and is intended to deal with the consequently higher levels of uncertainty and complexity. One human side effect is that successful project participants are usually able to handle higher levels of uncertainty than those who regularly work in a functional structure. If we agree that uncertainty increases as one goes up in the functional organization, by definition anyone

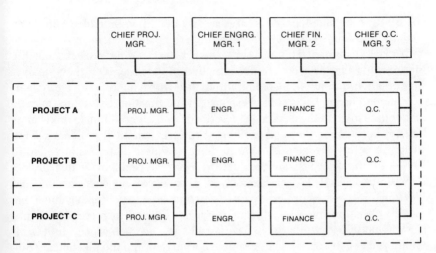

Figure 4-4 Project-matrix design.

who has successfully completed projects is trained beyond his present functional responsibilities.

> Successful experience in operating under a matrix constitutes better preparation for an individual to run a huge diversified institution like General Electric—where so many complex, conflicting interests must be balanced—than the product and functional modes which have been our hallmark over the past twenty years. (Davis & Lawrence, 1978, p. 132)*

FUNCTIONS VERSUS PROJECTS (STABILITY VERSUS CHANGE)

The uncertainty and complexity in the upper levels of large multinational organizations such as General Electric are not necessarily found only in these larger structures. Managers in medium and smaller companies have similar problems and uncertainties to deal with. Their markets are also impacted by competition, highly skilled personnel leaving and entering the company, and product life cycles decreasing for both small and large organizations.

The predictability and coordination of the in-place functional organizational structure provides few guidelines if an environmental change has not been accounted for in the structural design. The creativity needed for new product development, diversification plans, reorganization studies, and alternative investment analyses (as typical examples) is usually *not* part of most groups that are functionally (as opposed to project) oriented. That should be

---

*Reprinted by permission of the *Harvard Business Review*. Excerpt from "Problems of Matrix Organizations," by Paul R. Lawrence and Stanley M. Davis (May–June 1978). Copyright © 1978 by the President and Fellows of Harvard College; all rights reserved.

apparent, since functional structures are, by definition, intended to provide a *continuing* framework for coordinating participants' actions in accordance with prior decisions. This may also seem to be true for projects, but the difference is that projects are *not* intended to go on forever; they are *one-time* creations intended, by definition, to deal with a nonrepetitive problem.

When a change occurs, the response of the functional structure may be inadequate to meet the challenge (i.e., the response or solution to this change has not been found yet) and a different (project-oriented) problem-solving structure is then developed. That problem-solving structure is intended to have a limited life; just long enough to solve the problem. If that project structure reaches its goals it is dissolved, because it is no longer needed and it is much less confusing and expensive to operate a structure in a functional model. Occasionally, when the problem is expected to reoccur, but in a slightly modified way that may have novel and uncertain aspects to it, simplified project structures can become incorporated into the functional structure as a type of special unit intended to handle these problem modifications.

An example could be a research and development department that is organized functionally, with internally developed project organizations intended to achieve specifically assigned research tasks. Assuming that an initial task, that of developing a quiet air compressor for municipal street repair work, is successfully completed by the "quiet compressor" project, the tasks of developing *a line* of "quiet compressors" for other applications could be assigned either to the functional development group or to a mini-project team for development that could be set up.

The size of the task is not the only major rationale for using project structures. It is the nature of the task itself. One-time, creative tasks with high uncertainty and high situational ambiguity require project structures. Continual, logical, and relatively repetitive tasks with lower uncertainty and lower situational ambiguity require functional organizations. If solving a new problem is important and is a first-time effort, it generally requires project structures. Using that solution over again requires only the less costly functional structure. And in one organization in which both new and repetitive types of tasks must be handled, such as in the technical department, it seems reasonable to provide both types of structures. In effect, it is a continuation of the no "one best way" to organize, even within the department.

However, this no "one best way" is not intended to suggest that there are no recommended repetitive ways to set up project structures. There are, but they depend on the situation that exists, and each organization can develop methodologies that fit their situation best. It is even possible to have a standard operating procedure to be used in organizing projects. The model of project organizational structure that follows and the processes used to operate it are only one alternative design. The model utilizes concepts that I have proposed in many technical industrial organizations. Your structural design for projects might include this model, but only as a starting point, and since your situation is unique:

- Modify the model to fit your situation.
- Test it.
- Revise the model, improving it.
- Test it again, and so on.

A project structure is intended to be a totality designed to handle tasks involving great uncertainty. However, as the unknowns *decrease*, structural modifications should be made until, at a logical point (and only you know where that is), the organizational structure parallels the functional design and becomes part of it. A partial list of the project design sequence which a project structure is intended to cope with could be:

1. Establishing objectives (i.e., purpose, time, cost, resource limitations)
2. Defining internal task dependencies (i.e., logical development of sequential and concurrent activities)
3. Work scheduling (i.e., performance versus time)
4. Estimating (i.e., costs expected to be incurred as a result of work scheduling)
5. Deciding on work assignments and authorization (i.e., who does what and how work shall be organizationally controlled and measured)
6. Procuring services, materials, and equipment (i.e., requisitioning and purchasing methods)
7. Attaining required resources (i.e., personnel, funds, facilities)
8. Setting up information systems (i.e., reporting and measuring data)

But this design sequence or list of project tasks should be modified to suit your particular organizational needs. The organization of the structure (or repetitive human interactions) could include these four major areas:

- Charter
- Work breakdown system
- Operating procedures
- Close-down method

They are all covered here instead of being described in other sections of this book. They might be more appropriate elsewhere (such as in the financial information systems of Chapter 6) but because of the overall interrelationship between project structures and the operations that they are intended to accomplish, I have placed them here.

CHARTER

Projects have to start somewhere. Someone or some small group has to perceive a new problem and/or opportunity that is disturbing the organization or satisfies some requirement of the market or can provide an opportunity, if solved, for the organization to improve itself. Project goals may include marketing, sales, profits, quality or any other organizational, economic, and/or social goal. Since projects themselves may be repetitive structures, the *charter* is intended to be the starting point that defines *how* to start a project.

Charters are different for different organizations, but they have to exist either explicitly (as I am suggesting here) or implicitly in the minds of the technical managers. Without them, the organization starts from ground zero each time. That would be wasteful, because it's unnecessary. There may not be "one best way" for all, but there are commonalities to *each* organization that can be defined for that particular organization. The charter is the starting point. It outlines the requirements for the initial planning of any project. It is similar to a contract that defines the relationship between two separate companies or between the company and a customer, since it defines relationships. In the case of an internal charter, it defines the relationship between the project and the functional organization. Since this example is intended to be used internally, within the organization, there is no requirement for the legal terminology an external contract would have to have. The charter contains:

1. *The basis:* The importance of this project or its priority, comparing it to other existing projects in the company.

2. *The subject:* Describes the project—what has to be done.

3. *The price or cost* and how determined: For example, the price may be defined as the number of engineering hours and the various categories of engineering skills needed, in addition to how they shall be measured:

   • By percentage
   • At certain time periods, say through weekly time sheets
   • At certain events, say when some technical test is completed

4. *Standards:* Definition of acceptance criteria. How will we know when the project is finished? Do we have any final test needs?

5. *Changes:* How to handle changes. What happens if it just "misses" the final test and therefore fails to pass? What do we do then? What if personnel are transferred out? What happens then? What if the scope of the project is modified? And so on.

6. *Specifications:* Listing of resource people and facilities.

7. *Protection clauses:* Jurisdiction within the organization, dispute resolution procedures.

The suggestions above strongly indicate that the development of a charter for a project is not a fixed one-time process. It is iterative and a part of the development of the project itself and, of course, the structure through which it is expected to operate. The structure will be modified in the future as projects are completed. The first time that you, as a project manager, go through the charter process to design your project structure, you will probably have to do it alone, because *no one else* can really do it. Some apparent questions pop up immediately:

> How do you know the price or cost at this point, before the detailed planning needed for the work breakdown structure?

> How can activities be estimated without the concurrence of project participants, who have not yet been recruited?

Creativity is difficult enough to forecast, but without the concurrence of the individual who is expected to perform that creative act, it is practically a waste of time. It's very difficult to schedule inventions without including the inventor. That concurrence and those of the other managers assigned to your project can be obtained during the work breakdown system development which follows. Conversely, those other managers might suggest modification of the initial charter before they will agree with forecasted tasks, costs, and accomplishments.

WORK BREAKDOWN SYSTEM

The work breakdown methods are the initial planning and control mechanisms intended to forecast project progress, measure actual achievement, and point out potential problem areas. The complexity of these methods matches the complexity of the project that it is expected to help control. As the project becomes less uncertain and ambiguous, the method complexity decreases (and the costs of operation decrease) until the most straightforward

CONDITION 1: FIRST SHIP

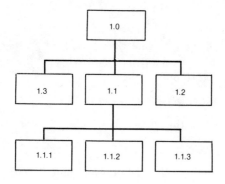

1.0 Spaceship
1.1 Nose
1.2 Body
1.3 Tail
1.1.1 Life Support
1.1.2 Steering
1.1.3 Nose Housing

Etc.

CONDITION 2: SIMILAR SHIPS

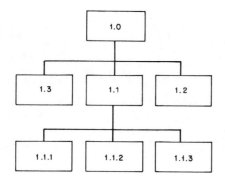

1.0 Spaceship
    Proj. Mgr.
1.1 Engineering
1.2 Finance
1.3 Quality Control
1.1.1 Mechanical
    Engineer
1.1.2 Electrical
    Engineer
1.1.3 Civil Engineer

Etc.

CONDITION 3: PRODUCTION

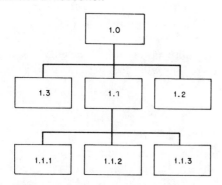

1.0 The Company
1.1 Production
1.2 Engineering
1.3 Quality Control
1.1.1 Spaceship
1.1.2 Turbines
1.1.3 Generators

Etc.

Figure 4-5 Work breakdown structures.

method or the standard functional level-of-effort budgeting method emerges. The work breakdown methods are therefore a family of planning and measuring methods, not just one particular method.

To illustrate, let us assume that the project in question is concerned with building an interplanetary spaceship under three different conditions of situational uncertainty (see Figure 4-5). The first or most uncertain situation shall be that the ship is the first one we have designed and built. This situation has the highest amount of uncertainty and ambiguity. The second or less uncertain situation occurs after the first ship has been built and others of similar configuration but slightly different destinations in space have to be built: medium uncertainty and ambiguity. The last situation occurs when many ships have been built and they then become part of the organizational product lines: lowest uncertainty and ambiguity.

Condition 1 is primarily product-oriented. It provides maximum control over product components, but it has the highest labor and management costs. It is the most differentiated and therefore incurs the highest cost for project integration and coordination. With the first ship delivered, condition 2 can be used for the design of the work breakdown for the next ship or group. This condition is task or functionally oriented but still structured in a project mode. There is less differentiation. Participants' skills are used across all parts of the project, thereby incurring lower integration costs which are major concerns of the project manager as the chief integrator. Risks are more definable because the work breakdown is becoming more person- and skill-oriented. When the spaceships become part of the organization's production lines, condition 3 should apply. At that time, the work breakdown is almost the same as a functional budget. It has the lowest integration costs (and therefore needs less project management) because it is the least differentiated method of control. In this last case, the risks are no higher than for any other product, and the uncertainty for all products is similar. Now the spaceships are just another product.

TABLE 4-5 CHANGES AS PROJECTS MOVE TOWARD FUNCTIONAL OPERATIONS

| | | Performance | | |
| | | Repetitive decisions | Cost | Management tasks |
Time and purpose	Structure			
Long range and innovation	Project	Work breakdown	Very costly	Highest differentiation and highest integration
Middle range and market response	Functional and project	Budget and work breakdown	Middle	Less integration
Survival or slow growth	Functional	Budget	Lowest	Coordination

Although the work breakdown method is primarily an administrative and budgeting tool, it could also reflect changes in the structural design as a project changes. Projects move, with time, from high uncertainty to lower levels of uncertainty as they approach completion. The work breakdown is supposed to be changed with this change in structure, becoming more simplified as uncertainty decreases. Table 4-5 is indicative of this general change in work breakdown methods as projects move into production phases. It reflects the three conditions of work breakdown noted before and the equivalent changes toward a functional structure in the organization.

PROJECT PROCEDURES

Project procedures can be categorized for descriptive purposes into three general areas, those that deal with people, those that deal with costs, and those that deal with accomplishment. Of course, the three areas are not separated in the actual operation of projects.

Those that Deal with People: Dual Reporting: Dual reporting is a procedure that is intended to reduce the negative results of the built-in conflict patterns when technical personnel that are administratively in functional groups are assigned to projects. Projects have limited lives and this often causes role conflict. "Which manager do I respond to, my functional boss or the project manager?" Or perhaps even more typically, "When I am assigned to several projects, how do I respond to the conflicting requirements of several project managers?" Dual reporting is one tool used to resolve this type of conflict. The management process is based on these definitions of structural responsibilities.

1. *Functional manager:* Prime responsibility is to train the person assigned to the project and to ensure that his or her administrative needs (i.e., vacations, time sheets, attendance, etc.) are satisfied.

2. *Project manager:* Prime responsibility is to provide financial resources, operating direction, and support for the people assigned to the project.

Therefore, under dual reporting, project personnel report to the project manager for day-to-day operating guidance and to their functional manager for training and administrative support. The performance review process of personnel assigned to projects under the dual reporting procedure is therefore relatively straightforward. At appropriate time intervals (e.g., six months, one year, the person's scheduled progress review, or whenever someone is moved off the project) the functional manager to whom the person reports administratively prepares a man-loading sheet for that person.

It shows the names of the project managers to whom the person reported during that time period and the *amount of time* spent on each project during the total time interval. For example, if the review is a six-month review, there are approximately 1000 hours to be accounted for. The functional manager produces a loading sheet for each person in the group, showing where they were assigned during that time period. The sheet shows the names of the project managers and how the 1000 hours that the person expended during those six months were distributed among those managers. One copy of the sheet for each person is then sent to the appropriate project manager. The project manager reviews the performance of each person *with that person* and tells that person the score, which is a function of his or her performance during that prior time period.

For example, George worked for project manager A for 100 hours, for project manager B for 800 hours, and for his functional manager for 100 hours. Assuming that the potential scores can be between 1 (absolutely terrible) and 10 (absolutely wonderful), if George gets a 3 from project manager A, a 9 from project manager B, and an 8 from his boss (since he worked in the functional group during that last time period in addition to working on various projects), his weighted score, which is reported to his functional boss, is:

$$3 \times 100 = 300 \quad \text{(For project manager A)}$$
$$9 \times 800 = 7200 \quad \text{(For project manager B)}$$
$$8 \times 100 = \underline{800} \quad \text{(For functional manager)}$$
$$8300$$

Weighted score is (7200 + 300 + 800) divided by 1000 hours = 8.3

The scores are summarized by the functional manager. The final score of the person is related both to the personal evaluation (i.e., performance?) and the time spent (i.e., importance?) on a particular project. Since this mechanism includes time spent in a functional area, it is an overall evaluation form. This example is indicative of the way dual reporting can work with several project managers and a functional manager. Each project manager rates independently of all others. Therefore, all the project manager sees is a sheet with an employee loading and a space for his or her own evaluation. Since each rater gets a separate sheet, it is obvious to that rater how much weight the evaluation will have versus all the other evaluations that are to be done.

The advantages are:

1. There is higher coordination of inputs received by the functional manager in determining the adequacy of functional training received for the needs of project managers. If the person is scored low, it probably

means that he or she is uniformly regarded as inadequate and needs training.

2. Dissatisfaction is less likely to be smoothed over and high performance is less likely to be overlooked, since all three people—the person, the project manager, and the functional manager—are involved in the evaluation process.

3. Functional managers are still responsible for the continual, long-term growth process of the person.

4. Project managers are responsible for evaluating the person's on-the-job performance.

5. The person is more likely to observe a direct connection between performance on projects and compensation: More connection equals more effective manipulation of rewards that are connected to motivation.

The disadvantages are:

1. Administrative costs go up to operate this evaluation process.

2. It is possible that no person will ever receive an extremely good or an extremely bad score. That is unlikely, because a logical response from the functional manager could be: "Why didn't you tell me about this before the review process? I could have (either) commended the person for his excellent work sooner, assuming that he had scored well, or put him into training, assuming that he had scored poorly, if I had known sooner."

Dual reporting supports coordination between project and functional areas and control of participants when they are assigned to different projects. It is probably one of the more important control mechanisms for the project manager, since people are the most independent variable. Control over expenditure of funds is next in importance, with control over accomplishment third.

Control Over Funds: Financial Networks: The work breakdown method is the mechanism that assists in:

1. Planning how the work will be done

2. Defining the interfaces among various work packages

3. Fixing responsibility for finishing these various work packages by assigning them to various project operating managers

Financial networks are set up, using hierarchical numbering systems (see Figure 4-5, showing the condition 1, condition 2, and condition 3 numbering system) that indicate which package is related to which. When the person informs the project manager that a task is to begin, the project manager can inform the financial control center (or, if you wish to give it a more utilitarian name, the accounting department) that a particular *charge* number is now open and can be charged. Time and material charges can only be accumulated against those numbers in the work breakdown that are *open*. When the particular task that has been assigned to the particular number has been completed, that number should be *closed*. This prevents charges from being made after a task is completed.

When there are no open numbers in the project, the only place that people can place their time charges is in the functional overhead accounts (unless they can find another project available with an accommodatingly open number, and sometimes they do). With only two alternatives for charges—project or functions—there is an automatic disclosure of progress and/or problems, since there is a plan for each of these alternatives. Projects use work breakdowns, and functions use budgets. When a charge is completed, something has been accomplished, like the reaching of some tangible milestone. A system that allows charges without responsibility for deliverables is like offering a full checkbook with all the checks signed and an unlimited source of funds. However, when the project manager can open or close any part of the hierarchical work breakdown method, she or he has a major part of an effective financial control.

The advantages of this type of financial control are:

1. Expenditures can be matched against critical work packages of the project.

2. Underexpenditures of particular packages that are closed upon completion of tasks can be reallocated within the project by the project manager. (This happens rarely, of course.)

3. Overexpenditure of funds that affect the total project *cannot* occur without the prior approval of the project manager (i.e., no surprises), since each charge number has only a limited amount of funds or time in it. The limitation on spending is ensured by the limited resources available under that particular charge number.

4. Any disputes between the project and the functional managers quickly surface if the functional overhead accounts become overexpended and project people have to charge overhead accounts and these charges have not been budgeted for. This can occur if the project is not "ready" to accept the person's planned charges; for example, when there are delays in other parts of the project.

The disadvantages of this type of financial control are:

1. This control cannot stand alone, but must be coordinated with a system to measure progress. Expenditures by themselves do not indicate anything; only when they are matched against progress (or lack of it) is this tool useful.

2. The administrative costs in opening and closing project work packages are not high, but they are not free either, since all controls cost something.

Going a step further, any system that merely matches financial or time expenditures against plan is inadequate for effective project control because the main task of the project is to achieve the *overall end* goal on time and within budget. Reporting that the expenditures are on target with the plan does not indicate that the *end point* will be (1) on target, (2) higher than target, or (3) lower than target. Hardly an acceptable state of affairs! The project manager is concerned with only one goal: the estimate at completion. And that is what he or she watches. It is easy to calculate (although sometimes difficult to get because people may be hesitant to cooperate). The formula is:

Actual + estimate to complete (ETC) = estimate at completion (EAC)

The "actual" is reported by the person expending the time or funds through the regular financial or accounting routines. The "estimate to complete (ETC)" for each component of the work breakdown method that is open for charges is reported by that same person with the same frequency that the "actual" is reported by the accounting department. By adding both numbers to obtain the "estimate at completion (EAC)" and *plotting the sum* as time goes on, the project manager can compare it with the original EAC and determine when the variance between the original EAC and the present EAC requires corrective action. (See Figure 4-6.) The ETC focuses attention on the end goal. There are other concepts such as estimate of percent completion. This seems to me to be less effective than the ETC because I always got the impression that the percent completion emphasized what had been done rather than what was yet to be done. But, it's a matter of preference; either concept could work.

When the chart of the EAC is flat, the person is probably not estimating the remaining tasks in front of him or her but is simply reporting arithmetical differences. It's virtually impossible to predict the future with no error, and that's what a flat EAC curve indicates. When this occurs, it requires looking into by the project manager. The shape of the EAC curve is quite revealing. If it has a sinusoidal curve of diminishing amplitude with a midpoint around the original EAC, it probably is as good as it will ever be.

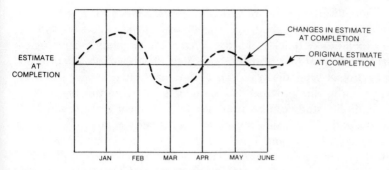

Figure 4-6 Estimate at completion over time.

Controlling Accomplishment: The Design Review: If it were possible to plot uncertainty as the ordinate (if we could measure it, it wouldn't be uncertainty) versus project elapsed time as the abscissa, it has been my experience that the curve would generally be in the shape of a decreasing S. Many of the factors that cannot be forecasted with any certainty at the beginning of a project become fairly well defined during the initial phases. In some cases that shape may not apply, but the subjective curve, S-shaped or not, can be estimated versus elapsed time of the project. If no one else can, the project manager must "guesstimate" where the major problems will probably occur during the project life. On the other hand, plotting a curve of cumulative expenditures of cost versus time in a similar fashion would probably result in another S curve, but reversed (see Figure 4-7.)

The starting phases of most projects are generally less costly than the middle and ending phases. It is almost impossible to spend a lot of money or time in the beginning phases, if for no other reason than it takes a certain amount of time to recruit and apply personnel, issue purchase requisitions, and turn those requisitions into purchase orders that eventually result in invoices to be paid from project funds.

Assuming that there is a similar vertical scale for uncertainty and cu-

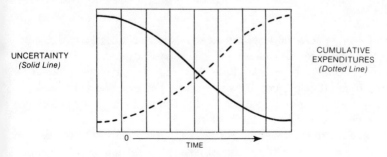

Figure 4-7 Uncertainty and expenditures.

mulative expenditures versus the same horizontal scale of time, it becomes apparent that after some point uncertainty will quickly decrease and cumulative expenditures quickly increase. *That is the optimum time to review the project for the first time* (usually about 20 to 30 percent into the project). Sometimes project expenditures, prior to the first design review, are called a feasibility study. The progress and learning achieved in this preliminary phase or feasibility study can be used to redefine project goals, time schedules, and budgets. The design review process is fairly completely detailed elsewhere (Jacobs, 1979) and I suggest that you become familiar with it. When the design review is used as a control on sequential reaching of project goals, it can be matched with the financial controls of opening and closing various work packages.

The advantages of design review are:

1. There is an ability to evaluate progress against goals at a predetermined interval.

2. Financial exposure is limited to the time before the design review, not the whole project.

3. It is possible to match cost and achievement.

The disadvantages are:

1. The cost of bringing all the project work packages to a halt during the design review process, restructuring them if necessary, and starting them all over again is high.

2. If the uncertainty curve does not have a typically decreasing S shape, because everything is dependent upon some crucial end test, the design review may not be particularly helpful. (Suppose you won't know if something works until the first field test, which cannot be scheduled until the project is almost over? An example is the first atomic bomb test in 1945.)

Handling Changes: The Impact Statement: No project has ever been completed without changes. The causes could originate anywhere from company management ("Sorry about that, Pete, but your best engineer has to be transferred to another project that's in a crisis. I'm sure you understand."), from the market in general ("Sales says that competition has produced the same product with half the tolerances you are estimating, and at a lower cost."), or even with the eventual customer ("Sorry, that's not the color we really wanted for our spaceship, so just change it."). Changes can even be caused by the project itself ("The design review indicates that 30 percent more time and funds are needed to meet the final specifications."). No matter where they come from, there has to be some mechanism built into projects to account for them. That mechanism is the *impact statement*.

The sequence for handling changes is similar to the sequence for setting up the original project:

1. Determine the new or revised estimate to complete.

2. Complete a new charter if necessary.

3. Reconstruct the work breakdown.

4. Write the impact statement, which explains the change and its impact upon the project. What is *now required to reach* the project goals and how will that be done?

The form is not as relevant as the content. Something has changed, and if the forecasted estimate at completion is affected to the extent where the project will not meet its targets, the project manager's responsibility is to present those data as quickly and as accurately as he or she can. When there are insufficient resources to change conditions within the *existing* budget, the time to issue the impact statement is *now*.

The project manager is no different from any other manager in the decision-making situation; he also has to absorb uncertainty in making decisions, but he can only absorb a limited amount. That limited amount is generally defined as the contingency resources available within the project. When the change is great enough to exceed this, a new estimate at completion is needed. When the amount expected to be spent is greater than the existing budget, regardless of the reasons, it is vital that the next upper levels of management, with greater amounts of resources, be advised and that either those resources be made available or the scope of the project be reduced to meet the resources that are available. I have found that impact statements are a very valuable defense against sudden changes imposed from any source, but especially from your own management. When the cost of changes is known, those changes may not be imposed after all.

Eventually, however, the project will be completed. Planning for completion should begin *before the expected completion date*.

PROJECT CLOSE-DOWN: DOING IT RIGHT

Closing a project requires almost as much planning and skill as starting one. The close-down has its own special needs, and these needs have to be addressed here. There are generally three main areas to be considered:

- The client-user-customer—the outside
- The functional needs internal to the organization—the inside
- The project itself—the project summary

The Outside: A list of open items is drawn up by the project team, listing the tasks to be completed in order to reach the project goals. This list is proposed to the client-user-customer as all the things that still have to be done. If accepted, a new work breakdown is developed to accomplish the tasks on the list. In effect, the scope of the work has been defined and a new project (for close-down) is set up. By checking the amount of funds available in the cumulative expenditure curve (difference in the y axis of the curve) and the amount of time to complete the open items (difference in the x axis of the curve), we know how much money and time are still available to complete the project. At this point, we may have to issue an impact statement requesting more of both.

The Inside: This is a subjective list of tasks that should be completed to take care of the administrative aspects of the project. It is a list that I have used. It is not the only possible one, but it does provide suggestions for you to use when you draw up your own list applying to your particular situation. My list includes:

1. *Personnel:* Review who you, as the project manager, will need, for how long, and when they will be reassigned back to their functional areas. When they are reassigned, write any dual reports needed, since you might not be working with them when the next six-month or year-end review is due. Just do it for the elapsed time from the last review to now.

2. *Test reports:* Complete the documentation. The tests were done, but where are the reports?

3. *Capital assets:* Did you buy anything for the project? What are you going to do with it when the project is completed?

4. *Inventory:* Did you have anything left over? What will you do with it? One project manager used to notify the customer that there was X amount of material left over. If not advised otherwise in ten days, he would scrap it and send the value received to the customer. It's an idea.

5. *Documentation:* Have the blueprints been updated? What about spare parts lists, manuals, and other documents? Remember, nobody ever

seems to need that stuff until you have just been transferred to the next bigger project.

6. *What else?*

The Project Itself—The Project Summary: This is a brief outline of the project history: what it was expected to do, the changes that occurred, and the results achieved. Since all the other project documentation is available, such as minutes of meetings and drawings, the summary is not supposed to be voluminous. The files are already bulging with all those materials. Therefore, a simplified two- or three-page outline summarizing all those wonderful achievements would be just fine. Remember how helpful it would have been, when you were starting the project, to have a summary of similar projects to review. Now's the chance to leave one for the next project manager.

DEVELOPING THE "HOW TO" OF PROJECTS: THE PROJECT OPERATIONS METHODS

The development of functional organizational structure is glacially slow when compared with that of the project organizational structure. That could be because environmental uncertainty has a much more gradual impact on the total organization than on projects. Rather than take the time to evaluate the design elements and assign weights to them, as recommended for functional organizational structures, we are forced by rapid change in projects to use a more radical and faster method of design. This allows fast response to uncertainty. The project plan should consist of the following three steps:

Step 1. Forecasting: The project manager (and possibly a small planning staff) is responsible for the initial documentation of the project charter, which includes a list of tasks and responsibilities and the work breakdown method. The selection of the project personnel (actually only the first line of management under the project manager) then takes place through negotiations between the project and the functional managers concerned (e.g., controller for financial persons, chief engineer for engineers). When that has occurred, the first meeting of the project takes place under the chairmanship of the project manager and the whole group evaluates a preliminary list of tasks that the project manager has developed as an initial program plan. Then the appropriate people on the project accept the responsibility for each task.

For example, if one task is to "perform a market review of the product potential," the marketing person on the project would probably accept that task, perhaps with some assistance from engineering or finance. The important point is that each task is "owned" by someone on the project. When a person accepts tasks, he or she initials the preliminary task list, thereby accepting the responsibility for developing the budget to match this preliminary list. If the group disagrees with the list, the first order of business is to develop a list with which all its members agree. A revised work breakdown is mutually developed and costed, and a revised charter written. An organization chart for the project is drawn up and people initial their own jobs on the chart as an indication of preliminary acceptance of the tasks assigned, the work breakdown, and the charter.

Step 2. Measuring: At this point, a cumulative spending curve can be drawn by the project manager. This will be the basis for determining the design review schedule and the schedule of dates (including general subjects) of future project review meetings. Since uncertainty should be decreasing with elapsed time, but not on a linear basis, this is the opportunity to request a schedule of accounting reports, which will also decrease in frequency of publication as the project moves along. Weekly reports on project costs when the project is 25 percent finished are vital, but they are almost useless when the project is 75 percent finished. The dual reporting mechanisms, the work breakdown release dates, and the estimate at completion reports should then be in place.

Step 3. Initiation: With all of this complete, a project organization chart is drawn up showing names and responsibilities of people on the project who have some responsibility for the expenditure of project funds. Each participating person then initials the chart as an indication of agreement, at least with respect to the work done so far. After the project manager and the appropriate managers over him or her sign the chart, the chart becomes a notification to the company controller that project funds have been approved and can be spent by the project manager as parts of the project work breakdown are opened or closed. It is both a responsibility and a funding authorization. When impact statements are issued, this chart must be reapproved.

SUMMARY

We started our review of some ideas and concepts about the second component of the organizational model, the structure, by dividing theories of organizational design into those that were proposed to be universally applicable and those that were based on the contingencies of the situation. Although the universal theories were developed earlier, in the late nineteenth and early twentieth centuries, they are still used successfully in many technical organizations.

One of the better-known classes of universal theories is that of classical management theory. This has the logical and consistent approach of the production-oriented industries in which it was developed. Fayol managed a large coal mining company and Taylor, a steel mill. Therefore, these designs still have applications in industries that are relatively slow to change and whose strategy involves satisfying fairly constant market needs. Human-relations theory is a bit more modern, having been developed in the 1920s and 1930s, but it has many of the characteristics of classical theory. It is also universal and states that there is a "best way" to organize. It deals with smaller groups or functions within the overall structure.

On the other hand, contingency theory is relatively recent, more flexible, and oriented primarily toward responding to the variables in human behavior. It has no "one best way," since it recognizes that all organizations and situations are different, but it does provide guidelines for "If A happens, B is probably appropriate." It attempts to evaluate the particular situation in terms of some of the psychological aspects of the people who work in the organization. The informal organization is no longer an enemy to be crushed or manipulated. It is a major consideration that interacts with the formal organization. There are many ways to organize, depending upon the environment, the people, and the times.

If the universal theories can be considered analogous to the deterministic physics of Newton, contingency theory can be considered analogous to the uncertainty of the atomic physics of Einstein. As we learn more about organization design, we find (just as the physicists do when learning about the physical world) more complexity and many alternatives as we get closer to our own organizations, rather than the relatively simple and rigid pyramidal structure of classical theory or the manipulative structure of human-relations theory that are supposedly applicable to all organizations.

This progression into complexity seems almost inevitable when changes in the environment and in organizational people are considered. Technical organizations are the producers of innovation, and that is the support of future organizational growth. Organizational participants are no longer interchangeable workers, they are the human capital that produces innovation and consequent growth. The formal structure has adapted to this change or increased complexity. The informal structure which interacts with it is at least equally complex. For example, we now have research that shows that people in different departments really do think differently from each other. The design engineers are not being difficult (according to the contingency models of Lawrence and Lorsch), they really have different ways of looking at things than the sales engineers. The structure must be responsive to these differences if the organization is to grow.

Functional structures can be either discipline- or product-oriented, and each design has its advantages and disadvantages. Designing functional structures includes different criteria than designing project structures. The principal variable is that of time, with functional structures considered to be "immortal" and project structures to have a limited life with limited, defined goals. Conflict levels and differentiation levels (or differences in attitudes among departments) become higher when the structure must handle rapid change. Both functional and project organizations have opportunities and problems in coping with environmental change. There is, however, one general rule for all types of organization structures: No matter how well they are designed to fit the perceived contingencies, they are never completely finished. With time, the fit between the structure and the external and internal demands decreases, making redesign a repetitive activity similar to the normal budgeting process.

The next chapter covers the third component of our model: the technology. It is defined in that chapter as the technology of production, which interacts with the first two components, people and structure. In some ways, our analysis will increase the tasks of learning how to manage technical organizations, since it adds another variable to be considered. It is an important variable, and one that cannot be overlooked if your technical organization is to adapt to its situation. In general, technical management is not a simple study, and since the effect of technology upon technical organizations grows more important every day, we should understand it thoroughly.

SUGGESTED ANSWERS TO CASE QUESTIONS

1. This *could* be a change in scope of the program but as of this moment, it is not, since there has not been a new direction, just a request for an evaluation. Therefore, Eric should lay out the alternative recommendations that he could make in an engineering report format. That format is quite straightforward:

   • Description (to and from; date; identification)
   • Recommendations
   • Everything else

   We can assume that Eric (as is typical of most project managers), generally is competent when it comes to his project. There is no reason he can't come up with several suggestions including *"ballpark" costs and benefits*. For example, he could probably tell Milt in general terms what it costs, how long would it take, and what the probable (risk?) effect on the existing project would be if:

   a. The casing were redesigned to be corrosion proof.
   b. A fail-safe were built into the turbine to preclude its operation unless it were covered by an appropriate shelter. External sensors could be designed in.
   c. The existing design was sold by the company for a limited time, and then only to highly industrialized customers who would permit the company's service personnel to inspect periodically. At a reasonable later time, a new turbine for more severe external conditions could be redesigned.

There are other alternatives that I'm sure you have considered at this point. The idea is that there are probably many alternatives, and before valuable time and money is spent, these should be documented along with recommendations for action.

2. A request for minimizing cost and delay is, if not explicitly stated, an implicit part of every impact statement. In this case, Eric should provide a cost-benefit analysis that forecasts the potential cost to implement any of his recommendations versus the time that he is allowed to start the implementation. Milt may delay a decision. If so, he should be aware of what that delay could mean in terms of additional rework, design, or what have you.

3. Since there has been no decision to commence the corrosion design work at this point, nothing should be done about Alan. If he is unavailable when the work is to be done, either alternative sources must be found in or out of the company (outside design and test labs) or the work will be delayed until he returns. If Milt does decide that the corrosion work is to go forward immediately, I suggest that Eric deal with that problem and possibly include a trip to San Diego for Alan as part of his impact statement. In this example, the person responsible for the decision to spend additional funds is Milt. Eric is responsible for seeing that Alan completes the job within budget, to schedule, with Clyde supervising the actual work done. The trip to San Diego is contingent upon the work outcome. In project operations, responsibility is rarely a single, clearly defined sequence.

4. Since Maryann now has information that could be important to the success of the project, if I were Eric I would request that she be added to the team, lay out any additional tasks for her, to preclude this problem and other field service problems from happening, and have her inputs disseminated throughout the turbine project.

5. If I were Milt, I would not have gotten Eric involved until I had satisfied my own level of uncertainty. For example, how bad is this problem in the field? What can our potential losses be? What alternatives do I have for a change in the existing "Blue" turbine project? What would corporate headquarters want? Eric has not been given enough information to come back with a specific recommendation. As noted above, since he had no idea which alternative was suggested, it was advisable for him to come back with several from which Milt could choose. If he had thought of the one that would satisfy the upper levels of management, that is fine, but he might not have, and then his work would have been wasted. If Milt had done his "homework," it would have saved everyone a lot of time. This is a case of poor absorption of uncertainty.

Do you agree? What other answers do you think would apply?

Case Study
THE CASE OF THE RIGID STRUCTURE OF CORVIS MANUFACTURING

CAST

George Mulvaney: Sales manager

Mike Casey: Manufacturing manager

Albert Halloran: Chief engineer

Don Corvis: President

The Corvis Manufacturing Corporation had been started about fifty years ago by Will Corvis, Don's father, as a producer of high-quality children's toys and dollhouses. Will had just retired, and although Don had been general manager for ten years, this was his first staff meeting as the president after his father's retirement. All four attended.

Don: According to our agenda, we have several problem areas to review. As I see it:

Decrease in sales of "missy" dolls is continuing.

Manufacturing costs for our new dollhouse assembly line have not decreased to meet the forecast when we approved the budget for it last year.

We might be sued because one batch of doll dresses we shipped last year was not treated for inflammability, and we have several reports of children being burned last Christmas when the clothes caught on fire.

Who has some answers to these problems? Anybody ready to start?

George: As I see it, Don, the problems are all caused by ourselves. Let me give you some background that you might not have had as general manager, since you weren't handling sales then. Although we have the best sales force in the field, cheaper prices and some product innovations from competition overseas are killing us. More lifelike, flesh-feel plastics and cheap imports are affecting our volume. We need new, less expensive, and more attractive products, and we've got to take more advantage of our long relationships with our customers to give them everything they want from one source of high-quality, low-price products. If we can provide all their needs, they won't have to go to imports. Imports are an administrative bother, even if they are occasionally less expensive. We've got to expand into allied areas. Dolls and dollhouses are not enough. We've got to sell the whole market of children's recreation: dolls, toys, books, and games. Give me some new products.

Don: OK. Sounds good so far. Anybody else?

Mike: Look, fellows, we've been trying to keep costs as low as we can. The new dollhouse assembly line was expected to lower costs, but if you keep changing the designs, there's no savings left because of the increased costs we incur when we break down and set up a new product. We can't break setups and produce cheaply too.

Don: OK, what do we do now? Sales wants new and different products at lowered costs and manufacturing can't produce at the costs we need unless we keep our products fixed long enough to get some economies from long runs.

Albert: Our engineering department is set up to design toys the same way that we used to. Before we used any new materials, we always tested them thoroughly in our own materials lab. Every plastic, fabric, and metal was checked against our standards to ensure that no problems would happen in the field. The one time that we delegated responsibility to manufacturing to get materials to meet our fireproofing standards, a batch of nonstandard doll clothing slipped through. This company has always been organized on a product-line basis and engineering has assigned technical people to each line. Dolls and dollhouses are almost separate departments. Why don't we become a bit more flexible and change the way we're organized?

QUESTIONS

1. What would the organization look like if George Mulvaney were running it?
2. What would it be like if Mike Casey were running it?
3. Are these managers responding to the questions that were asked? Why? How would you explain it? Correct it?
4. What can be done to resolve the questions on the agenda?
5. If you were Don Corvis, what would you do that would affect communications? The organization structure? Do you have any theories that might be applied here?

SUGGESTED ANSWERS TO CASE QUESTIONS

1. If George were running Corvis, the organization would probably include a small-run, very flat structure that would be organized in a project orientation for rapidly changing products such as doll's clothing and dollhouses. It might also have a relatively fixed production line for the dolls themselves and other products that did not change very rapidly. This would be similar to the classical pyramidal structure.

2. The organization would be set up on a pyramidal basis, using the assembly line as the central core. Production would be primary, there would be high centralization of authority, and direction would be top down.

3. No, because they are only responding with other problems. None of them has any specific operational hypothesis that can be proposed, with the best one selected for testing in order to correct the defined problems. It seems that the situation-manager interaction influences both parts. The situation is influenced by the manager and vice versa. Therefore, it is quite difficult for George to put on Albert's or Mike's "hat." Don has to be able to "listen" to what his differentiated people are saying and attempt to integrate the answers into a framework for everyone to use.

4. Don Corvis might do several things to improve communications. One quick solution in the meeting is to request that each of his managers tries a little experiment in communications. Suggest that each one tell another manager what that other manager has just said. In other words, "I'll tell you what you told me and you tell me if I'm correct. If I can do that, and you agree that I have it right, at least I understand it." Another technique would be to have each manager prepare a position paper outlining his suggestions, in an engineering report format, before the next meeting. By that time, everyone will have had an opportunity to review and compare the other people's positions. It seems that rapid change is coming to Corvis Manufacturing, and it might be appropriate to set up a project team to handle several of the new product developments. For example, a project team might try to develop a new plastic for doll bodies or an improved method of developing and changing production lines or install multipurpose lines that can handle different products under computer controls.

Some theories are suggested in this chapter. Would you use any of them? How? Do you agree with my answers? Why?

REFERENCES

Argyris, Chris. Being human and being organized. In Erwin P. Hollander & Raymond G. Hunt (Eds.), *Current perspectives in social psychology* (2d ed.). New York: Oxford Univ. Press, 1967, pp. 573–585.

Beer, Michael. A social systems model for organizational development. In Thomas

G. Cummings (Ed.), *Systems theory for organizational development*. New York: Wiley, 1980, pp. 73–114.

Chandler, Alfred D., Jr. *Strategy and structure*. Cambridge, Mass.: MIT Press, 1962.

Fayol, Henri, *General and industrial management* (C. Storrs, trans.). New York: Pitman, 1949.

Filley, Allan C., & House, Robert J. *Managerial process and organizational behavior*. Glenview, Ill.: Scott, Foresman, 1969.

Galbraith, Jay. Environmental and technological determinants of organizational design. In Jay W. Lorsch & Paul R. Lawrence (Eds.), *Studies in organization design*. Homewood, Ill.: Irwin/Dorsey Press, 1970, pp. 113–139.

Gulick, Luther. Notes on the theory of organization. In L. Gulick & L. F. Urwick (Eds.), *Papers on the science of administration*. New York: Institute of Public Administration, 1937.

Jacobs, Richard M. The technique of design review. *Proceedings of the Product Liability Prevention Conference of the American Society for Quality Control*, 1979, PLP-79 Proceedings.

Katz, Daniel, and Kahn, Robert L. *The social psychology of organizations*. New York: Wiley, 1966.

Kotter, John P. *Organizational dynamics: Diagnosis and Intervention*. Reading, Mass.: Addison-Wesley 1978.

Follett, Mary Parker. In Henry C. Metcalf & Lionel Urwick (Eds.) *Dynamic administration*. New York: Harper, 1942.

Krupp, Sherman. *Pattern in organizational analysis*. New York: Holt, Rinehart and Winston, 1961.

Lawrence, Paul R., & Davis, Stanley M. Problems of matrix organizations. *Harvard Business Review*, May–June 1978, 131–142.

Lawrence, Paul R., and Lorsch, Jay W. *Organization and environment*. Boston: Harvard Univ. Press, 1967.

Likert, Rensis. *New patterns of management*. New York: McGraw-Hill, 1961.

Meir, Richard L. Explorations in the realm of organization theory, decision-making and steady state. In W. A. Hill and D. Egan (Eds.), *Readings in organization theory—a behavioral approach*. Boston: Allyn and Bacon, 1967, pp. 471–477.

Miller, Jeffrey G. Fit production systems to the task. *Harvard Business Review*, January–February 1981, 145–154.

Mintzberg, Henry. Organization design: fashion or fit? *Harvard Business Review*, January–February 1981, 103–116. Excerpts reprinted by permissison of the *Harvard Business Review*. Copyright © 1981 by the President and Fellows of Harvard College; all rights reserved.

Mooney, James D., and Reilly, Alan C. *Onward industry*. New York: Harper and Row, 1931.

Morse, John J. Organizational characteristics and individual motivation. In Jay W. Lorsch and Paul R. Lawrence (Eds.), *Studies in organization design*. Homewood, Ill.: Irwin/Dorsey Press, 1970, pp. 84–100.

Osborn, Richard N., Hunt, James G., and Jauch, Lawrence R. *Organization theory: an integrated approach*. New York: Wiley, 1980.

Roethlisberger, Fritz J., and Dickson, William J. *Management and the worker*. Cambridge, Mass.: Harvard Univ. Press, 1939.

Scott, W. Richard. Organizational structure. In A. Inkeles, J. Coleman, and N. Smelzer (Eds.), *Annual review of sociology*, (Vol. 1). Palo Alto, Calif.: 1975, pp. 1–20.

Shannon Robert E. *Engineering management*. New York: Wiley, 1980.

Shockley, W. On the statistics of individual variations of productivity in research laboratories. *Proceedings of the I.R.E.*, March 1957, 45(3) 281.

Silverman, M. *The technical program manager's guide to survival*. New York: Wiley, 1967.

Taylor, Frederick W. *Principles of scientific management*. New York: Harper and Brothers, 1911.

Thompson, James D. *Organizations in action*. New York: McGraw-Hill, 1967.

Weber, Max. *The theory of social and economic organizations* (2d ed.; A. M. Henderson & T. Parsons, trans.). Glencoe, Ill.: Free Press, 1957.

Woodward, Joan. *Industrial organization: theory and practice*. London, England: Oxford Univ. Press, 1965.

FURTHER READINGS

Arrow, Kenneth J. Control in large organizations. In Erwin P. Hollander & Raymond G. Hunt (Eds.), *Current perspectives in social psychology* (2d. ed.). New York: Oxford Univ. Press, 1967, pp. 573–585.

Barnard, Chester I. *The functions of the executive*. Cambridge, Mass.: Harvard Univ. Press, 1938.

Drucker, Peter F. *Toward the next economics*. Westport, Conn.: Greenwood Press, 1981.

Fairfield, Roy P. *Humanizing the workplace*. Buffalo, N.Y.: Prometheus Books, 1974.

Gibson, James L., Ivancevitch, John M., & Donnelly, James H., Jr. *Organizations: behavior, structure, processes*. Dallas, Tex.: Business Publications, 1976.

Haberstroh, Chadwick J. Organization design and system analysis. In James G. March (Ed.), *Handbook of organizations*. Chicago: Rand McNally, 1965.

Lawler, E. E. *Pay and organizational effectiveness: a psychological view*. New York: McGraw-Hill, 1971.

March, J. G., & Simon, Herbert, *Organizations*. New York: Wiley, 1958.

Pugh, J. S., Hickson, D. J., & Hinings, C. R. An empirical taxonomy of work organizations. *Administrative Science Quarterly*, 1969, 14 115–126.

Sherif, M. Superordinate goals in the reduction of intergroup conflict. *American Journal of Sociology*, 1958, *43*, 345–356.

Silverman, M. *Project management: a short course for professionals*. New York: Wiley, 1976.

Stein, Barry A., & Kanter, Rosabeth Moss. Building the parallel organization creating mechanisms for permanent quality of work life. *The Journal of Applied Behavioral Science*, 1980, *16*(3), 371–386.

Strauss, George, & Sayles, Leonard. *Personnel: the human problems of management* (3d ed.). Englewood Cliffs, N.J.: Prentice-Hall, 1972.

Walker, Arthur H., & Lorsch, Jay W. Organizational choice and product vs. function. *Harvard Business Review*, November–December 1968, *26*(6).

5
TECHNOLOGY

Case Study

The Case of the Pregnant Computer

CAST

George Jackson: Design team leader

Michael Jacoby: Chief engineer

Mary Hughes: Chief, product development

Sam Disko: Data processing manager

SCENE: THE JOHN TEXTILE WORKS

It was late on a beautiful spring afternoon in South Carolina and Mike Jacoby decided to take a five-minute break to smell the roses. Just as he stood up to lean close to the open window in his office, Sam Disko walked in unannounced, as he usually did.

Sam: Hi Mike. Got a minute? (He proceeded to go ahead without waiting for Mike's answer.) I'd like to talk to you about some things I've noticed about computer usage in the company in general, and specifically right here in the engineering department.

Mike: OK Sam. I've got a meeting in half an hour, but pull up a chair and let's talk. What's the problem?

Sam: Well, you know that my department is responsible for procurement and operation of all computers in the company. We make sure that when they are "born" in this company they will work the right way. We service all the departments, but my fellows tell me that they are having a lot of trouble with your engineers because your people don't want to follow the right processes.

Mike: Tell me about it.

Sam: Well, you know how we design a system. After a request for our work is approved by you, I send in a systems analyst to work with your people to do an information flow diagram and a rough layout of what the programs are supposed to do. When your people sign off on the layout, I give it to our programmers, and when they get finished, they run trial data and present the results to your people.

Mike: OK, what's the problem?

Sam: I've never been able to get a system signed off by your people. They always say that we misunderstood, or that's not what they really wanted; they wanted something a little different. They never really tell me what they expected to get. I've been here over five years and I've never gotten a straight answer yet and I'm getting tired of it. Oh, we finally get the computer systems to run, but only after a lot of wasted time and meetings. It's wasteful and inefficient.

Mike: Well, let me look into it and I'll get back to you tomorrow.

After Sam left, Mike called George Jackson and Mary Hughes. He briefly told them what Sam had said and asked them to discuss it with him the next day at lunch. The next day, at lunch:

Mike: Well, that's the story. What's to be done about it?

George: Mike, that guy Disko and all of his people are just a big pain in the neck to me. For example, all I wanted four months ago was a simple modification to our design program so that the graphics display would be able to check our piping diagrams for the process plant. I wanted to be sure that we didn't have any more fiascos like we did last year when a designer put two pipes in the same place because he didn't have a complete assembly drawing to look at first. By the time Disko's people came over, the last piping design was almost finished, and by the time they seemed to understand what we wanted, they had wasted four days of my best designer's time in these endless meetings that they seem to like to have. We finally got the job done without them.

Mary: Mike, my problems with them are a little different. I am never sure how to design the fabrics to meet the special requirements from marketing, since fabric design is still pretty much of an art, not a science. So since I wasn't sure what I was looking for, I couldn't exactly tell

Sam's systems analyst what I wanted. Finally, I signed the approval form just to get rid of him. I then took a course in BASIC (a computer programming language) for several weeks and started to program my own terminal. When Sam found out about it, he complained because he said that it wasn't up to his standards and my attempts at programming—I think he called it a simulation program or something like that—were interfering with his mainframe.

I then asked him (you must remember, because you approved the requisition) for a stand-alone microcomputer and when I finally got it, I found that any programs I wrote for it had to be approved by Sam's group. They had programmed my stand-alone before I got it to accept only those programs that they unlocked from the mainframe. They said that this was necessary to prevent unauthorized use of the computers, and they wanted an approval form signed every time I tried to use the terminal. They called it the computer delivery process, just like it was having a child in the hospital or something. I finally just stopped using the darn thing! It's too much of a bother to get those guys to do anything. They use more paper forms than anybody else I know and I just don't have the time to fill them out if we want these new fabrics out. So I just do my designs the old way, by trial and error on the drawing board, and then down into the experimental weaving room to try them out.

Mike: OK. Everybody has problems and we'll have to deal with them in a while, but right now Sam is complaining about your groups. What can be done about it?

QUESTIONS

1. Does the company policy for centralizing computer procurement and operations accomplish its intent? What is that intent?

2. What would you suggest if you were Mike?

3. How would you optimize the use of computers in the firm and still prevent their unauthorized use? What are unauthorized uses and why should they (or should they not) be prevented?

4. How should George's group be structured: tall, flat, or pyramidal? What about Mary's group? Why?

REVIEW AND INTRODUCTION

As I have defined it here, technology has achieved an importance equivalent to the other two components—people and structure—only within relatively recent times. Its interactions with the other two components used to be as a junior rather than an equal partner. Changing inputs into outputs (i.e., the most general definition of technology) was almost a secondary consideration, since management was primarily supposed to spend its time controlling the people and the structure. Technology was expected to flow from these other two components. At the beginning of this century, it had a very limited role, since it consisted of only the production methods and/or changes in products. Technology was useful for making the product faster, at less cost, and with fewer inputs of human effort and raw materials. It was occasionally extended to include product innovations, but this was rare. The more modern definition, that of changing inputs into outputs, is relatively new and includes many more aspects of management decision making.

In the past, this practice of limiting technology improvements to production or product changes often affected the people component negatively. For example, when competition installed specialized assembly equipment, the company quickly followed to maintain competitive or lowered labor costs. It was a very rational and logical sequence. Technology, which was applied through work simplification, changed workers from craftsmen into machine feeders and maintainers—in effect, adjuncts to the machines. A familiar but overemphasized example of the use of this type of technology was on the automobile assembly lines, and at the time it was quite satisfactory. Much of our industrial strength was built on it. During that period, the people did not affect technology as much as technology affected the people.

The other major component, structure, seemed to be less involved with technology since there was supposed to be a "one best way" to be organized regardless of any technological needs. For example, when the company's marketing strategy determined the structure, that structure accepted the technology as a given and product innovation occurred only when market changes required them. The technology was a secondary, or even minor, part of the company strategy. If the structure was well-designed in accordance with the "one best way," the technology could be assigned to the production engineers or manufacturing managers, who would do the job of eliminating unnecessary motions and simplifying the work. That, then, would reduce costs. (This process was still being taught in universities long after it became obsolete. It might even be taught today. In fact, I learned it when I was exposed to an introductory course in time-and-motion systems. We were supposed to find the best way to organize the work and were told that the structure and the technology would then be optimum.)

There have been changes since those times. Decreases in product life cycles, with quickly changing market needs, now require company responses involving more than the production processes through which the organization transforms inputs into outputs. Production methods and new-product development have been joined by management decision-making processes.

Typically, techniques used by managers in information processing and management sciences are added to the most general definition of technology.

The transformation of inputs into outputs involves more than tangible processes. The mental processes are (and perhaps always have been) as important. Technology has therefore become as important as people and structure. We will try to optimize it, as we did the others, using our general contingency approach. In Chapter 2, we covered some of the thinking and decision-making processes available to the manager as an initial discussion of technology. In this chapter, we will continue with other processes through a by now familiar path of description of applicable research findings about technology and its interactions with the other two components, followed by prescriptions and specific recommendations.

As noted before, when technology was defined primarily as the methods of production, it was almost a secondary part of the company that wasn't supposed to have much of a direct effect on its economic success compared with the effects of people and structure. The search for the one best way to manage people and the structure limited and defined the smaller role played by technology. Research has changed all of that and expanded the technology variable from the limited actual production processes to include all the managerial functions and information handling methods used. Now the management operations and the knowledge and sequencing of activities—the work flow itself—were considered part of technology.

Technology, as redefined, was no longer contained entirely within a relatively unified section of manufacturing departments. It became an expanded cluster of working and management operations, tools, and materials. It included both the measurable organizational and manufacturing processes and the nonmeasurable managerial actions. Those interactions were quite complex, since they interlocked and affected each other.

In order to understand this, we will use a familiar but artificial analysis method. We will untangle the human and production interaction by assuming all other things to be equal and investigate only one interaction at a time. If we are to understand how this technology of changing inputs into outputs works, we'll keep everything else fixed and analyze the interaction with people. Then we'll do the same thing for structure. That means we'll be able to look at this component from several viewpoints.

TECHNOLOGY—INPUTS INTO OUTPUTS

In previous chapters, we said that classical management theory considered people to be relatively interchangeable. While they were an important commodity, they were still only one of the resources that were needed to produce revenue. The pyramidal, hierarchical organizational structure that was developed as the best (and only) way to manage this commodity was intended to clarify and help simplify the work that people did. Therefore, although the design of the structure was a company problem, the solution to that problem was well-marked and clear, since the resulting structure generally reflected the society of that time. That society had some very interesting historical roots that, interestingly enough, involved the technology of production.

SOME HISTORICAL ROOTS

The industrial revolution, which occurred in the late eighteenth century, took place in a social and economic environment that rewarded technological achievement in manufacturing very handsomely. That achievement was in producing new and more effective products in more efficient ways. Owner-managers who could purchase or develop the technology were able to gather labor and materials into one location (using a new invention called the *factory*) and transform those resources in a predetermined, repetitive fashion into articles of value. These articles were then sold to an accepting and hungry market, resulting in profits for the entrepreneur. Cottage industries, in which small producers used the labor of their families and close friends and manufactured parts and products in their homes, could not compete with the efficiency of these centrally organized factories.

Moreover, the political scene completely supported free or laissez-faire trade of manufactured goods and developed a fascinated interest in the financial benefits that accrued to these entrepreneurs (Bronowski, 1965). The development of accurate machine tools and cheap energy sources to drive them defined the primary mass production technology, which suited the resulting patriarchal organizational structure. Productivity increased when work was simplified by being divided up and machines were used to speed up the simplified, divided work.

A classical example of those times was the productivity of pin makers, which would be increased more than 200 times through task specialization and effective tooling. The master pin maker who straightened the wire, cut it off, headed it, sharpened the point, and put it in a paper was outclassed by technological change. Dividing the job into approximately 20 operations performed by several workers increased the worker's relative skill through job simplification and minimized the nonproductive planning tasks of preparing or setting up. It also allowed skills to be learned quickly and to be repeated. The concept of specialization, implemented through a change in technology, probably contributed to the success of those technologically adept pin-making companies and the demise of other companies that did not change their technology (Smith, 1977).

Then, the improvement of technology was a cost-driven activity. The idea was first to develop the production tasks and methodology, then to develop a disciplined manufacturing structure to manage it. The goals were clearly order, logic, and acceptance of authority. The primary objective was optimal performance within rigid contraints of a pyramidal type of hierarchy, with communications only in vertical directions. Orders were sent down and information was sent upward through the structure. People's jobs were clarified, spans of control were limited, and responsibility and authority were clearly placed together. Uncertainty resolution was the task only of the owner-manager and the structure was intended to minimize costs in using available technology to carry out that owner-manager's wishes completely. This system was believed to be ideally applicable to all factories and organizations.

About the beginning of this century, there were occasional protests and suggestions from some researchers that were intended to modify this approach. However, these researchers were not considered to understand the "correct approach" for designing company structures, considering the technology. Since the technology of mass production was relatively new and hierarchical-designed structures worked well with it, one wasn't supposed to argue with success. The design worked then and sometimes still does, but recent research indicates that it has a limited application. It works best with mass production technologies such as those with relatively fixed, well-understood manufacturing processes.

The prime example of the technological (or classical) approach to the design of organizational structures was that of F. W. Taylor (1911). That design exemplified the efficiency, control, and pragmatism of the production machine applied to the design of the organization. Science was applied rigidly in the factory. Authority was centralized and vested in planning specialists The new technology of mass production was optimized, using task specialization and relatively complex production methods. The individual worker's tasks were made simpler, but the planning and coordination that the manager had to do became more complex. The worker no longer had any scheduling or methods flexibility, since that was all placed in the manager's hands. These new management responsibilities were difficult, but positive results were achievable with a machinelike, disciplined approach. The plannning tools used to direct this technology were limited but they were all that was required.

I can still recall seeing many manufacturing operations for which planning the best machine feeds, speeds, and loading and unloading operations for a machine operator was supposed to be done with a stopwatch and a small table of standards. It was the day of the time-study man, the efficiency expert, methods engineers, piece-rate setters, and job classification specialists. Taylorism flourished because the technology of the time suited it well. The typical organizational structure supposedly included the best technology of production, but there had never been any testing or research of this design, since there had never been any large-scale evaluation of organizational effectiveness. The obvious answer when an organization succeeded was that it was well-organized. Failure had an equally obvious answer—the company was not well-organized. It almost seemed to be a matter of faith or of ideals, with little empirical backup data. These were the mainstays of management texts for many years.

Taylor's work meshed nicely with the description of bureaucracy proposed by Weber (1947), his predecessor. That description covered the specialization of skills and tasks, specific limits on discretion embodied in a known set of rules, professional impersonal behavior, and a complex hierarchical structure concerned with administrative control. This model of the organization was based in the intuitive and deductive logic of managers attempting to develop order in managing technological organizations. They believed that the machine model of the organization was surely the best, since the machine-driven technology had obviously produced the best social and economic arrangements.

However, in more modern times, when this machine model was tested by researchers who observed the success patterns of companies that used it and compared them with the patterns of nonusers, they found that the nonusers sometimes succeeded very well also. Some companies with well-documented internal procedures and systems and clear, unambiguous organization charts were even declining in the face of successful competition from companies that were very loosely organized. Both kinds of companies, of course, had the same technology. Is it possible that something was missing that the researchers were not aware of?

TECHNOLOGY AND STRUCTURE: IMPROVING MODERN ORGANIZATIONS

One important study of recent times (Woodward, 1965) attempted to determine what the real relationship between structure and technology was. That study used a simplified definition of technology: the methods of production—including more than manufacturing. It included organizational structures and management techniques. Those methods had changed radically from the ideal of mass production that came out of the industrial revolution.

The research findings relied on data obtained from approximately 100 manufacturing firms in southern England. The original purpose of the research was to determine if there was any relationship between organizational structural design as the independent variable and business success as the dependent variable. There were comparisons of measurements of the organization's structure, such as the chief executive's and the first-level supervisor's span of control, levels of management, the ratio of indirect to direct workers, and the ratio of staff to line personnel against business success.

Initially, the researchers found nothing. There was little or no correlation, even when comparisons were made on size of the firm, production control techniques, and organizational change. No empirical structural relationship emerged to predict business success. Even classical management theory predictions were not supported, since firms succeeded both with and without it. Firms classified as being less successful also had all kinds of organizational structures. Empirical research had not supported the logical machine model or any other model at this point.

According to Woodward's description, the research team decided to re-evaluate the data it had gathered after thinking through some of the inherent design concepts that classical theory possessed. The bases for this machine theory were developed in a *technical* setting, but *independently* of the technology used. Taylor and his contemporaries were engineers who practiced successfully in manufacturing industries. They had generalized on that experience as if it were applicable to *all* administrative organizations. The similarities between the background of the people who had developed classical theory and the technology that they had used had been overlooked. The research team felt that this was a major point to be considered.

The members of the team then redefined their independent variable of organizational design in three separate technological methods of production. When they reclassified their data about the 100 manufacturing companies into these three categories, they found that economic success occurred when there was a closer match between the different kinds of organizational structures and particular production technologies. The companies that were economically successful had structures that were very similar to each other *within specific categories of technology* and those that were less successful did not. There were three categories:

1. *Unit or small-batch production:* the manufacture of articles to customers' individual requirements. This is the oldest and simplest form of production. Since each article is produced to order and is modified to suit some customer's needs, the processes used are sequential and noncomparable to each other. The unique requirements of the customer's specifications defines the machinery used and the nature of the product. In a factory environment, the manufacture of tools and dies, custom machinery, specialized test equipment, and very large industrial turbines are typical of this unit or small-batch technology. The worker is not a machine adjunct. His skills are central to the manufacturing process. He plans and controls his own work.

2. *Large-batch and mass production:* the technology with which Taylor and his associates were familiar. It produces standardized products in sequential manufacturing with little deviation, except with extensive setups and reworking of production equipment. The customer gets a product that is essentially the same as all those that are produced. Some examples are bathroom fixtures, moderately priced clothing, and most types of automotive vehicles. The worker is both a machine adjunct (i.e., assembly line) and a work controller (i.e., maintenance).

3. *Process production:* the flow manufacturing process of the refinery, the pharmaceutical plant, and chemicals. The factory can be considered to be one large manufacturing tool and the product can be varied but only within very narrow limits. A refinery, for example, can turn out various kinds of gasolines but it can't be converted into producing something other than petroleum-based products, such as maple syrup. The worker is highly skilled but those skills are primarily used to control the process. The technology of the process has been defined by the plant designers, not the plant operators.

The three-way classification included twenty-five firms in process technology, thirty-one firms in mass production, and twenty-four in unit production (Woodward, 1965, pp. 52–65). Woodward found the results shown in Table 5-1. Organizations with structural designs that varied greatly from the above median figures for their category of technology were less successful than organizations whose structures were close to the medians.

Ratio of direct to indirect workers gets lower as more technologically complex production processes are used. (A refinery uses more complex technologies of production than a plant in large-batch and mass manufacturing, and that plant, in turn, is more complex than a tool-and-die shop.) Process industries employ many maintenance people and relatively few production people. A trip through an operating refinery can be an unusual experience for the uninitiated. There seems to be a great jungle of hissing, stationary

TABLE 5-1   TECHNOLOGY VARIABLES FOR UNIT, MASS, AND PROCESS PRODUCTION

	Unit	Mass	Process
Ratio of direct to indirect workers	9:1	4:1	1:1
Chief executive: levels of management	3	4	6
Costs in wages, %	35	30	12.5
Ratio of manager to total personnel	1:28	1:16	1:7
Ratio of staff to industrial workers	1:8	1:6	1:2
Span of control: first-level supervisors	1:24	1:50	1:15
Span of control: chief executive	4	7	10

machinery and silver-coated pipes with no people visible, except for the occasional worker adjusting a valve or tightening a pipe connection. It is only when you open the door of the control room that you find the workers—watching dials and adjusting controls (and there are very few of them).

Levels of management and span of control of the chief executive increase with technological complexity. This indicates that the unit technology company has an organizational structural pyramid that is flat, while the process industry has a narrow, high structure. Unit production executives are more closely involved with the day-to-day operations typical of a customer-oriented shop environment. Process production executives manage more by committee and are more concerned with working group activities that are intended to minimize the production downtime of the particular process.

Ratio of managers to total personnel and percentage of costs in wages show the effects of increasing specialization as technological complexity increases. The generalist worker who sets up and controls his own operations becomes less prevalent and the number of managers coordinating the lower skills of workers increases.

Span of control of first-level supervisors is the only internal variable that appears not to follow the rules. Small spans are an indication of the breakdown of the working force into small groups in both unit and process operations. According to Woodward, the relationships between the first-line supervisor and the highly skilled generalists of unit production or the highly specialized workers of process production are more intimate and informal than in the mass production or large-batch technology firms studied.

The implications for successful design seem to be:

Unit production: If your company produces special or unique products to meet the specific requirements of a customer, it should have a flat organizational structure (i.e., few levels between workers and top management and a wide span of control), few line managers, and a small staff. A high proportion of costs will be in wages for very skilled generalists who independently plan and execute their own work (e.g., the systems development and programming section of your computer operations).

Process operations: If your company produces continuous printing of any kind, chemicals, hydrocarbons, pharmaceuticals, or fabrics in facilities that are operated on a multishift basis, it should have a tall organizational structure (i.e., many levels between the workers and top management, with a narrow span of control), a high proportion of line managers to workers (this may mean fewer workers but the same number of managers as a comparable size unit production company), and a fairly large number of indirect and staff personnel.

Mass and large-batch production: This is the organizational model familiar to Taylor and is very close to the bureaucratic model of Weber. Its internal social atmosphere differs considerably from that in both unit and process production models. This model is used in auto production, food, clothing, and consumer tools. Production processes should be routine, efficient, and mechanical, with workers tending machines that are paced by engineering standards (Gibson et al., 1976). Work control and supervision are separated from the worker, and any deviation in standard working methods is not acceptable. The first-line supervisor is primarily a disciplinarian in this model and a problem solver in the other two models. The organizational structure is the familiar equilateral pyramidal shape of the classical theory, with spans of control and intermediate levels of management somewhere between those in unit and in process production.

This and other replicating research (Zwerman, 1970) provided an important link between the technology of production and the structure of the organization. It also partially explained why the mechanical models of classical organization theory sometimes does and sometimes does not work well. Taylor and the other pioneers of organization design were right when they suggested that their personal successes could be extrapolated to other companies. They were not aware, however, that the technical and management organization structures applicable to steel mills and other large-batch or mass production manufacturing companies might not fit all production technologies, especially those in unit and process industries.

Woodward's research showed that there is no direct connection between operating success and organization design unless the specific manufacturing technology (as an intervening variable) matches the structure correctly. When the structure matches the technology of production, the correlation between structure and economic success become clear. The technology component is no longer a built-in kind of process; now it is an active component or variable that must be considered in the organization's design. Management discipline, logic, or intuition is not enough.

But our new definition of technology is larger than that of merely the methods of production. It includes *all* the processes of changing inputs to outputs, and that involves innovation and product change. Do these also affect the way the organizational structure is designed? There is research that provides some potential answers.

Technology Defined as Innovation

Burns and Stalker (1961) defined technology slightly differently and produced results that explained some other structural designs of the firm. They were concerned with companies in which the products have considerable variety and may be changed during develoment processes. Instead of the manufacturing processes, innovation was their independent variable and organization structure the dependent variable. Their data were based on the reported perceptions of management personnel and on the communications patterns within the total organization.

They found two general types of structures. One kind was *mechanistic* and applied during stable conditions. The other structure was *organic* and applied during conditions of change and innovation. The typical mechanistic structure included rigid descriptions of functional specializations, precise definitions of duties, and a well-defined system of command through which orders flowed downward and information flowed upward. This is the epitome of the middle (mass production) model in Woodward's categories and the classical theory of Taylor.

On the other hand, organic structures have less formal job definitions, with internal communications that flow up, down, and horizontally to wherever they seem to be needed by the participants. These structures are more collegial or consultative, less directive, and very adaptable to changing environmental conditions. They are, however, more expensive to operate because of the additional efforts at coordination required. The two technological extremes of Woodward, unit and process production, seem to be in this organic mold.

These alternatives of organic and mechanistic models help to explain the metamorphosis of the loosely organized, smaller technical firm using flexible unit production modes with a great number of product innovations into a tightly defined, more rigid larger company using mass production modes with fewer innovations and product changes. The high costs required with rapid change and innovation and unit production methods are reduced through being channeled into more fully defined organization charts, responsibility diagrams, and production standards. It's a different technological world and the firm must change its organizational structure if it is to continue its success.

Rate of Product Change

This research seems to have provided different kinds of answers to the design questions in the technology and structure interrelationship, but are they really different or just part of a larger theory? If we think about the total organizational structure as being a monolith or a strictly uniform design for each company, we definitely have a conflict between theoretical results. But, as you well know, most technical organizations are not like that. They could have both flexible (i.e., advanced research and development) and rigid (i.e., mass production) departments *within* the same overall corporate structure.

When this happens, the structure cannot be uniform. Different structures are designed to fit each situation *within the overall company*.

We reviewed some of the ideas of Lawrence and Lorsch (1967) in Chapter 4, but those findings can be applied differently when considering the variable of technology. Lawrence and Lorsch were also concerned with how technology affected departments *within* the organization differently. To review briefly, their data involved observations and the results of questionnaires administered to the personnel of firms in three kinds of industries: plastics, food, and container manufacturing. These represented fast (plastics), medium (food), and slow (container manufacturing) changing technological, scientific, and economic environmental conditions. The three industries did not exactly fit across Woodward's ideas, since all three of them used a type of process technology (Lawrence & Lorsch, 1967, p. 191).

However, the researchers discovered differences in responses for the *same kinds of technical groups* between companies when they asked "What kind of organization does it take to deal with various economic and market conditions?" and "What are the differences in cognitive and emotional orientation among managers in different functional departments . . . or groups?" (Lawrence & Lorsch, 1967, p. 11). They selected three functional departments of the company: science, technical-production, and marketing groups. Even within the single production-oriented process technology of Woodward's classification, they found that effective organizations that had lower differentiation and minimal integration needs between internal groups were typically concerned with high volume, prompt delivery, consistent product quality, and producing a relatively unchanged product at a minimal cost.

The most successful company in this case was the one involved in high production of relatively fixed product designs of cardboard containers. It had very little product change and was almost mechanistic in its operations. Those with higher differentiation needed special integration functions, and although they also were concerned with the same criteria of volume, delivery, and quality, they were more involved in rapid product change and innovation. Typically, it was the plastics manufacturer that was very organic.

Differentiation was defined as behavioral subsystems that were consistent within their own groups but very *different* from the behaviors of other people in other groups. *Integration* was defined as the qualities of the state of *collaboration* that exists among departments required to achieve unity of effort by the demands of the environment. In these cases, the technology was the rate of change in products that each company manufactured to satisfy the requirements of the environment. See Table 5-2.

The container industry's customers wanted minimum innovations and proven containers that could be processed at high speeds on automatic packaging lines. They did *not* want technological or product innovations that could make their existing packaging obsolete. Improvements were secondary

TABLE 5-2   DIFFERENTIATION AND INTEGRATION IN THREE INDUSTRIES

	Container	Food	Plastics
Degree of differentiation	Low	Medium	High
Major integrative device	Direct managerial contact	Individual integration	Coordinating department

to continuing production, as is, and lowest prices were important. In the food industry, there was competitive pressure for some innovation and change in the foods produced. Selling prices didn't make a great deal of difference to the market and those prices had little internal relationship to volume or costs. Therefore, moderate change could be paid for through the increased costs that this rate of change would cause. But the most product change occurred in the plastics industry, which provided specialty plastics tailored for specific customers. Although production was process-oriented, with relatively few workers needed to monitor automatic and semiautomatic processing equipment, the product life cycle was very short. The development of plastic materials, moreover, was more of an art than a science, since there was a poor understanding of cause-and-effect relationships. The customer environment definitely wanted high innovation, was willing to pay for it, and operated under major conditions of uncertainty.

We can use these research findings by applying a little self-analysis to our own technical situations. Since differentiation was defined partly by the questionnaire reports of the organizational participants, it had to reflect the subjective perceptions of those participants. Therefore, if high uncertainty is perceived in the environment by your technical organizational participants, perhaps your structure should have high differentiation and the best organizational design would be the organic model of Burns and Stalker. Conversely, if low uncertainty is perceived, the structure should have low differentiation, with a mechanistic or classical theory model.

In these examples, changes in technology such as production methods, product innovation and adaptation, and the information processing responses of organizational participants all impact both the structure and the people. The classical management theorists were correct in their limited design concepts of the bureaucratic, mechanistic, and relatively static organizational designs. They erred, however, in suggesting that these concepts were universally applicable, since they could not account for variations in success with companies that were not organized "correctly." The error probably came about because the classical management theorists were familiar with only one technology.

A similar error of supposed universal application occurs in some modern laissez-faire–oriented mangement where there is an implicit suggestion that all structures should provide free and open environments that support people in developing their job potentials. It is possible for organizations to use the same general rule that there must be a match among the components of people, structure, *and* technology, with this match agreeing with the differing requirements of the external environment, and yet come up with different structures, depending on how they have defined the match of components and environment. The problem is the expectation that either of these designs is really universal. This is not so, of course. For example, an inappropriate match might occur when the structural freedom of the small research and development company prevents the development of a separate mechanistically structured group that is needed to manage the disciplined, logical manufacturing systems required to mass produce the new and wonderful widget that this free and open organically structured R&D group has developed.

It's a difficult design task to complete because the managers must, in this example, turn their backs on the free and open organization in which they succeeded in developing that widget and build a closed organization to produce it. That means hiring different kinds of people, setting up different organizational structures and policies, and using different technologies to produce with and make decisions. Success in the past may have little relationship to success in the future.

SUMMARY OF TECHNOLOGY AS INNOVATION

High uncertainty, innovation, and unit or process production technology seem to require organic, very flexible structures with highly differentiated people. Those firms in the middle range or mass production technology require mechanistic structures and more collaborative people who can coordinate both with each other and with other groups within the firm. These summary statements begin to unravel some of the tangles in the interaction of technology with the organization. But let's see if these recommendations have been supported by research findings.

TESTS IN ORGANIC FIRMS: THE RESULTS

Some research was done by Walton (1979) on using open, organic structures with worker-defined working conditions. He gathered data from about three dozen technologically varied companies, generally in process manufacturing technology, and found no clear *long-term* indications of success in applying organic structures. His major definitions of organic structures were:

Self-managing work teams: They developed their own goals and norms.

Whole tasks: The reverse of task specialization. Simple operations organized into meaningful wholes for operators.

Flexibility in work assignments: Movement from one set of tasks to another; systematic rotation of tasks.

But early gains in organizational performance were reported almost uniformly. For example:

1. Higher production efficiencies were reported. These were derived from less wastage of materials, less downtime, or more efficient methods.

2. Quality improvements were significant.

3. A reduction in overhead was common, due to a leaner supervisory and staff structure, with less paperwork.

4. In several cases, a more rapid development of skills produced promotable people at a more rapid rate, increasing the number of operators who were promoted to foremen outside of their own department.

5. Turnover and absentee rates were generally lower.

In fact, Walton found that almost every organization that employed this structural design reported *initial* gains. However, with the passage of time, other patterns began to emerge. Some plants returned to conventional mechanistic patterns within a short time, others regressed from the ideal organic structure toward these mechanistic patterns after a few years of successful evolution, and a very few persisted, although in a limited fashion. As an explanation for this, Walton (1979) pointed out that "new demands may also tax the system's ability to perform and survive, producing a return to more conventional patterns."

Two representative types of these new demands were based on technology and competitive pressures. The new demand based on technology happened when the markets expanded rapidly and new products had to be developed quickly to satisfy that expansion. The organizational structural responses to the increased technological inputs in this example were the undertaking of quality corrective actions in field service in order to fix unsatisfactory products and the provision of additional engineering and development support. These structural responses were always tied to closer supervision. It's a familiar response. It involved typical questions like: "What are those guys in advanced R&D doing? Dammit, we needed this new widget on the market last month. I want a written report every day on their progress!" The increased speed of innovation (or technology) seemed to force an inappropriate return to mechanistic close controls. Creativity seemed to be acceptable but only when it met a predetermined schedule!

The other new demand occurred when companies came under severe or long-term competitive pressures. Under these conditions, upper management began emphasizing cost reduction and *near-term results,* insisting upon discipline and compliance with their programs. Authoritarian decision and *"do it now"* commands don't support group efforts and cooperative, organized decision making. The uncertainty of the technological environment was quickly converted to the primary and certain goal of survival first, regardless of the obvious negative impact on creativity and innovation. While all of this might seem a bit harsh, it's important to remember that innovation can only follow if the firm survives to support that innovative effort.

This research indicates that the organic and loosely constrained organizational structure works well only when technology and external economics are supportive and when there are few extreme pressures to contend with. In effect, a lifeboat is no place to hold a participative, free discussion on future organizational planning. Neither is it the place for highly differentiated behavior patterns, since rowing a lifeboat requires a great amount of co-operation (or integration, if you wish to use a more precise term). A technologically innovative and organically developed decision to use either gas turbines or nuclear energy to propel the lifeboat is also inappropriate, since you only have oars and the delivery schedule on these other items is fairly long, even if the vendor knew where in the ocean he had to deliver them. Therefore, the naval tradition on lifeboats is justifiably very authoritarian, with task-oriented, closely supervised groups. Most technical organizations that are in economic trouble and wish to survive, therefore, should tend to be able to adopt a mechanistic structure that almost instinctively demands integrative or collaborative behaviors from people.

Again, there is no "one best way" to organize, since *time,* in terms both of survival and of growth or development, must be considered. Even if the structure matches all the manufacturing requirements of technology (Woodward) the environmental uncertainty (Lawrence and Lorsch), and the innovative, organic structures that support progress in product technology (Burns and Stalker), the technical department must consider both the organic elements to produce creative product improvements and the mechanistic elements to produce quality, high-volume–low-cost products. Organic structures and highly differentiated behaviors work well when there are sufficient resources within the organization to protect it from day-to-day problems (Walton, 1979). And we can also find a need for the mechanistic, closed, well-defined organization structure with its well-trained, cooperating professional people for the broad middle range of technology. It also still seems to be the best design to use when the organization is in immediate trouble.

Summary: Technology and Structure

Table 5-3 summarizes the relationship between technology and structure.
Why should we be concerned about developing an organic, relatively

TABLE 5-3 CLASSIFICATION AND DESCRIPTION OF TECHNOLOGICAL STRUCTURES

Classification	Description
Unit and small batch, organic	Few procedures. Short, squat structure. Few in staff. Few levels of management. Workers are generalists.
Mass and large batch, mechanistic	Rigid procedures. Triangular structure. Intermediate staff. Large span of control of first-line supervisors. Decrease of percentage of costs in wages. Classical theory model. Workers are semiskilled.
Process, organic	Process-controlled procedures. Tall and thin structure. Many levels of management. Small groups. Greater proportion of staff to workers. Workers very skilled and specialized.

differentiated structure if we can predict that this free and responsive design requires changing into a mechanistic, rules-oriented organization if an economic or technological downturn occurs? The reason is that growth depends on organic structures, and perhaps some of these can be protected within the organization if a mechanistic fire-fighting approach is needed for short-term survival. Two longer-term potential answers to this question also apply to technical organizations. They are:

1. The organic, highly differentiated structure promotes innovation, high innovative productivity, and prompt flexibility in responding to changing market demands. In short, it promotes business success *in a relatively benign economic environment*.

2. As a general rule, both our economic environments and the product-process technologies are moving away from mechanical, high labor content into electronic, high knowledge content for products. Two major reasons are the increased skills of the knowledge workers and the gradual taking over of the drudgery in work by automation and computers.

The industrial revolution required the assistance of precise, mechanical machine tools and the replacement of expensive muscle energy with inexpensive mechanical and electrical substitutes. It was a revolution that drastically changed people's lifestyles. Presently we seem to be heading into an equally drastic kind of new industrial revolution. Our products, our work methods, and our industrial systems are changing. The proportion of our work force in manufacturing is dropping, but the work it does is still quite critical to business success. Machine-serving labor, which is measurable, has been replaced by knowledge and information work that is very difficult to measure. The latter type of work typically requires organic structures, because the person is, in effect, determining what to do and how hard to work at it.

It is possible that we are involved in as revolutionary a change in technology as that which occurred in the early part of the industrial revolution. Then the worker moved from the farm and the cottage industry into central, more efficient work places called factories. Production and growth were a function of a new kind of industrial discipline located in a factory, and they were paced by the machine. Now the worker seems to be moving from the factory into the laboratory, the computer area, or, if the worker has a computer at home, the living room. (Back to the cottage industry?) The same kind of industrial discipline would not support the innovation and creativity required to meet new and changed market demands. We may be facing an information and technology revolution that is changing the way we work and live as drastically as the industrial revolution did in its day.

While speculation and prediction can be important guides to use if we want to modify our organizations (and to improve our own position within that organization), the existing environment confronts us with today's problems to be solved before tomorrow can be planned for. If we are in mass production (or threatened with severe or negative changes in the economic environment), the mechanistic, logical, systems-controlled, relatively undifferentiated organizational structures are optimal. Those structures in the short run still account for much of the employed working force, even if they probably are not the way we will be organized in the future.

Classical theory defines the mechanistic conditions. That theory allowed little deviation to account for the needs of people in the organization, since those people are considered to be almost interchangeable resources. Taylor's workers were uniformly (according to him) motivated only by wages. Their technology (work methods) was controlled by the job foreman and deviations were frowned upon. Even the managers or the professionals (of Weber's definitions) had specific behavioral standards imposed upon them. The technology, as they defined it, determined how and what to do and the measures by which they would be compensated. *Then,* the job defined the worker.

Recent research tends to support some of these ideas, but not all of them. Where the worker defines the job, mechanistic concepts no longer support optimal productivity. The sequence of organizational design when considering technology could then be that we should first develop the structure and systems to survive (i.e., keep the ship afloat first) and then begin to differentiate and grow. The differentiation occurs in those departments that are expected to produce the innovation accelerating the company's growth.

Information workers and human capital require organic structures, since the technology that they use is either unit production ("I've got to get this computer program debugged by tomorrow morning.") or process operations ("What kind of automatic setup and run programs do we need to keep that machining line operating even if we're manufacturing different products on them all the time?") The relationship between technology and structure is complex, because productivity then depends on nonprogrammable behaviors of the people. That is very difficult (if not impossible) to measure. With organizational economic success, the demand for more organic and less mechanistic structures *within properly defined organizational subunits* increases. How will this occur?

REVIEW OF ORGANIZATIONAL MODELS AND TECHNOLOGY

In much of the management literature, the mechanistic model is held up as an example of American know-how and productivity. The technology is not considered to present the major problem; the nature of the problem-solving tasks is. Effective programming of tasks and elimination of uncertainty are supposed to help in solving these problems. The mechanistic model of the production line (mass production or large batch) is the result and the prime example of the attempts to program tasks. However, from the worker's viewpoint, it has had a mostly negative effect. This is how it is described in the rigidly controlled, technologically advanced, manufacturing environment of an auto plant:

> A typical assembly line worker can learn his job in 30 minutes. He gains no skill that might qualify him for a better job. He has little incentive, investigators claim, to acquire the instincts of workmanship, to take pride in his job, or to feel a sense of purpose. The point of all this is that my experience working in an auto plant has convinced me that alienation of workers as a personality characteristic is in the eyes of the social scene beholder. (Widick, 1976, p. 71)

This writer seems to believe that alienation comes with the job, not the psychology of the worker. The job is the controlling factor. The "beholder" (in our example, you) could feel this way if you believe that these two conditions exist:

1. All workers react this way to repetitive jobs and have these same feelings.

2. These feelings, in some workers, really can affect the organization's productivity.

It may be a bit surprising to find that the first condition does not always apply.

> I would characterize this work as highly repetitious (this author is writing about a food processing plant, not auto work, but the idea is the same), with little

room for autonomy or growth and no position to rise to, but processing. Here there is also a lack of that critical job satisfier, the freedom to walk around and schmooze. Yet because workers, like the rest of us, are not of one mold, one worker said, "I like packaging because I do not want to think about the work anyway" (Schrank, 1978, p. 233).

Some people apparently may not like their work for reasons not associated with the organizational model of the job. We carry our mental models of the world around in our heads and they are very different from each other. Not all of us are the creative, innovative, I-can-hardly-wait-to-get-to-my-job kind of people, but no matter whether you are creative or not, it now appears that you, as the worker, are the major variable in increasing productivity.

Brookings Institute economist E. E. Denison (1978) estimates that the new technology accounts for 38 percent of the factors affecting productivity. In other words, if we assume that technology, with its 38 percent effect on productivity, is kept constant, the worker controls more than 62 percent of the increases or decreases in productivity! If we then assume that the worker also controls the technology (which she or he does in technical organizations), it would appear that *almost all* the productivity changes are controlled by the worker. Therefore, the people in technical organizations are the determiners of organizational success. Of course, this line of reasoning has taken some liberties by comparing actual production with technical workers as if they were the same. If they are not so now, they will be in the immediate future, since the production workers of the future will probably be programming a computer that will then control the plant.

HOW DOES THE COMPUTER CHANGE THE ORGANIZATION?

Technology changes both the formal and the informal organization, impacting on the behaviors of the people in them. In mechanistic classical organizations, most of the response of workers has been negative: "Dissatisfaction with work seems to be a function of technology. The greatest dissatisfaction is reported on jobs with short job cycles and relatively little challenge in industries, especially the automotive industry" (Fairfield, 1974, p. 39).

Although other research (Hulin & Blood, 1968) shows no relationship between this dissatisfaction and productivity, there *is* a relationship between self-interest and productivity. Therefore, if the technology can increase that self-interest, and dissatisfaction thereby decreases, productivity will increase. This assumes, of course, that there is an inverse relationship between self-interest and dissatisfaction. This is being attempted in the same industry that was described negatively before: the auto industry.

Selected according to seniority, they (the workers) studied the slimmed-down V-6 (engine), its automated machinery, the high quality standards that would apply, and how workers would be given responsibility for assuring that quality. . . . In a stroll down the new line, Mr. Wilson (assembly line worker) and Mr. Hayes (maintenance electrician) pointed out numerous changes their three work groups had made in the equipment. . . . Under the traditional concept, they would have told us. . . . "This is the way it is; you live with it. . . . I'm no longer a job rat under this concept. . . . I feel like I'm part of the product (Hayes, 1981).

This description doesn't seem to be one of an auto assembly line, but it is. Since the worker now appears to control many of the technological elements of productivity, there is a change in the psychological environment of the assembly line that seems to increase his or her control and self-interest in keeping the line moving. The worker is no longer an adjunct of some assembly process, but a controller of it. It is a change from a labor commodity resource into a type of semiartisan similar to those working in unit production or to the manager of the flow controller computer in the process industry. It has also enlarged the ranks of management, since the worker has become a manager of his or her own area of responsibility. First-line supervision changes from a primarily disciplinary job in mechanistic production to small-team coordinator in unit and process production.

While mechanistic structures are economically most successful for mass production and large-batch manufacturing, they can also be quite successful for people if there is a change in technical methodology to remove deadening and repetitive tasks. This would provide workers with nonrepetitive, controlling jobs which should be a source of work satisfaction and self-interest tied directly to an increase in productivity. Installing computer-controlled or automatic production machinery accomplishes this to some extent. The nonhuman kinds of work have been assumed by a machine. In that example, the mechanistic procedures could be handled by automated production devices. In some of our larger manufacturing companies, the installation of production robots has begun. The organization then becomes more mechanistic in production technology (that's good) and more organic in its human structure (that's good, too). We begin to move toward multiapproach, much improved, and highly differentiated organizational designs.

> Technically and eocnomically, process production is at the opposite end of the scale from unit production, but from a social organization point of view, this inner system was surprisingly similar to the social system of the special-order production firms studied. There was the same identification of formal and informal systems (Woodward, 1965, p. 161).

The important relationship between the variable of worker behaviors in the more human-oriented unit or process production technologies and organizational success was that the workers controlled their work inputs. To a large extent, they determined productivity and seemed to be able to *manage* their own jobs very well. If we can assume that there is also a relationship that applies between these variables for mass production technologies, those machinelike organizations should be concentrating the necessary discipline and logic of production on the shop floor, where it belongs, through increased uses of computers and robots. They should then be able to provide more decision-making capabilities in worker behaviors to allow those workers to handle unforeseen contingencies more freely themselves.

In other words, with that computer-robot technology in place, workers will be able to work more "processlike," thereby exerting psychological control over the work flow. Taken to a possible conclusion, the mechanistic organization would be that existing only on the manufacturing shop floor involved with the actual metal cutting or what have you, but the human (i.e., managing) organization would be organized in a process mode. It would be like a two-layered structure, the bottom being mechanistic and the top organic. The top layer would be closer to the refinery-type organization workers who manage the work flow and control it. They are definitely more than machine adjuncts.

Organizing to satisfy the needs of technology could be achieved by emphasizing more automation or computerization, accompanying an organizational structural redesign into this two-layered, very "differentiated" style. It is possible that the general categorization into unit, mass, and process technologies in industry could be moving toward a consolidation into one overall design with automated manufacturing technology (organized mechanistically) that can generally operate in any of the three modes overlaid (or differentiated) with an organic management structure.

This seems to be happening on some machine shop floors, for example. Introduction of computer-controlled general-purpose equipment allows accurate machining of *different* products on the same machine, since the computer programs control setups and operations. Those programs, of course, were developed in the organic group of the factory's manufacturing engineers. While we still have some distance to go before we can expect to see general-purpose manufacturing companies that can turn out anything from steel to gasoline or even any kind of machined product, there will be *decreasing* differences in overall organizational structures, first among companies in the same industry, such as all steel mills, and then between various industries such as steel mills and refineries. But that is going to take some doing and will probably be in the more distant future.

Even though organizations within the same industries, on an overall basis, will become more similar on an overall basis in the future, they will still be differentiated internally, as they are today. People are different and each manager affects the organization as well as it affects him or her. The internal differences perceived by the participants (i.e., engineering is different from sales) can determine how well the firm will succeed. Therefore, although I forecast that steel mills and refineries will become more organizationally alike because technology allows production to be completed in any of the three modes mentioned, they will still be very different inside because of department differentiation. And because of differences in their participants.

In this case, technology (or the capacity to process information quickly and efficiently) supports differentiated social behaviors within each department. Computerization can therefore promote more organic kinds of structures for human beings inside while supporting more mechanistic structures on the shop floor and more consistent external interactions with the market or external environment. (It seems to be a good thing all around.)

However, before using this general prescription for organizational change into similar "outsides" but different "insides," which is intended to help the human side of the group, we should be concerned with the effect on the quantity and the quality of the product output. Would more open organic structures, using the technology suggested above, help or hinder those outputs?

MECHANISTIC INTO ORGANIC STRUCTURES

The relationship of machine-oriented, organizational designs manufacturing large quantities of satisfactory products to excellent company profits has always seemed to be fairly clear. At least it was before the research of Woodward, Lawrence and Lorsch, and Burns and Stalker. Now it seems that this relationship is no longer very clear, and many organizations are not optimally structured to fit their technology. Is it possible that they have been operating at less than optimal effectiveness? Why have they been able to survive this long? Has the relatively recent decline in our national productivity been the result of poor use of technology? Do other countries use it better? Let's analyze what has really been happening to our so-called mass production operations. Have they ever worked as well as they could have or at least as well as our research-based predictions of the optimum relationship between a mechanistic structure and mass production technology indicate they should? Should we have been using organic structures all along, and would that have helped our products to achieve higher quality levels? The answers are quite surprising.

It seems that mass production has never really worked as well as it should have. There have been problems (Skinner, 1979, p. 214). Some of the obvious ones were:

1. Poor worker reaction and cooperation in serving as machine adjuncts

2. Many management-dictated changes in product mix, which negated the advantages of single setups and long runs

3. Many changes in product design, which affected the way that they were made

The most important of these was the first problem, poor worker reaction. When behavior at work, or to put it another way, when organizational behaviors were in conflict with the requirements of the formal organizational structure, the informal structure fought with, rather than supported, the formal structure. This might have indicated that the human discipline required in mechanistic structures in mass production was lacking and conflict was therefore a constant problem. According to some recent data about the metal-working industry that uses this structure quite a bit, it seems that we really don't have real mass production. Therefore, a rigid mechanistic structure could be inappropriate and the poor worker reaction could indicate a need for a looser, more organic, structure. For example,

> Complex metal products . . . occupy a central part in metal working. They come in myriads of types, shapes, and sizes and are made in untold millions. But contrary to popular belief, most of them are *not* mass produced (i.e., over 100,000 units per year). (Moreover) . . . over one half of the mechanical parts are made in small batches, less than fifty pieces at a time, and in terms of value, mass production accounts for less than 30% of the total. Small and medium batch manufacturing is the predominant mode of production and is characterized by the employment of general purpose—as opposed to special purpose—machinery. . . . Cutting time is on the average less than 30% of the time that the work piece is in the machine. The remainder is spent on loading, unloading, idle motions, measuring, operator rest, etc. . . .
>
> If anything, the situation has become more difficult because the number of different product types is increasing, doubling every twenty years, so the batches grow even smaller in size. (Barash, 1980, p. 38)

And if you consider the total time (not only the time at the manufacturing station) that the work is typically on the factory floor, more than 95 percent of that in-process time the part is waiting, tying up costly floor space and working capital (Kops, 1980, p. 110). Our metal-processing facilities, which are typically supposed to be mechanistically organized as if they were using mass production technology, are *really* using semiunit production technology. Therefore, according to the research that we have reviewed, they should be organized in a more organic fashion! Surely this would be a radical change for most companies, and I can predict that it will not occur quickly. Most organizations change their structure slowly. (The only fast changes that I have seen are those caused by rapidly dropping sales. That always seems to tighten the mechanistic design rather than loosening it.)

My feeling is that this reluctance to move quickly into organic structures is based on the behavioral patterns of the very top management personnel in mechanistically oriented companies. They are familiar with more deliberate, mechanistic thinking and seem to be uncomfortable with supporting change intended to allow more decision making at lower organizational levels in a wholesale manner. I remember once being told that delegation and decentralization are alright within limits but "there's no way that we are going to let everyone run his own job since that would be anarchy." This appears to be a common frame of reference typical of a patriarchal and mechanical organizational structure.

There also seems to be a general match between the organizational structure and the people that will stay in it. Do their perceptions constrain the mechanistic structure or does the structure constrain them? Which is the cause and which the effect? Perceptions and structures eventually seem to match, even if it means the creation of an opposing informal structure, as we have learned from Burns and Stalker (organic versus mechanistic) and Lawrence and Lorsch (differentiation versus integration), but which are really the independent and which the dependent variables?

The existing research is still incomplete. Answering this question might provide clues to changing those perceptions (and/or structures) in order to move quickly toward the more organic designs complex technology seems to require. According to Woodward (1965), economic success was higher for those firms whose structures matched the technology that they used. Apparently, many of our "mass production" metal-working shop organizations are mismatched with their technology. Can it be because of a mismatch in management perceptions? Are those top managers' personal theories wrong? "Popular perceptions and images lag behind objective reality. Perceptions and images, personal and collective, shape the behavior of individuals and groups when changes in their systems are proposed" (Kops, 1980, p. 101). Generally, organic structures seem to be the better design for technical organizations when changing technology is a major factor.

THE REAL MECHANISTIC STRUCTURE

Do better mechanistic structures matching mass production technology exist? We know that technology affects product quality, quantity (i.e., productivity), and organizational structure. Within recent times, quality comparisons between American and foreign products have often been decided in favor of the imports. Japanese cars and cameras, for example, have typically been considered to be the best in their class. This has on occasion been attributed to the Japanese social system, training, or many other (typically organic) factors, depending upon the academic background of the writers of articles about them. While these factors may be important, I suggest that there could be another approach, one that more effectively relates the technology of production to the organizational structure. When that structure supports quality and discipline, it is really mechanistic and the worker services the equipment well by becoming, truly, a machine adjunct.

According to a recent report, the modern Japanese factory is not really a prototype of the factory of the future, as many of us might believe. "Instead, it is something more difficult for us to copy: it is the factory of today running as it should be." (Quotes and concepts in this paragraph are from Hayes, 1981, pp. 57–60.) Automation consists primarily of adequate materials-handling equipment used in conjunction with standard processing equipment. There are monitoring systems that allow Japanese workers to oversee the operation of more machines than their U.S. counterparts, and production schedules are based on capacity measures derived from actual performance data (not, as often seen in the United States, from theoretical or obsolete standards). There is no expediting, and no overloading is allowed. Work is measured out to the plant in careful doses instead of being, as one U.S. manager put it, "dumped on the floor so the foreman can figure out what to do with it."

If these are some of the reasons for success, it appears that for the last twenty years or so in this country, we have had the wrong goals for our mass production organizations. Those mass production organizations have selected other goals, such as mass distribution, advertising, financial controls, and new-product development as more important than the day-to-day emphasis on the basic production of reliable, low-cost, defect-free products to a predetermined schedule. They seem to have selected failure rather than success as one of those goals.

In international markets, we have increasingly been displaced by others who understand that difference. Others believe in manufacturing fundamentals such as decreasing manufacturing changes and increasing product quantity runs, increasing maintenance personnel to keep equipment operating optimally, and continuing training of workers to achieve a technically updated, informal organization that supports the formal organizational goals. Repetitive and machinelike tasks are assigned to robot machines or computers. The classical theorists would have recognized this type of organizational structure immediately, since it is quite appropriate under these technological conditions. If companies professing to use a mechanistic structure really wanted to develop one, models are available.

With respect to the so-called differences in culture that require lifetime employment in Japan or the wide distribution of robots in the Japanese factory, similar lifetime employment is not strange to the U.S. scene. In most large companies, about 30 to 40 percent of the management and "knowledge" work force has almost lifetime jobs, and the use of robots, as we define them, is not very widespread in Japan. Their definition of robots is different since it includes many of the automatic materials-feeding devices and transfer fixtures that we do not dignify with that name; they use a lot of them. The technology does not limit the organizational structure when it matches it, but it will when there is a mismatch. The mechanistic structure is alive and very useful when properly applied. That proper application seems to be the major problem. The question really could be, "In your company, what should the proper application be?"

THE COMPUTER AND TECHNOLOGY

We have reviewed how productivity represented by structure interacts with the technology as much as any other factor, and although we have cited several definitions of technology, none has explicitly covered the effects of changes in processing *information* as much as they have covered the changes in *production* methods affecting products and organization *design*. There probably have been more changes in processing information than anywhere else. The abacus still works, but the technology of the computer is no more like it than the foot-driven lathe of the industrial revolution is like the modern high-speed automatically controlled chucker.

The technology of processing information is now as vital as the technology of processing products or creating innovation. Improvements in production processing or product innovation are important, but if it were possible to separate out the effects of all changes on the organization, we would probably find that changes in information processing technology have become one of the most important. These changes have supported a lot of the moves from mechanistic structures into organic designs by relieving people of many of the short-cycle, repetitive work patterns that they follow so poorly and computers do so well.

While many of the definitions of technology used here have included the computer or its predecessors, since there always has to be some method of repetitively processing information supporting the particular organizational structure, the combination of information processing controls and automatic manufacturing equipment that results in process technology is still not commonplace. In my opinion, this lag contributes to less than optimal technical production organizations. We use ineffective human "computers" to perform the repetitive coordinating and information processing tasks. For example, consider the shop floor:

> In a country where 60% of the average foreman's time is spent expediting, looking for material, and in general, firefighting because of poor scheduling . . . there is an opportunity to improve productivity that is largely untapped, and for that matter, poorly understood. . . . Before computers, scheduling in a manufacturing company was simply out of the question. (Wight, 1980, pp. 93–94)

It seems that before computers, the informal, unstructured organization would often keep things going by moving lots around, finding lost work, and changing schedules constantly. Even technologies that were intended to support mass production industries needed internal miniorganic organizational structures to keep going. The informal organization served that purpose. The inefficiencies that continually cropped up on the shop floor were then overcome by human corrective action that was *not directed* by the organizational hierarchy. These actions may seem to be relatively minor—finding a part there, setting up a machine there—but the aggregate cost of using this very valuable human effort is quite high. It is now possible that the valuable, creative human being can be released to assume the nonrepetitive tasks she or he is best suited for and the nontiring computer can take over the minor repetitive tasks or the drudgery of keeping track of things. Even though the possibility has existed for some time now, it has not been acted upon as rapidly as it should have been because, in my opinion, based on observations in many companies,

1. Top management has an underlying fear that it might not be able to understand this new technology. Therefore, any movement in this direction has been kept very slow and deliberate (i.e., because of lack of understanding).

2. The concept of having several different kinds of organizational structures—which this technology of computerization would support—could be a bit disturbing to traditionally oriented top managers (i.e., because of fear of the unknown).

For example, it is unusual for top managers to take the time to attend training sessions on computers (or anything else, for that matter). Therefore, when the advanced engineering department (organic structure) wants to use computers for innovative design alternatives and suggests that each designer have his own microcomputer or a connection from his home to the company's central processing unit, but the manufacturing and production departments (mechanistic structure) want to use them only for tightly disciplined and centrally controlled production processing, top management no longer can set a companywide personnel and computer policy. Each group has to be treated differently, and many conservative top managers have not had (and may never accept) the training to handle this change.

This type of change places a much greater administrative load on the top operating management. When the change is extended to the organization, even the most basic types of uniform rules—such as vacation, working hours, and compensation areas—may have to be evaluated on close to a one-to-one basis. The amount of uncertainty increases, since general, long-standing rules which are typical of decisions included in company policies can no longer apply companywide. However, there is a positive aspect to making all managers' jobs (not only those at the top) more flexible. The organization is now more responsive than it was, and a lack of those management decisions that are intended to cope with this increased flexibility (or even a poor decision) is more easily detected.

For example, computer operations (running the programs) require tight discipline very close to that of a well-managed mechanistic organization. They require a close collaboration between the mechanistic human structure and machine processes. On the other hand, the actual programming of the computer (not its operations in production) is closer to the creative efforts of the organic human structure. This is no longer a change between two different departments; it is now a change between two sections of a seemingly similar computer-based organization. Management efforts and the costs of administration are increased. However, whether the structure is organized mechanistically, as in mass production technological manufacturing (similar to the organization of some Japanese factories), or organically to help produce the design alternatives from which innovations and new products will come, the necessity for computerization will continue to grow in all organizations. The cost of this change will happen whether a company installs computers or not.

Consider the two alternatives of computer installation versus noninstallation for the simpler mechanistic structure. In addition to equipment purchases, installing computers typically involves the expenditure of funds for training people to perform highly skilled, machine-supporting, self-managed maintenance tasks in the shop office, and for managers to develop rigid rules that ensure production optimization on the shop floor. The mechanistic organization will have no more changing lines to suit temporary shortages, a lack of extra parts, or any of the many small, but heretofore necessary, jobs to keep the plant operating. The computer is able to plan the purchasing, production, and shipment of a higher-quality, uniform product; if something goes wrong, the cause is very apparent to the supporting human organization. On the other hand, not installing computers means spending almost the equivalent, if not more, money and time, but in a less obvious way. Decreasing quality levels, increasing labor costs, and other effects that result from the mismatch of technology and structure and people slowly cut into markets and profits. Eventually, competitive edge and perhaps the whole market is lost.

The cost equivalency is more difficult to visualize with an organic organization. Here the role of the computer is to increase the effectiveness of the human operator. When that operator is the organization's human capital (see Introduction) and he or she has always controlled productivity, it is more difficult to determine when, or if, that productivity has been increased. In that situation, the motivation plan is the first task to be completed if any resulting organization structure redesign is to achieve results, since that plan is as differentiated as the people to whom it applies.

In this case, the computer is a tool of the person, not of the production process, and that person will increase throughput only as a function of self-interest which the correct motivation program will promote; when motivation is high, computers do increase output in organic organizations dramatically. In both mechanistic and organic structures, one conclusion seems inevitable: computerization increases costs but it also provides great benefits either way. When there is a deliberate investment, it can pay off in greater future returns. When there is no deliberate investment, one is made by the market anyhow, since there is an increased cost to compete that can eventually drain the vitality of the firm, causing its probable demise.

COMPUTER APPLICATIONS IN MANAGING

Using computers in manufacturing technology is an obvious application for a mechanistic organization. Other uses involve application to some of the repetitious activities that do not result directly in products. In some special applications in manufacturing technology, the computer sequences equipment and work stations, stores all the programs for different parts, analyzes system status, takes corrective action if needed, and makes operational decisions if it has been programmed to do so. These are only the obvious beginnings of a multitude of management tasks which heretofore had to be performed by people.

Among these tasks are computer aided design/computer aided manufacturing (CAD/CAM) for automated drafting, design, manufacturing sequencing, documentation, and engineering development. The primary job is to assist the user in creating and/or revising engineering drawings. Secondary tasks typically include preparation of accurate bills of materials, flow diagrams, manufacturing process descriptions, purchase orders, machine tool instructions, instrumentation layouts, and the multiple, and often exasperatingly detailed, administrative jobs that require great accuracy. (Did you ever wonder who wrote the assembly instructions for that not-so-easy-to-assemble airplane model that you just bought, or those instructions that didn't tell you which way the handlebars were supposed to be put on the new bicycle until you were finished with the job? Someone didn't work out the small details correctly.)

Another computer application is the word processor (WP) computer that can record documentation, manuals, and internal memos; allow revisions almost instantaneously; correct spelling; and absorb all the deadeningly repetitive work of the person who had to generate a bill of materials, for example. In these tasks as in most others, all the computers have essentially the same design:

1. A logic unit for data manipulation

2. Interactive display for visual and keyboard interaction with human beings

3. Data storage and retrieval, usually disc or tape

4. A hard-copy output device that could provide something permanent on paper

5. The software or the program that controls and directs the machine to manipulate the data that you, the operator, put into it

Using computers, in effect, could make each operator like the highly trained worker of the process technology mode. The machine (or the process) is programmed to produce the product (the flow of information) and the skilled computer operator maintains the process machinery on a day-to-day basis. Similarly, computer systems designers and programmers themselves resemble the artisan-workers of the unit technology mode. Each product (system or program) is developed to meet the customer's (user's) needs and is unique with respect to all the other systems or programs produced. The organic structure with independent manager-workers in both unit and process technologies is again the optimal with the advent of computerization.

I have described and minimally prescribed how technology affects the technical organization but potentially larger effects are expected in the future, and these have not been well-covered in the literature. They include changes in resources, both inanimate (such as materials) and animate (such as the work force).

CHANGES IN RESOURCES

It seems reasonable to assume that the world's resources are limited and that we will run out of specific raw materials in the foreseeable future. Historically we have developed new sources of energy and materials as the old sources have been used or the costs of extraction risen to the point where it was economically unfeasible to continue. The proportion of product costs represented by materials and energy has always been low compared with other costs, such as labor, overhead, and administration. Therefore, most of the economic pressures for technological innovation have been concentrated on labor and time saving, rather than materials and energy saving. On those rare occasions when a particular material becomes scarce (and higher priced), technological advances were able to support ingenious product redesigns and substitutions of more common materials. However, when there is a general concern about the finite limit of all of our nonrenewable resources and not just a few materials, the importance of conservation and recycling of materials rises to rival the historical importance of cost reduction through labor savings with automatic machinery.

Computer-assisted design of new products can balance all the cost components and perform multiple trade-offs within seconds, doing the job that the technical designer would require months to do, if, indeed, the time were available. Weight of materials, production usages, and energy consumption for production can also become part of the design data banks, prompting a reexamination of every machine part for machinability, processing costs, and relative availability before inclusion into the product design. Directions for production and standards for material and energy use become tighter and more effective, which decreases costs. These additional funds can then be used to support the gradual (with all deliberate speed?) reorganization of the structure to fit this new technology, which requires an organization designed for continual processing.

On the other hand, the ability to search the computer memory quickly for all the ways to produce a machine part and then select the best method at a minimum cost of materials and energy decreases the need for repetitive planning at the work station. This situation is often associated with the technology of unit production. The artisan-worker in that production mode can then manage more equipment, increasing his or her area of responsibility to include the many small but important maintenance problems that are not programmable. Manufacturing engineers can use their time to develop improved processes while the computer issues the routing sheets and orders the materials. The structure has become more optimal by becoming disciplined toward production but organic toward workers.

THE WORK FORCE

Hazards

The application of technology to reducing hazardous and unpleasant working conditions can be related to changes in the social environment (i.e., the law) and the shortages of skilled people who will work under these conditions. Changed social attitudes, management techniques, militant unions, and federal and state laws, such as OSHA regulations, have dramatically affected production methods and the appearance of the factory floor. The prisonlike, inhuman aspects depicted in Charlie Chaplin's movie *Modern Times* have largely disappeared. At one time the factory was considered to be a monster that "alienated" workers. This attitude (which is still seen in some of the writings of social scientists, who probably never have worked in a modern one) was born in the early nineteenth century with the shift from "artisanry to . . . the more time disciplined, task oriented, large scale organization of machines, workers, management, and materials we have come to equate with 'The Factory' of industrial society" (Miller, 1980, p. 100).

The advent of classical management theory in the early part of this century, with its insistence on discipline, directed planning, and hierarchical structures, did little to dispel these attitudes. Human-relations theory tended to treat the differences among people as dysfunctional and attempted to fit the human situation to the work through management attention to workers' informal organizational structures. Militant unions demanded better and safer working conditions and governmental regulations on pollution control and environmental protection have continued the pressures to use technology to reduce hazardous and unpleasant working conditions. But foundries will always have fumes, coal mines will be dirty, paint spraying is dangerous, and the production of nuclear power has its own set of problems.

In many cases, the application of technology to the reduction of these hazardous working conditions results in unexpected changes to the organizational structure. Automating a process to eliminate the need for continuing human intervention can also eliminate the need for heating the work place in winter and cooling it in summer. Dangerous punch presses that now feed themselves and distribute their products to other machine tools no longer require hourly workers to operate them. The work environment (and the structure) has changed to support fewer, but more highly skilled, maintenance engineers who program and maintain this new "factory." The factory has changed, and the nature and quality of the work to run it have changed accordingly.

Scarcities

There is a steadily increasing preference of workers in industrialized countries for employment in the service industries instead of manufacturing . . . in recent years, the percentage of the work force in manufacturing has begun to decline—from 30% of the work force in 1947 to 24.9% in 1968. The U.S. Bureau of Labor Statistics projects that in 1980 it will decline further to 22.4% and a Rand Corporation forecast projects that by the year 2000 only 2% of the labor force will be employed in manufacturing. . . . This migration of the labor force puts a direct social pressure on the manufacturing industry to continue to produce and remain productive as it gradually loses its labor force, which is slowly but surely moving out of manufacturing into the more attractive jobs in the service industries. (Merchant, 1980, p. 75)

Perhaps an appropriate question would be whether technology has to increase because of fewer workers in manufacturing or whether there are fewer workers because technology is displacing much of the need for manual labor. Is there really "a steadily increasing preference of workers" or do workers prefer to work in service industries because there are fewer manufacturing jobs available? Regardless of the direction of cause and effect, there seems to be a direct relationship between the two. There are fewer workers qualified to operate the new technology and a potential scarcity of them in the future.

Present indications are that this scarcity will become more critical as the requirements for higher skills in workers increase (Tesar, 1980). And when changes in technology require changes in the organization, operators will not as easily be replaced with others who have more skills. These others will not be there. Highly skilled people will have to be developed internally, by being trained to accept the changed work patterns and independently organized responsibilities of more organic structures.

SUMMARY

The existing mass production model of the "factory," with manufacturing technology, including semiautomatic machinery, large-batch processes, rigid performance standards for labor, and a management structure that resembles an equilateral triangle, will become less viable unless a management discipline different from those that presently exist in "factory" firms can be applied. Organizations that have structures more closely modeled on either a unit or process technology will be able to grow by taking advantage of the economic, social, and environmental changes *as they occur*. These more organic structures will be able continuously to adapt to change.

With respect to technical organizations, while it is true that all may have similar general goals involving the technology of innovation, it is equally true that the management process that directs, assists, and controls the achievement of these goals will be changed internally, since technology, such as computer-assisted design, will affect management as much as it will affect the "factory." Both will become more organic.

The next chapter deals with a part of the organizational model that is both the buffer between the external environment and the organization and the flexible cement that binds the structure as a whole together: the information systems. I will, as usual, describe before prescribing.

SUGGESTED ANSWERS TO CASE QUESTIONS

1. No, it does not accomplish its intent, which seems to be to provide the best computer service to different groups. The design group in this organization seems to be almost mechanistic in operation, and it should have its needs serviced by Sam Disko's people. That servicing should be on an overall systems design in which the computer is programmed to support the designers, not

allow piping tolerance errors, and, in general, perform the administration work needed. Mary's group should have its own microcomputers and Sam should be able to provide members of the group with training so that they can do their own programs, perform their own simulations, and, in general, use the equipment as just another design-assisting tool.

2. If I were Mike, I would suggest that Sam assign several of his programmers to the design group on a project basis. By reporting to the users, the programmers would be part of the design team, not outsiders who have only a limited functional interest in them. I would suggest that Sam start training sessions for Mary's group, since it would be doing its own programming tasks.

3. I would consider them differently in each group, depending upon the internal needs of that group and its goals. (Can you give specifics that apply to your own organization?) Unauthorized uses of computers are those that are prohibited by law (e.g., fraud or unauthorized use of proprietary data or access to restricted financial data). These kinds of data can be protected through the use of *passwords* (i.e., the computer programs for these areas can be accessed only if the user has the correct authorization codes).

4. George's group seems to be almost process-oriented and therefore would probably have a very tall structure, while Mary's is unit-oriented and would probably be very flat. Do you agree?

Additional discussion question: How would you use computers in the technology of manufacturing processing in unit mass or process production? Include examples.

REFERENCES

Barash, M. M. Computer integrated manufacturing systems. In L. Kops (Ed.), *Toward a factory of the future*. (PED Vol. 1). New York: ASME Winter Annual Meeting, November 16–21, 1980, pp. 37–50.

Bronowski, J. *Science and human values*. New York: Harper & Row, 1965, p. 67.

Burns, T., & Stalker, G. *The management of innovation*. London, England: Tavistock Publications, 1961.

Denison, E. E. U.S. productivity problems stir up attention on the federal scene. *Professional Engineer*, September 1978, 20–21.

Fairfield, Roy P. *Humanizing the workplace*. Buffalo, N.Y.: Prometheus, 1974.

Gibson, James L., Ivancevitch, John M., & Donnelly, James H. *Organizations: behavior, structure, processes*. Dallas, Tex.: Business Publications, 1976.

Hayes, David. Workers and bosses get ahead by getting along. *New York Times*, July 5, 1981, pp. F4–F5.

Hayes, Robert H. Why Japanese factories work. *Harvard Business Review*, July–August 1981, 59(4), 56–66.

Hulin, Charles I., & Blood, Milton R. Job enlargement, individual differences, and worker responses. *Psychological Bulletin*, 1968, 69(1), 41–55.

Kops, L. The factory of the future—technology of management. In L. Kops (Ed.), *Toward a factory of the future* (PED Vol. 1). New York: ASME Winter Annual Meeting, November 16–21, 1980, pp. 71–82.

Lawrence, Paul R., & Lorsch, Jay W. *Organization and environment*. Cambridge, Mass.: Harvard Univ. Dept of Research, 1967.

Merchant, M. E. The factory of the future, technological aspects. In L. Kops (Ed.), *Toward a factory of the future* (PED Vol. 1). New York: ASME Winter Annual Meeting, November 16–21, 1980, pp. 71–82.

Miller, R. R. The transformation of the factory of the future. In L. Kops (Ed.), *Toward a factory of the future* (PED Vol. 1). New York: ASME Winter Annual Meeting, November 16–21, 1980, pp. 99–108.

Schrank, Robert. *Ten thousand working days*. Cambridge, Mass.: MIT Press, 1978.

Skinner, Wickham. The impact of changing technology on the working environment. In Clark Kerr & Jerome M. Rosow (Eds.), *Work in America: the decade ahead*. New York: Van Nostrand, 1979, pp. 204–230.

Smith, Adam. *The wealth of nations*. New York: Modern Library, 1977. (Originally published, 1776.)

Taylor, Frederick Winslow. *Scientific management*. New York: Harper, 1911.

Tesar, D. Mechanical engineering R&D. *Mechanical Engineering*, February 1980, 37.

Walton, Richard E. Work innovation in the United States. *Harvard Business Review*, July–August 1979, 88–98.

Weber, Max. T. Parsons (Ed.), *The theory of social and economic organization*. Glencoe, Ill.: Free Press, 1947.

Widick, B. J. *Auto work and its discontents*. Cambridge, Mass.: MIT Press, 1976.

Wight, Oliver W. Tools for profit. *Datamation*, October 1980, 93–96.

Woodward, Joan. *Industrial organization: theory and practice*. London, England: Oxford Univ. Press, 1965.

Zwerman, William L. *New perspectives on organizational theory*. Westport, Conn.: Greenwood Publishing, 1970.

FURTHER READINGS

Fayol, H. *General and industrial management*. New York: Pitman, 1949.

Follett, Mary Parker. *Dynamic administration: The collected works of Mary Parker Follett* (Henry C. Metcalf & Lionel Urwick, Eds.). New York: Harper, 1942.

Griffin, Faith. *Computer aided design and computer aided manufacturing industry review*. New York: Special report issued by L. F. Rothschild, Unterberg & Towbin, July 25, 1980.

Homans, George C. Discovery and exploration in social science. In Edwin P. Hollander & Raymond G. Hunt (Eds.), *Current perspectives in social psychology* (4th ed.). London, England: Oxford Univ. Press, 1976.

Krupp, Sherman. *Pattern in organizational analysis*. New York: Holt, Rinehart & Winston, 1961.

Leavitt, Harold J. Applied organizational change in industry. In James G. March (Ed.), *Handbook of organizations*. Chicago, Ill.: Rand McNally, 1965, pp. 1144–1170.

March, James G., & Simon, Herbert. *Organizations*. New York: Wiley, 1958.

Perrow, Charles. A framework for the comparative analysis of organizations. *American Sociological Review*, April 1967.

Scott, W. Richard. Organizational structure. In A. Inkeles, J. Coleman, & N. Smelzer (Eds.), *Annual review of sociology* (Vol. 1). Palo Alto, Calif.:1975, pp. 1–20.

Seaman, John. Distributed data processing. *Computer Decisions*, September 1980, 71–82.

Skinner, Wickham. The factory of the future: always in the future? In L. Kops (Ed.), *Toward a factory of the future* (PED Vol. 1). ASME Winter Annual Meeting, November 16–21, 1980, pp. 83–98.

PART THREE

TYING IT ALL TOGETHER

6

INFORMATION SYSTEMS

Case Study
THE CASE OF THE CONFUSED MANAGER

The company was the hydraulic controls division of Monlith Industries. Tony Ogard, the new general manager, was sitting in his office. He was trying to make sense of the reports that he had received from the division controller through the interoffice mail. He couldn't understand what they were all about, and the instructions that accompanied them didn't seem to be consistent. He picked up his phone and asked Sam Greenshades, the controller, to meet with him that day. Later in the afternoon, at the exact time scheduled for the meeting, Sam came in and sat down.

Sam: Before you start on the analysis of these sheets, let me tell you that I had nothing to do with designing them. That's the way they want them up at corporate headquarters, and you know that I report to the corporate controller as well as to you. It's not easy, having two bosses, but I'm trying to satisfy you both. Now that that is out of the way, what can I do for you?

Tony: Very interesting speech. I find it encouraging because you seem to believe that something is wrong too. But let's go on and maybe we can sort this thing out. My reports show that our inventory in some of the smaller valves has increased over 100 percent since last month, and some have dropped by about 10 percent. Why should I care about percentages? An increase in a valve that sells for $20 to $40 indicates a 100 percent improvement, and one that sells for $300 whose price increases by 15 percent equals a change of $45, but that seems to mean less because of the low percentage figures.

Another thing that I don't understand is the detail in these reports. They cover everything that has happened in the division for the last month. Not only that, but when I phoned George Wheatley up in engineering and asked him about some variances in his overheads, he said that he'd get back to me as soon as he got his reports from you. Apparently he gets them after I do.

Sam: Well, that's how the previous general manager wanted things. He said that he wanted to keep on top of everything, and how could he do that if the people who reported to him had more details than he had. He also wanted percentages because that's the way the president of the company reports to the stockholders and he felt that being consistent in his reports to the president was very important.

Tony: I also have this package of instructions from corporate finance that lists in detail how to do my business plan and then the budget to put the plan into operation. Apparently this is a one-time-per-year operation. As part of the instructions, I found this memo from the executive vice president telling all the general managers that they spent too much money last year and that they've got to cut down by at least 20 percent this year. What's going on?

Sam: Well, we do follow an annual budgeting plan because we are publicly owned and we have to report to the stockholders on how their company is doing. The budget is often used in the president's address that goes along with the annual report. You know the sort of thing: "We'll be doing big things in this division or that division next year." Of course, for the last few years we've been hit by large increases in raw materials costs, so we haven't been able to hit those optimistic forecasts. When that happens everybody gets all excited. The general manager before you (who, by the way, is now a group vice president) used to fire people when they didn't meet the budgets he set out for them. He said that he wanted to run a tight ship and he did. He got things going.

Tony: I understand what you're saying, but I noticed that this division is being charged for the corporate expenses that have been allocated to it. Do they ever get cut? Can you find that out for me?

Sam: Look Tony, you're new around here. Questions like that are not answered very well at corporate. If I called up my boss at headquarters and asked about it, I'd never hear the end of it. They say that they provide us with coordinating services and other valuable things and we should be grateful that they're around, especially when we can't meet our budgets and we have to go to them for money.

QUESTIONS

1. What should Tony do about designing an information system that can help?

2. Any suggestions for the frequency of the budgeting process? Should it be done more often than once a year? How? What would the additional administrative costs be?

3. What defense do managers have against budget cuts and/or directed budget expenditures that fit into the decision-making process as it is laid out?

4. What should Sam do? How can he do it? What would the results be?

REVIEW

We have covered the three components of the technical organizational model, developed some ideas about their interactions with each other, how to define them, measure them, and use them in our own personal management theories. We have also covered some ideas about how economic and social environments typically affected the technical organization. Of the three components, *people* was the most independent and important. Moreover, it was also the most difficult to understand, since it provided only a partial set of observable variables or behaviors from which we could develop personal theory. We were not even sure that the partial set we observed was typical of all the behaviors people exhibit.

The second component was *structure*. This was the combination of the repeated interactions among people in the organization. Those deliberately approved by management were known as the formal structure, and those that were primarily social and/or culturally based, the informal organization. The intended design of the formal structure as we know it has a history rooted in the major social changes of the industrial revolution, when there was a change from cottage industries—where workers and their families lived and worked in the same building—into a new arrangement where only the workers congregated for a part of the day in a place known as the factory.

The organizers and the managers of this new arrangement believed that there were universal designs that were the best possible to use for the formal structure. These designs were modeled on the social mores of the then-existing master-servant culture. They were the "one best way" of classical theory. We found, however, that this one best way is a fallacy. More flexible situational or contingency recommendations are better suited to the modern, changing, innovative organizational structures of today. There is no longer only "one best way" but many alternative ways. Structure is based on many of the variables of people, with a few of its own variables added in. These latter semi-independent variables include growth, the passage of time, and changes in the economic, social, and political environment.

Structure is less independent than people because it depends on people for its existence. When people leave the organization, they continue to exist, but the organizational structure does not. Structure is semi-independent. It is both the *repetitive* planned (i.e., the formal organization) and unplanned but acceptable (i.e., the informal organization) behaviors of people in an organizational situation. The informal organization can occasionally exist in a partial form outside the usual work environment, so it, therefore, is semi-independent. Since it is a group activity, structure is probably more predictable than an individual is.

The third component, *technology,* is perhaps the youngest in terms of being studied by management researchers, and the least independent of the three. It does not exist separately outside of the organization. It has changed from a variable that was supposed to be an integral part of the organization's structure or purpose to a variable that is of almost equal stature with the other two components. An overall definition of this component is the methods by which inputs are transformed into outputs. We first evaluated it in Chapter 2 in terms of decision making, since that was the area discussed there.

In Chapter 5, we used other definitions, such as the methods of production, changes in product and innovation, information handling, materials utilization, and the use of humans in the organization. Some of the prescriptions offered for inclusion in your personal theory entailed a possible speedup of the evolution of the technical organization into a two-layered functional group, comprising groups that were production-oriented and those that were problem-oriented. Production-oriented groups were proposed to be very mechanistic, or machinelike—logical, and effective in delivering products and services at the lowest cost with the highest quality that the market wants. However, those groups should have upper management organized in an organic or creative manner in order to handle the unpredictable problems and rapidly changing environments that the mechanistic structure cannot respond to very well. In general, that management includes the problem-oriented parts of the organization and should be organic, more susceptible to change, and should be able to respond effectively and quickly to rapid variations in the environment.

Both mechanistic and organic types of organizations will use computers more. The mechanistic organizations will use them for the repetitive decision making and the control of variances from goals the organization has established, and the organic organizations for the fast response in supporting alternative product or process designs in development and modifications to the administrative tasks necessary to bring those designs to market.

However, in addition to interacting with the organizational model separately, these three components—people, structure, and technology—also interact with each other and the organizational environment in both repetitive and nonrepetitive ways. The information systems that assist in both the interorganizational and the intraorganizational communication process are the tools that are used in those interactions.

WHAT ARE INFORMATION SYSTEMS?

Information systems are the patterns used to communicate both between the organization and its environment and among various internal functions. It includes both those patterns sanctioned by the formal organization and those approved by the informal organization. There seem to be two major reasons for information systems: to communicate decisions and to modify human behavior.

These purposes are intertwined in the firm. They typically include formal internal communications such as budgets, drawings, material requisitions, time sheets, standard operating procedures, internal memos, and minutes of meetings, in addition to all the nonmemorable trivia that often avalanches into the in basket (and then, if you are well-organized, goes expeditiously into the wastebasket). They also include the external documentation through which the company communicates with its environment, such as purchase orders, financial statements, employment applications, and the press releases that forecast ever-increasing sales and profits.

The broad purposes and uses of all these communications as they pass through the many information systems make our usual procedure of research description followed by general prescription very difficult; the variety (and the consequent descriptions) would be almost infinite. Each system is a function of many variables, typically the operations of the particular organization, the people in it, the environment in which it operates, and its past history.

However, one formal group of systems predates all others in every organization and most of the others probably depend on it. The group is the accounting systems. Since these are fairly important, they are the first that we try to describe and prescribe. We want to understand how they *should* work and then develop one that supports our unique methods of decision making. Many of the concepts used in this design are common to most effective accounting systems. Therefore we start, as usual, with definition.

ACCOUNTING SYSTEMS AND TECHNICAL MANAGEMENT

One of the better definitions is:

> The accounting system is the major quantitative information system in almost every organization. It should provide information for three broad purposes:

1. Internal reporting to managers, for use in planning and controlling routine operations,
2. Internal reporting to managers for use in making non routine decisions and in formulating major plans and policies,

   . . . The data raise and help answer three basic questions:

   a. Scorecard questions: Am I doing well or badly?
   b. Attention directing questions: What problems shall I look into?
   c. Problem-solving questions: Of the several ways of doing this job,   which is the best?
3. External reporting to stockholders, governments and other outside parties. (Horngren, 1972, pp. 3,8)

These definitions cover most of the intended uses of accounting systems. Unfortunately, too often the emphasis is on the third purpose, reporting to stockholders, governments, etc., and it operates to the detriment of the first two purposes. I suggest that this occurs because the developers of the information systems usually are *not* the same people as the users of these systems. In too many cases, the development of the accounting system is, by default, left to the people in the organization who have historical, rather than decision-assisting, purposes in mind. The "scorecard" questions are the most important to them, and systems designed to answer those types of questions are typically not very useful to operating managers when they are designed by people other than the managers themselves.

The wrong scorecard questions include "What were the mistakes that I made last month?" rather than the right decision-assisting questions, such as "Why is there a variance between my plan and the actual results and does this variance indicate that a change in my theory is necessary?" These latter types of questions, in *attention-directing* and *problem-solving* frameworks, are those that guide the future. In general they are intended to help in answering the question, "What can I do better *today* to influence results tomorrow?"

The future is where we are all going and our expectations about it, as much as our behaviors in the past, can change what we will do when we get there. Of course, our expectations are a combination of our past conditioning and our present information. Accounting information systems that provide guidelines to assist in making that dimly seen and changing path to the future easier to trod are the ones that should be developed. While tomorrow's potentials are not unlimited (after all, we are always limited by physical, mental, and material conditions), more of these potentials are available than we can ever conceive and use. Therefore, we need all the help we can get to choose the best one when we are making decisions—especially under conditions of uncertainty. As William James said,

   . . . the mind is at every stage a theatre of simultaneous possibilities. . . . The mind, in short, works on the data it receives very much as the sculptor works on his block of stone. In a sense the statue stood there from eternity. But there were a thousand different ones beside it, and the sculptor alone is to thank for having extricated this one from the rest. (James, 1890/1950, p. 288)

Getting the data that are helpful to us in making decisions *today* that affect our behaviors *tomorrow* is the main determinant of effective information systems. Accounting systems are one important subset of all information systems, so they should also be designed with an eye for the future. There are accounting systems that have been designed to assist in looking forward, but they are always developed from inputs supplied by the *users*. They produce outputs to help in extricating the particular statue that the manager thinks is best from all the statues that are possible in his or her organizational stones. Accounting systems that serve this purpose will invariably include a most valuable device. It is called a budget.

BUDGETING AND RESPONSIBILITY ACCOUNTING

A budget is not a conceptually complex communication system. It is described easily; the difficulty comes in the prescription and implementation. It has two general parts: a forecast and a measurement of the actual situation. The forecast is intended to be initially a qualitative, and then a quantitative, plan of action that reflects some previous qualitative inputs and decisions. It coordinates and assists in decision implementation. I like the definition of a forecast as the selection of a particular alternative *today* from those that are perceived. The alternative selected is intended to achieve some future goal. It may be obvious to point out that no one can predict the future. Therefore, a forecast is not the same thing as looking into a crystal ball. It is only a selection of one of a number of alternatives perceived at the present time by the person who is doing the selecting. (I am dealing only with single estimates here, not probabilistic ones that are more complex but won't add much light to our present discussions.)

When the forecaster and the user of the forecast are different people, there is going to be less than optimal forecasting. This happens even under the best conditions; both people rarely perceive the same range of alternatives. That is almost impossible. There are also differences in interpreting which alternative out of this unlikely condition of agreed-upon range of alternatives is the best. If you have ever been involved in a budgeting process, I'm sure you'll agree that this happens a lot.

Another difficulty that occurs when the user is not involved in the forecasting process is the probable development of less than optimal plans because there is less motivation to apply the information to affect the future of the organization. Few of us are objective enough to prove that we, as the user, were wrong when we insisted that the forecaster didn't select the best alternative. In other words: "The preparers (forecasters) focused on their technical responsibility and the users focused on the performance of functional tasks. This disparity has led to the neglect of the important task of fitting the forecast to the function required of it" (Makridakis & Wheelwright, 1978, p. 647).

When the forecaster or selecter of the alternatives and the user is the same person, the inferred motivation would probably be to select the best behavior (or sculpt the best statue from those perceived in the block of stone) and thereby make the optimium forecast to fit the function or job required as interpreted by that user. But this occurs only when that user is reasonably sure that his or her plans will be accepted by the approving levels of management. (One learns very quickly if there is a repeated pattern of budget cuts and, of course, adjusts the original budget accordingly beforehand.) Another way to support the selection of optimum budgeting behavior is to provide that all accounting and other information systems be developed in accordance with *responsibility* accounting concepts.

With responsibility accounting, "revenues and costs are recorded and automatically traced to one individual who shoulders primary decision responsibility for the item" (Horngren, 1972, p. 158). It fits into the ideas on decision making in prior chapters, since it explicitly states that if you can't control it or change it, it shouldn't belong to you, so why bother with it? For example, some budgeting systems assign an overhead or burden rate to a direct-cost activity. If the managers of those activities, such as project A or engineering design section B, have no control over the costs in that overhead, the forecasting and the measuring of actual overhead costs is a waste of time. The manager-users of A and B can't change them anyhow.

The overheads, however, do belong to someone. Every cost can be changed somewhere in the organization, and that is where it belongs. Using our example of overhead, if you take one of the least variable elements such as real estate taxes on the plant or offices, they can be assigned to the chief executive officer. She or he is the person who can decide to change those taxes by suing in court for tax relief. The decision maker, in this case, is the chief executive officer and it follows that those taxes should be forecasted and become part of the chief executive officer's budget in this example. That follows the concepts of responsibility accounting.

We now move on past the part of the ubiquitous accounting system which is the budget and the accounting systems themselves into a model of the larger information systems, which serve two general purposes: tying the organization together and supporting its interactions with the environment. The general model of the information system prescribed here includes some of these accounting-oriented parts and is presented as a sort of general hypothesis of forecasting, measuring, and managing that can be modified to fit your situation. It follows the ideas of responsibility for one's plans and is intended to be a cooperative rather than a competitive system, since "a competitive system can almost always be beaten by cooperation" (Haber-stroh, 1965, p. 1184). It is a system that supports the future orientation of managers, providing the data today to help make the decisions intended to solve uncertainty-based problems. It provides managers with the ability to respond to environmental and internal change by allowing repetitive prob-lems to be classified and resolved through a prior decision matrix, as we noted in Chapter 1. Managers can then do what they should be doing: solving nonrepetitive problems by absorbing uncertainty.

BACKGROUND: DESIGNING THE INFORMATION SYSTEM

The scene is probably all too familiar to the experienced manager. A com-munication problem arises, and in attempting to solve it, the manager gets the feeling that it has all happened before, that the same problem has been solved dozens of times in the past. It may have appeared in different dis-guises, but beneath the changing surface there lurks the same tired beast that should have been put to rest long ago. Too many of our management tasks fit this description. They have to be done over and over again simply because their real causes are not recognized, and therefore there is no effective means of preventing their recurrence. The only reliable way to avoid having to solve the same problem repeatedly is to establish a system that permits the manager to compare the present problem with past solu-tions, determine if they fit, and allow only really new problems to occupy his or her time.

In a generic sense, the same type of situation seems to occur in companies in which an otherwise enviable record of growth and prosperity is all too often marred by periodic convulsions resulting from attempts to solve sup-posedly new problems involving communications or some other major area. Those problems are nothing more than a variation of a recurring theme. What is generally lacking in these instances is a management system capable of defining and forecasting these repeated problems and then documenting the solutions in an easily accessible matrix.

A company grows and prospers in direct relation to its ability to accept change. Changing markets, sources of supply, consumer tastes, and organ-izational needs all require flexibility and fast response. The marginal com-pany, or even the one that grows only as fast as its markets, is one that responds to change only when the mismatch between organizational actions and those needed by the changes in the environment is very obvious. The response is usually a major "crash" program of some kind. As examples, lost markets or dropping profits can trigger responses in the form of "establish-ing a field office to improve communications about quality problems" or restructuring the sales and marketing reporting systems. These are often nothing more than creations of the moment or, at best, merely stopgaps next crisis occurs. These triggered reactions treat the symptoms without curing the disease, and the underlying disease in each case is slow response to change.

A more effective and prompt response cannot occur with information systems that are historically oriented because many of them assume that the future is an extension of the past. This is not true in all cases, and especially not in technical organizations. When information systems can provide timely, accurate, and relevant data to support management response to constant change, providing guidance through indications of variances that help in separating novel and repetitive problems, the first step in solving problems under conditions of uncertainty is successfully taken. That first step always involves definition. If the problem can't be defined as new or old, important or trivial, it is treated as new. This wastes the most valuable assets the technical organization has: its managers' time and efforts. Definition of new and important change, therefore, is a major criterion in designing information systems. This type of change for most organizations usually begins in the organizations' environment.

ENVIRONMENT

Companies don't exist in vacuums. Events that can have the greatest effect on company growth often occur outside of the particular company's organizational model. The promptness with which the managers within the company detect and respond to events or changes in its environment can determine organizational growth patterns. Conversely, if the company's internal environment is isolated from or nonresponsive to these outside influences for too long, the chances of growth are minimal and eventually its chances of survival are affected.

As an example, when there are great differences between perceptions of managers and customers about the value of the company's products measured against those of the competition, those differences are going to affect the company. If the effect is decreasing sales, it may be due to a lack of organizational response. This could be quite destructive. But a change into a slightly different scenario can occur when customers change their evaluation of the company's products as greater than those of the competition. This scenario might be considered to be quite satisfactory, but it too requires an organizational response. If there is no equivalent increase in production when present levels of inventories are gone, customers might be willing to buy from others and potential income could be lost. Although lack of increased production is not quite as destructive as turning out less desirable products, the company still suffers because of this lost potential. Response to change depends on an understanding of and taking of appropriate management action as a result of information exchange and understanding between the organization and its environment and within the appropriate parts of the organization. In other words, managers should be responding almost continually to both external and internal sources of information that are new and important to preclude major organizational dislocations.

The foundation of control is based on an act of reciprocal communication. Not only must the managers communicate easily with the external environment, they must also be able to communicate easily internally within the company. This includes communications up and down throughout the company hierarchy and horizontally across it. It requires cooperative systems connecting cooperating departments. Responding to an increased or decreased sales demand for a product often cannot be done by only one department of the company. Engineering, sales, manufacturing, quality, procurement, and so forth must be able to coordinate their responses. Therefore, perhaps the most important single ingredient in the company's ability to plan and direct its growth (or response, as the case may be) is adequate systems that provide effective, relevant, and timely information to minimize uncertainty and support decision making. The results of the decision making must also be communicated easily.

COMMUNICATIONS SYSTEM

The design of the information system seems to be as important as defining effective organizational components. It cannot be a haphazard response. Even though internal departments (and the components) are "different" from each other, they must be able to feed information to each other and receive it in such a way that changes can be accepted quickly. This usually means some type of information or communication system in which most interactions can be forecast and can be measured when they occur, and the variances can be fed back to each responsible manager for appropriate actions. There are design parameters for these systems and, as we have noted before, responsibility accounting is one of the most important ones. It provides that the user, if not the same person as the forecaster, at least has the right to accept or reject inputs that he or she has not approved. There are other important design parameters.

DETAILED SYSTEM DESIGN PARAMETERS

The following list cannot be complete because of the varying types of organizations that exist, the rate of environmental change, and the specific goals the company or department has set. It does, however, provide some important considerations for the design of effective information systems that you can modify as required.

The system must be a closed loop. Acceptance of change is a continuous process. Therefore, an information system must be operational at all times and oriented to respond to change. There must be provisions for management to alter the course of the company at any point in the loop without disturbing the continuity of the systems. In effect, when there is a change in course, at that time the manager must be able to compare the new direction with the old one in order to determine if she or he is on course. This requires some type of feedback of the difference, or error between the new course and the old one. The greater the difference, the greater the amount of management action is required. *The system must permit maximum use of all appropriate management systems, including those with different philosophies coming from both inside and outside.* Certain external institutional restrictions are placed on some company operations—e.g., antitrust, interstate commerce regulations, and community relations—and the information system should permit the company to operate well within these restrictions. It should also provide for free operation of different management methods and philosophies in different parts of the company. We know from our analysis of the way change is perceived in different parts of the company that some managers have perceptions that are consistent within their own groups but very different from those in other groups. As a minor example, the application of predetermined output standards as used in mechanistically organized production machine shops is probably not appropriate and would not be tolerated in the advanced engineering design department. There should be standards—just different ones for different groups. Similarly while there might be some commonalities due to legal, social, etc. requirements, all systems do not have to be exactly alike in every detail.

The system must have a predetermined set of measures of efficiency. The most familiar measures are revenues and costs. While the arithmetical difference betwen the two is usually called profit and is generally supposed to be maximized, sometimes profits are put aside in favor of making a long-term investment in new products, improving production machinery, satisfying changed political environmental demands for pollution controls, or any other organizational goal of the moment. The system should be able to include the measures of efficiency in achieving any of these or other goals.

The system must be applicable to all forecasting time periods. It should permit the use of data generated by the system for making short-term, medium-term, and long-term plans. This assumes that the detail with which the system deals shall change with the time period involved. I have seen budgeting systems that require weekly forecasts of activities when those activities are not scheduled to take place for several months into the future. That is a waste of time. In my opinion, the detail in forecasting generally should be inversely related to the time period being considered. The longer the time in the future, the fewer the details that are forecast. As an example, weekly forecasts may be appropriate for the first six months, monthly from seven months to a year, quarterly from a year to two or three, and semiannually for several years beyond that.

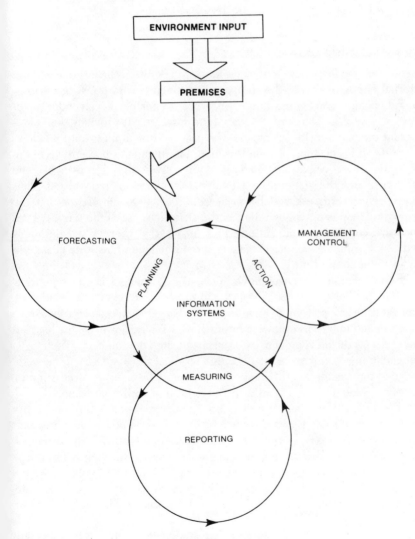

Figure 6-1   Management communication system.

Within these general parameters, an information system can be constructed that provides the kind of information managers need to make effective decisions. In other words, although the specific system may vary from company to company, the parameters that limit the theoretical design should remain fairly constant. The proposed general design shown in Figure 6-1 is built within the parameters mentioned above and can be adapted to many kinds of organizational structures. It is the adapting process of this general design to your specific situation that gets to be complex. This general system consists of one main closed loop that represents the total information system intersected by three subsidiary closed loops.

THE FORECASTING LOOP

The forecasting loop is the primary connection between the internal and the external environments of the company. Its function is to translate environmental changes and requirements—e.g., consumer needs or vendor qualifications—into the standards that apply to forecasts. It provides the guidelines for the planning for the future. The process from input to output follows a definite series of steps. Within this loop, management's perceptions of the environment are converted to company objectives, the objectives become standards, and the standards become the basis for operationalized plans against which progress will be measured. I want to emphasize that the forecasting loop is not a beginning in the strictest sense of the term. All the loops are continuous and closed. This means that there is a continuous feedback through them and managers should be able to institute change in any of them at any time.

The translation of environmental change into changes in the repetitive decisions behind the company policy and the possible reverse translation of changed company policy into environmental changes should be equally likely. It is intended to be a free-flowing interaction. Usually, however, the heavier flow of data is the movement of external change into the company. Therefore, being able to accommodate to this influx is one of the first tasks of the well-designed forecasting loop. It helps the managers involved in forecasting to perceive which data being received are consistent with prior standards and which are not. That is the purpose behind company statements of the premises on which objectives and plans will be formulated. These company premises are probably some of the better data-screening tools that the company can provide. They provide answers to typical kinds of very important questions, such as "What business are we in?" and "What business should we be in?" Answers to those premises are inputs to this loop and help to determine where the company will be going (i.e., which statue will be sculpted out of the block of marble).

For example, a railroad company that says it's in the freight business will have a different set of premises and competitors than one that says it's in the tourism business, even though both businesses require the tracks and locomotives of the railroad. The major freight business competitors would be trucking firms. Therefore, the railroad company setting up data in its forecasting loop would be very concerned with minimizing operating costs. One outcome could be the abandoning of trackage that does not immediately produce sufficient freight revenue.

On the other hand, tourism has different organizational premises for a railroad company. The major tourism business competitors would probably be all the buses and airlines. A possible outcome could be an expansion and upgrading of trackage that hadn't been used very much, provided that trackage went into very scenic areas of the country. Buses have to stay on major highways, and highways don't go through some of the most beautiful country scenery, and there isn't much to see from an airplane at 30,000 feet. In this obvious example, the selection of the company premises defined the nature of the organization and its direction. They were different, even though in both cases the physical assets were much the same.

The definitions of the company's premises, which help managers to perceive relevant data, are usually followed by definitions of company objectives. Objectives are more specific delineations of the goals that the company must reach to keep pace with change. They are the goals resulting from testing external and internal data against company premises, e.g., premises mean "freight" and objectives mean "abandoning trackage." They could include changes to the product to conform to changing consumer tastes, profit goals, etc. The next step beyond company objectives are the specific standards which must be determined next and against which interim progress toward goal achievement will be measured. They may be direct departmental subgoals of the overall company goals or they may be departmental subgoals that have only a general relationship to overall company goals. This last statement may hold with a high internal differentiation among departments. However, the standards are generally the sequence of steps that must be taken to achieve the objectives within the restrictions or premises imposed by the environment and/or the company management. They apply primarily to the internal departments of the company. Usually they are formulated first and are followed by the detailed plans for the operations of various departments which support standards.

These steps constitute the complete cycle of the forecasting loop: environmental input through limiting premises, selection of objectives or goals, detailed departmental planning of subgoals, and setting of specific standards. With this cycle completed, managers can segregate nonrepetitive from repetitive problems, anticipate the repetitive ones, and eliminate the reappearance of many of those repeaters through development of a decision matrix—sometimes known as company policy. New problems that have not been caught in this forecasting loop, that have no standards or plans for their solution, should be the only ones to appear in the future since they are not covered by standard solutions. The final product of the forecasting loop, the plan, is a forecast of the company's position at some time in the future and a yardstick of expected performance against which the company's actual performance can be measured. The next step is to put the plan into effect.

THE REPORTING LOOP

Putting the plan into effect means moving the output from the forecasting loop into the total information system. That total system then carries the forecast into the reporting loop as an input. In other words, the total system connects the forecasting loop (projects future performance) and the reporting loop (tabulates data on actual performance and compares that with the forecast). It then carries the comparisions of the forecast and the actual data, usually called the variances, as an output from the reporting loop into the next loop, called *management control,* as an input.

The comparisons or variances produced by comparisons of the first two loops give the manager the starting point for implementing any required change. The management control loop is in the head of the manager and consists of the mental processes used to guide decision making. The second or reporting loop usually includes the standard accounting system that interacts with stockholders, governments, and others in the outside world. As part of the "scorecard" system that was described earlier, it may be the end result for historical accounting reporting but it is only an interim step for any effective management information system. It provides the *actual* in the comparison of actual versus plan. As noted before, that comparison is also known as a variance.

Variances are important when they are used as a problem classification tool. Of all the assets that every manager has, the most limited is time. There always seem to be more problems to solve than time to solve them. Variances can help the manager select which problems to work on first. A general rule is to select the largest variances as the first (but not the only) indicator of the size of the problems. As an example, it is probable that an unfavorable variance of $50,000 in standard manufacturing costs is going to be tackled before a favorable variance of $5000 in materials received.

This is true even though both variances may prove to be equally unsatisfactory if a complete investigation of both is made. The excessively high costs of manufacturing may be due to unavoidable labor spent on a power failure in a loading device. The favorable variance may be due to a vendor price reduction that has not been included in the system yet. Favorable dollar variances don't necessarily indicate favorable management consequence. That example of a favorable variance of $5000 in materials received may indicate a major problem in the documentation for the receiving department. Maybe the department really got it but doesn't know where it is. On the other hand, both variances might be predicting some terrible problems in the organization. Usually, the size of the variance is only the first criterion for problem selection. The direction, either favorable or unfavorable, is the next. However, these criteria are not the only way to categorize problems.

As another example, a small, continuing, favorable variance in the maintenance budget may be a major problem if it is an indication that the machines and the plant are not being repaired adequately. Conversely, large variances may not be as much an indication of potential problems occurring as they are of poor planning or setting of standards. The plan, the standards for the maintenance department, or the budget itself might be wrong and need correcting. A variance is only a *difference* between some predetermined standard and an actual cost; it is possible that both the standard or the cost may be incorrect.

There could also be as many problems lurking with no variances as there are with major variances. When compensating errors which are both plus and minus occur, they could cancel out any variance, keeping the manager in the dark about some problem building up until a disaster strikes. Going back to the original point about the limited amount of problem-solving time, I suggest that variances are useful first tools but not the only ones to use in establishing decision-making priorities. They tell where to look first.

Data that are generated in the reporting loop fall into two general categories: achievement and cost. Data on achievements are often self-explanatory. "We shipped 389 widgets yesterday against a planned shipment of 390. "The last endurance test was completed on November 21, three days before originally scheduled." Conversely, data on cost usually are not that clear and must be analyzed in detail, because they represent only the shadow of accomplishments or lack thereof. These accomplishments are the central concern of the manager. The costs themselves, with no explanation, are not very useful. The speed with which these kinds of data are presented and their accuracy are our next concerns. We assume relevance is built in when the forecaster and the user are the same person.

Accuracy versus Speed

Although information is the principal product of the reporting loop, the methods (i.e., technology) by which it is reported can be equally important. It takes time to gather data and to report them, even when using the latest computerization techniques. And even when they are reported, they occasionally include errors or even become mislaid. As examples, the labor reporting sheets showing the time spent by various technical personnel may get finished on time but then get held up for days in the company mail.

Typically, invoices for shipments are lying on somebody's desk longer than they should be, and as for the results of the latest market survey, that always takes "forever" (and "we didn't like the results anyhow").

You can make a choice between data accuracy and promptness. Following up all those labor sheets, invoices, surveys, or what have you will give you accurate data for management decision making, but in many cases the length of time after the fact that it takes for you to get it tends to make it stale. There are those who say that if you wait long enough, you'll find out everything about something that happened too long ago to be of any use to you.

When there is a choice to be made between accuracy and speed, I always choose speed. It is useless to know January's loss down to the last penny if that information is received the following May. It's too late to do anything about it then. I feel that inaccurate reports can be valuable if they are provided on a timely basis. However, they are only useful when these few rules are followed. You are informed beforehand that the report contains errors because it contains estimates and there is only *one* consistent source for these prompt reports. With only one source, repeated evaluations of the report's credibility and accuracy can be made over time. When the same person(s) makes the estimate week after week, both the report writer and the receiver can account for repeated errors that begin to crop up. Most inaccuracies are repetitive. After making the same labor cost estimate for three or four months, your cost accountant can include the errors that she or he knows will occur every week and correct for them. The reports then become very useful and timely tools. *Timely* means different things to different people, however.

Time Delay

The manager with primary responsibility for an activity should receive the reports about that activity *first*, and for a limited period of time should be the only one who has the report. This manager is in the best position to correct any shortcomings that the reports show, since they are supposedly within his or her direct area of responsibility. There is no need to give them to the manager's supervisor at the same time. In fact, doing so could be detrimental. The manager should have enough time to correct a situation.

Assuming that the problem falls into the manager's decision area (i.e., it's nonrepetitive), is within the manager's authority to solve, and seems to be important, why not allow him or her to solve it by providing enough decision time? If it's beyond the manager's decision area, the manager then has the time to push it up the line to a supervisor or down the line to a subordinate for action. On the other hand, if the manager does not or cannot solve the problem within a predetermined time period, condensed reports including this problem should be passed up to the next higher level of management.

A similar situation should occur at this next higher level. Let's assume that there are managers who are either unable or unwilling to correct a situation (rather unlikely, but we are designing the closed loop system that includes feedback, and no action is just as much feedback as positive action). The unsolved problem will eventually appear in a condensed form in the report of the chief executive officer some time *well after* the problem occurred. That chief executive officer is now aware of a string of management failures, beginning with the manager below him in the organization chart and continuing down to the manager directly responsible for the original problem. According to this concept, the problem is less important than the ability of managers within the organization either to correct a problem or to place it in the hands of someone who can. It's a person-oriented concept, designed in accordance with the rules of responsibility accounting. The amount of information varies almost inversely with the organizational level to which it is reported, and we see why and how that happens next.

Level of Information

The inverse relationship between the detail contained in a report and the level of management for whom the report is intended means that the higher the organizational structural level, the more condensed and consequently less detailed the report should be. When this rule is not followed, managers can get voluminous data that help very few except those at the bottom, who have to know the details in order to manage. You, as the technical department manager, are not really concerned with the status of a particular project. The project managers are responsible for that. Your concern is with the *total* of all the projects in your department. The project managers, in turn, should only be concerned with the overall performance of their projects, and not with the specific task assignments within them, assuming that they have someone who has that detailed responsibility.

This doesn't mean that all the data don't exist, just that they are "owned" by different individuals in different levels of detail. The same applies to groups, committees, and any ad-hoc teams that are established. Groups or teams do not own their assigned tasks and do not require detailed data on them. Only a specific person can be responsible. Either that person manages it or it shows up on the next higher level as a failure. But either individuals take action or no one does! The next design question could be how often these data should be transmitted.

Frequency of Reporting

The frequency with which reports of "actuals" are received is different for functional and for project operations. Functional organizations are supposed to be immortal; they go on forever. Therefore, the frequency of the reporting should be a constant for them. That frequency should be established by the shortest time span in which corrective action can occur; in other words, when a relatively constant, acceptable level of uncertainty applies.

For example, the engineering manager might be concerned with *weekly* reports on the status of all designs in the drafting department; she or he may want a *monthly* report on all purchase requisitions in the administrative department and a *quarterly* report on all performance evaluations for her or his people from the personnel department. In each case, the frequency of the reports is set to a minimum time span that provides sufficient time for that engineering manager to take corrective action. With other kinds of management tasks, paychecks are produced which are based on weekly records. Purchase orders are summarized and matched against invoices monthly. And people are reviewed semiannually for changes in compensation. Those types of reporting frequencies would probably be fairly constant as long as that particular organization functions.

The frequency of reports for projects is quite another matter. Projects are not "immortal"; they begin at some time and end at some other time, and the amount of uncertainty generally is inversely related to the life span of that project in a nonlinear way. Therefore, since uncertainty decreases with time and we want the reports to cover time periods of relatively equal uncertainty, it follows that the frequency of reports must *diminish* as time goes on. Initially, those reports, typically can be produced on a weekly basis, after a reasonable number of reports are issued, they can than be issued on a biweekly, then monthly, quarterly, and finally semiannual basis. The amount of uncertainty covered by each of these reports will be fairly constant for the project; since uncertainty is dropping, the time period between reports can increase. We are not concerned with cost, but with uncertainty. If we could measure and plot the curve of uncertainty versus time (We can't, of course, because uncertainty by definition is not measurable in the usual sense of the word.), it would probably be a decreasing curve, as shown in Figure 6-2.

For example, a project manager might request daily reports for the first few weeks of the project, weekly reports for the next few months, and, finally, a monthly summary report. But if these reports are expected to point out variances and are to guide the project manager in corrective action, why decrease the frequency of reporting? Is that not an implicit assumption that variances occurring at the middle or the end of a project are less important than those at the beginning? The answer should be yes. Variances in the middle and the end are not as important as (even though they may be the same size or even bigger than) those in the beginning.

In the beginning, there is greater uncertainty. Management input is much more effective during this phase than later on. When you are controlling

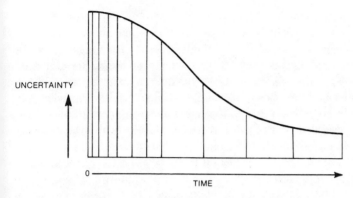

Figure 6-2   Project report frequency.

projects, if you have to make a choice, be more concerned with beginnings than with endings. If the projects don't get started and controlled right when they are born, they will never reach the goals set out for them as they mature, and may never mature at all. All projects that are started correctly do not end correctly but when they are started right, at least they have a reasonable chance of ending the way they are supposed to. The timing of variances is more important than their size when you're dealing with problems, and early reporting is therefore one of the most important design criteria of a project reporting system.

This change in the frequency of reporting can be done fairly easily with computer reporting. Just set the frequency of reporting at the smallest interval (say, one day or semiweekly at the beginning of the project) and then at some predetermined point such as the first design review have the computer programmed to suppress and summarize those daily or semiweekly reports and produce these summarized reports on a weekly basis. Do the same after the next design review, but make the weekly reports monthly reports. It's easy for machines to do; it's more difficult when people have to do it. People have trouble following the rigidly disciplined (and completely uncreative) path required for condensing data that the computers find easy to do. The reports are now available. Therefore, the next step is to use them correctly.

Correct Usage

Reports should be used for only one purpose, as guides for actions *today* that are intended to influence the future. The past cannot be changed, but the variances in reports can be used to correct those present activities that are affected and to preclude the same kinds of errors in the future. It's analogous to a "sunk cost" concept in economics. You can't unspend yesterday's money. When you hear someone say, "Why, we've spent so much time and money on this problem, we really should spend another two months or $2 million because of what we've got invested," you're listening to someone with little understanding of economics. The choice of spending the additional two months or $2 million *today* or of not spending them is the one that has to be made, and not the choice of whether or not to follow on what was spent before. Today is the setting for making choices, not yesterday. Yesterday may be useful to learn from but it doesn't help in absorbing uncertainty today. Management has to be existentially oriented.

The reporting loop encourages action to be taken at the point at which it is most effective. It also assists in fixing responsibility for such action firmly. The first-line manager of a function or an activity is the first person to receive a report on that function's performance. When a problem is brought to light in this report, it is his or her responsibility to solve it. If the reporting loop were not used properly, however, the first person to be informed of a problem would probably be an upper-level manager. Before that upper-level manager could correct it, he or she would have to seek out the source, losing valuable, irreplaceable time in the process.

THE MANAGEMENT CONTROL LOOP

The forecasting loop provided the plan against which the performance data generated in the reporting loop were compared. This comparison process is then carried by the total information system into the next loop—the management control loop. That is where decisions are made and actions taken. If actual performance does not match planned performance, there is a reported variance. In a simplistic sense, managers can only choose one of two available alternatives: *change the action or change the plan*. (I have not dealt with changing nothing, because that action results from a decision that there really is no variance and nothing has to be done.) In the first example, of changing action, the manager has decided that the plan is more correct than the actual performance and something has to be done to bring the performance into line with the plan. In the second example of changing plan, the actual performance is more correct and the plan has to be changed. The intent in either case is to match actual and planned performance in some *future* time period, based upon what we know *today*. Which alternative is chosen, of course, depends upon your diagnosis of the causes of the shortcomings.

Most of us have been involved with management control loops when it is always the actions that are supposed to be changed. ("We've got to work harder to get rid of that production slippage showing up on the charts.") But the other alternative is really just as viable. There are many reasons for changing the plan, some of which are:

1. The plan might be unachievable. ("That production slippage should never have been on the charts because there's no way that we can get those mainframe castings into the shop on time to meet that schedule. Who set that thing up anyway?")

2. It might be arbitrary. ("We're cutting the expense budget by about 25 percent this year, but I'm sure that you'll make it because you're such a great manager.")

3. It could have been inadequate because of unforeseen conditions. ("I know that we're in the petrochemicals business, but how were we to predict the tremendous increase in the price of crude oil during the early 1970s?")

4. Probably the most important reason of all is based on our definition of a forecast. That definition referred to is only a selection *today* of an alternative from all the alternatives presently perceived to solve some future problems. That selection may change as time goes on.

Since we cannot actually predict the future, when experience later indicates that the objectives of the plan cannot be reached, the plan itself must be changed. Changing numbers on paper is easy and plans that cannot be used must be changed into those that can be.

THE APPLICATION OF THE DESIGN

To show how an information system works and to clarify the terms used in the foregoing discussion, it might be helpful to examine the system in operation. This limited case study is intended to illustrate such an information system and show how it operates.

A major consumer electronics firm, closely directed by its founder for over thirty years, had grown from a small radio shop into a multimillion-dollar operation. Its products were well-known and reliable, and the company had always produced a respectable profit. In spite of its reputation in the marketplace, its profits had declined steadily in recent years. The founder died suddenly, and his replacement as president was recruited from a competitor. The new president brought along a new vice president for engineering as part of a general management reorganization.

The first task of the new team was to determine the reasons for the decline in profits. They found that the company had failed to keep pace with changing consumer tastes. The founder had always been able to do market research informally by communicating directly with his major distributors, but the company had grown too large and unwieldy for one man to serve as an adequate pipeline for marketing input as customers' needs changed.

An additional factor was that these favorite distributors no longer seemed to be able to supply accurate data. When the company's competitors changed their distribution systems successfully by using their own warehouse as retail stores to sell products at a discount, the company's independent distributors did not follow suit but moved away from the eventual customer by expanding their retailer-serving warehouses and moving out of the retailing end themselves. They became middlemen rather than the partial direct retailers that they had been before, since some of those prior sales were direct to consumers. They lost all direct contact with the eventual customer. The marketing data that they fed back were not as accurate as they had been.

This deterioration of the quantity and quality of marketing feedback data was not the only problem, but the new president knew that problems are often symptoms of underlying company diseases. He decided to treat the disease, not the symptom. He diagnosed it as the lack of an adequate overall communication system including marketing and internal operation. He therefore started the cure by requesting the vice president for marketing to evaluate the company's present distribution systems, set up market information-gathering processes, and, in general, determine the environmental needs upon which design engineering would develop the next year's line of products, since the product designs would be based on those needs.

In more typical situations, the establishment of a management communication system (it's the same here as information system) requires lengthy management training if the system is to be effective. We know that it is difficult to change organizational culture, and the information system is a past of that culture. The company culture is a strong in-place force that often resists change. Modification then requires either drastic action or long-term training. Drastic action could mean wholesale replacement of personnel or other draconian measures to make new systems work. Training is a more complex change tool. It usually includes setting different behavioral standards and rewarding personnel who exhibit those new behaviors.

In this case, however, the new president felt that immediate action was needed for a product redesign and cost reduction, with management training of all the management staff to increase their sensitivity to external change. In the long run, most of the burden would fall on the new vice president for engineering, who was familiar with the forecast-reporting-control system concept and could implement it in production. A request to the vice president for marketing initiated these changes as far as reviving the flow of current, accurate market information.

The Forecasting Loop

A market research program directed at past and potential users of the company's products was completed to determine what changes these users would like to see in the company's products. The company also set up wholly owned retail stores in strategic locations as sources of these marketing data. Since we are concerned here primarily with information systems within the organization, we'll deal only with the findings that applied to product design. The findings were that customers wanted:

1. Simplified controls with more push-button operation

2. Decreased external radiation in anticipation of proposed restrictive legislation

3. Longer product life with fewer defects over a five-year expected use, even at the expense of slightly higher purchase price

With this information, the president knew that new-product development was one change that had to be emphasized. However, the funds available were limited, since the past president had always insisted on high dividends being paid out (he was a major stockholder). Therefore, the development program had very definite restrictions on spending. He consequently placed these restrictions into the planning process by stipulating that although there was going to be funding for new products, those products had to generate enough profit to pay for their development costs in two years. To ensure successful use of the applicable market research data, he also required that no new markets for the company's products would be entered until the newly developed product line was first made available to presently existing markets. One suggestion that he made to gain the simplified control that customers wanted was that the new designs would incorporate the latest developments in component miniaturization.

Within these limitations, the president and the vice president for engineering established the company's objectives under the new program. First, they directed that two models be developed: a red and a blue. The red model would have miniaturized components, push-button controls, a shielded chassis for radiation protection, and modular construction with heat sinks for longer life. The blue model would essentially be an improved version of the red model with a slightly higher selling price; it used wood instead of plastic cabinetry and was intended for a different market. Estimated sales were 30,000 red units per year for three years and 20,000 blue units per year for two years. Management set a profit objective of 30 percent of sales after charging off development costs. Finally, it directed that all new models should be ready for pilot production runs by a fixed date in the next year.

Many simultaneous changes occurred all over the company to meet these new goals. For ease of understanding, we will follow just one change that affected the technical group involved in manufacturing operations. The vice president for engineering called a meeting of the various engineers in charge of mechanical, electrical, and manufacturing services. He told them of the company's new objectives and requested that they develop a plan to reach those objectives.

The chief manufacturing engineer, using the data generated by the chief of electrical and mechanical engineering services, established standards of manaufacturing upon which detailed manufacturing plans would be based. These standards included a breakdown of all product assemblies into manufacturing units requiring approximately the same manufacturing elapsed time. This was difficult, because the new miniaturized components took more time to handle in the assembly process. Additionally, they included the purchasing department's estimates of material deliveries from various vendors and the establishment of testing procedures and projected rejection rates from quality control. The manufacturing managers used this information as the basis for establishing detailed manpower loads for each product line, material requirements by time period, and designation of delivery points for raw materials.

With the publication of detailed plans, the forecasting loop in the information system was complete (at least at this point in the cycle). Up to this point, the information system had given management valuable information about potential market demands, provided specific targets for production and sales over an extended period of time, and produced a set of forecasted standards against which the company's operations could be compared.

The Reporting Loop

The plan was then put into effect. The two new models were designed, tested, and prepared for manufacture. Materials were ordered, production lines prepared, and work begun. Following the ideas of time delay, level of information, frequency of reporting, and correct usage, the reporting worked as follows:

TIME DELAY

The accounting department prepared daily production reports, which were submitted to department foremen. A weekly summary of these reports was sent to the plant manager and a monthly recap sent to the chief manufacturing engineer.

LEVEL OF INFORMATION

Daily production reports showed actual achievement and costs versus planned achievement and cost. The data in these daily summaries included production units, labor hours, and materials used. Weekly summaries were exception reports, since they included only summarized actual performance over or under planned performance and beyond acceptable limits of tolerance. The monthy recaps were similar to the weekly summaries; they, too, were exception reports.

FREQUENCY OF REPORTING

Since this was a functional operation involved in mass production, the frequency of reports was established at a constant rate but modified to fit the needs of the particular functional managers. Daily reports were needed for line foremen, but they were useless to the plant manager since he couldn't change the actions of the assembly personnel directly. Since the plant manager had to work through the line foremen, his report actually measured those foremen's performance as managers, not the performance of the assembly personnel. A similar analysis worked for the chief manufacturing engineer; he was concerned about the performances of the various plant managers who reported to him. He could do nothing about the line foremen directly and had no need of data on their performances but he could do something about the activities of his plant managers and therefore needed those recaps.

CORRECT USAGE

The major purpose of an information system is to give managers the data used in decision making. Reports save management search time and, if designed correctly, pinpoint the areas to be attended to. The timing of the reports is such that the manager immediately responsible for the department in which a problem exists knows about it first and has a reasonable time either to solve it or to pass it on to the appropriate upper management level.

In this example, a foreman noticed a slow daily increase in the number of labor hours expended in chassis production. He found that the chassis metal being used was harder than that specified in the print, a condition that caused unforeseen shortening of drill life and more time spent in sharpening tools. He took what he thought was the only course of action available to keep the line moving by decreasing the drill sharpening schedule, but he also informed the plant manager and requested assistance.

At the end of the week, the plant manager's summary report showed a slight decrease in chassis output. Seeing that the variance was becoming unacceptable to his overall plant performance, the plant manager requested that a team made up of purchasing and engineering personnel investigate the problem and report suggested corrective action. He also forecast to himself when the variance would exceed the allowable limit of his own level of uncertainty, at which time he would have to bring the chief manufacturing engineer into the problem. For now, the prompt action by both the foreman and the plant manager indicated that there should be sufficient time to resolve it without going further up the organization. During the investigation, the metal hardness problem did not show up on the chief's report.

Had the reporting loop not been properly designed in accordance with the principles of time delay, etc., this might not have been the case. In this example, the foreman directly concerned with the problem was informed first, and was the first to begin work on its solution. Without this proper design, the plant manager and the chief manufacturing engineer would probably have become involved and valuable time would have been lost while both of them tried to find out where the problem was coming from.

The Management Control Loop

The problem of the metal hardness, having been solved temporarily by shortening the drill sharpening schedule, was now in the hands of the purchasing and engineering team. Members of the team found that a design change had increased the strength requirements for the chassis at the last minute and the tooling, which was on order during that change, had not been altered to meet the new requirements. They recommended that production switch to carbide rather than high-speed steel drills since the metal hardness couldn't be decreased because of this engineering requirement.

This change in the manufacturing standards raised the cost of tooling slightly but decreased the production time to meet the prior standard, since carbide drills could handle the harder chassis metal as easily as high speed drills handled the softer chassis metal. Changing the drill material changed *performance,* since it brought man-hour expenditures back down to planned levels. Changing the processing print to show the new drill material changed the *plan;* i.e., raised the planned cost of tooling to match actual cost. This completed the management control loop and solved the problem that the information system had brought to light.

SUMMARY

The three loops in the information system—forecasting, reporting, and managing—operate continuously in every functional and project group served by the system. The foregoing example applied to manufacturing, but any technical department or group of departments could have been used in the illustration. Properly applied and maintained, the information system sup-

---

Note: With the permission of the publishers, some of the materials noted above were borrowed from an article that I wrote, published by the Society for Manufacturing Engineers in *Modern aspects of manufacturing management.* See references for details.

ports the continual adjusting to changing environmental and internal conditions to which all organizations must respond or die. It provides timely and accurate communication of all the managers in forms that they can use; and it also allows errors to be picked up equally on all organizational levels. It supports the ideas of exception reporting, responsibility accounting, and a continuously self-correcting system. The company is able to direct its growth along those paths that it considers optimal, having full opportunity for the incorporation of environmental and internal change. Change is no longer a problem but an opportunity, as is management in general.

SUGGESTED ANSWERS TO CASE QUESTIONS

1. In this case, it seems clear that the overall direction and general goals are top-down generated. The actual plan should be provided by the forecasting loop. Managers can be given direction as to general goals but they have an equal responsibility to tell how much they will cost and what kinds of resources are needed to achieve those goals. The units used (in this case actual dollars, not percentages) should be part of the forecasting and measuring loops. If the corporate staff likes to see percentages, perhaps one answer would be to give it a special report. Let Sam's group program a computer to convert actual results to percentages before sending the data upward. There's no reason not to give them what they want, so long as the operating managers can also have their own information needs satisfied.

2. Budgeting processes are primarily documentations of forecasts and measurements against those forecasts. Forecasting can be learned, and it should be used as such an opportunity. I would change the budgeting system from a once-a-year exercise to a *rollover* exercise. In effect, budget for a year with the first three months planned in weeks, the next three months in months, and the third and fourth quarters laid out as one number each. After three months, I would repeat (i.e., roll over) the budget and plan for another year. That is, I would add three months on the end, but this time the three months that *were* planned in months would *now* be planned in the greater detail of weeks, therefore again planning the next three months in weeks, the following three in months and the final two quarters as one number each. This way, a manager is required to inspect his or her own planning process at least four times while moving from a single-number plan for a quarter to a monthly plan, and back into the weekly plan. The numbers may change from rollover to rollover, but that can mean that conditions have changed and/or the manager has learned how to budget better.

Administrative costs might increase slightly, but not very much. As an example, do you remember how long it took and how painful it was to complete the very first budget that you ever did and now how relatively simple it is? Cognitive change or learning progresses very quickly, and although there is some increase in time that is spent planning, the returns are much greater than the cost because you, as the manager, are able to evaluate your learning four times as often with this method as with an annual method.

3. They have the best defense possible, which is a complete lack of responsibility for the numbers. ("We really tried to increase production with your directed cutback of machine maintenance, but some of that equipment is pretty old and those continual breakdowns just held us up.") This statement can be translated into "We don't have any responsibility for the numbers anyhow, because you told us what to do." When the details of a budget are top-directed, the manager receiving those numbers faces a no-win situation. If the manager meets the numbers, the director often takes the credit, and if she or he doesn't, the manager is blamed for the failure to meet them.

4. Sam should assist Tony in designing a new information system and then present it to corporate headquarters with the suggestion that if it wants he can convert the data to meet corporate's needs. Since there is no change so far as the corporate offices are concerned, he should be able to do it. The result would probably be better and less expensive output from Tony's division, since each manager would have a greater capability for managing.

Have you ever been in any of these situations? What did you do? What happened?

REFERENCES

Horngren, Charles T. *Accounting: a managerial emphasis* (3d ed.). Englewood Cliffs, N.J.: Prentice-Hall, 1972.

James, William. *The principles of psychology* (Vol. 1). New York: Dover, 1950. (Originally published, 1890.).

Makridakis, Spyros, & Wheelright, Steven C. *Forecasting: methods and application*. New York: Wiley, 1978.

Silverman, Melvin. Directed growth through management communciations sytems. In Ivan R. Vernon (Ed.), *Modern aspects of manufacturing management*. Dearborn, Mich.: SME, 1970, pp. 305–316.

FURTHER READINGS

Haberstroh, Chadwick J. Organization design and systems analysis. In James G. March (Ed.), *Handbook of organizations*. Chicago, Ill: Rand McNally, 1965.

Hall, M. F. Communications within organizations. In Walter A. Hill & Douglas Egan (Eds.), *Readings in organization theory: A behavioral approach*. Boston: Allyn & Bacon, 1967, pp. 403–415.

Mechanic, David. Sources of power of lower participants in complex organizations. In Walter A. Hill & Douglas Egan (Eds.), *Readings in organization theory: A behavioral approach*, Boston: Allyn & Bacon, 1967, pp. 196–206.

Stanley, C. Maxwell. *The consulting engineer*. New York: 1982.

LEADERSHIP

Case Study

The Case of The Expert Group Leader

International Chemicals, Inc., was one of the major suppliers of industrial chemicals and for many years had operated an internal applied research and development group that was closely associated with the marketing branch. Whenever a customer had a particular problem, that customer would call its local salesman, who in turn would contact the marketing group. This group would then get the problem defined and bring it back to corporate headquarters for the engineers, scientists, and technicians to solve.

CAST

George Jessup: Chief metallurgist

Mike Jensen: Group leader, marketing

Melanie Michaels: Metallurgist, technician

Bob Andrews: Vice president, research and development

It was early on a Monday morning when George Jessup received a phone call from Bob Andrews to come by his office when he had a chance. Later that day, George phoned Bob and asked if it was OK to come in then. Within fifteen minutes he walked into Bob's office.

Bob: Oh, hi, George, Glad you could come by as quickly as you did. I know that you've been busy running life tests on that new gas-metal interaction that the utilities were interested in to increase boiler life. The problem seems to be centered around your new technician, Melanie. She's been in charge of some test that the fellows in marketing wanted to get done quickly and according to what Mike Jensen tells me, she's absolutely refused to do anything for them. I'd like you to look into it and tell me what's going on.

George: I really don't have to do much looking to tell you about it, Bob, because she told me what happened just this morning. It seems that Mike had this rush job for a steel mill out in the midwest and he, as usual, just came into our department and asked the first person he saw to work on it for him. Usually, we try to respond because we know how important it is but Melanie is new to this firm so she tried to get him to define the problem a little more before she went to work on it. She asked him to write a functional specification so that she would know what to look for and she asked him which of the other ten "hot" jobs she was working on he wanted her to drop. That seemed to me to be a reasonable question.

Well, you know Mike, he doesn't respond too well to things like that so he came over to complain to me, as her boss. I, of course, told him that she was absolutely correct. She hadn't refused to do anything; all she wanted was some direction and priorities. I've been meaning to discuss it with you for some time, but we really should get some structure into what we're doing. If Mike can't even take the time to define the problem or give it a priority, how can we know either what he wants or how fast he wants it? While we certainly try, it's impossible to meet all the demands all the time. So far, he hasn't come back with an answer.

Bob: Yes, it sure sounds different when you tell it, but we've always been able to operate on any demand from marketing. Why have things changed, and what is to be done about it?

George: I don't really know, Bob, but I sure do agree that there have been changes. Why don't we discuss it at the next staff meeting?

Later that week, at the regular staff meeting, all four of the principals to this discussion were present.

Bob: Mike, let's find out what this problem really is. It sounds a whole lot bigger than just having your feelings hurt.

Mike: Well, it seems to be a combination of things. Marketing used to be able to get an inquiry from the field, run it through our experts here at the home office, and have an answer, and a satisfied customer, in a few short weeks. Now it seems that we need all kinds of paperwork, that none of the experts wants to work overtime, and it's almost impossible to do anything in less than six months. Our competition is eating into our markets because they're lean and mean. What's happened to us?

George: Why don't we look at our present situation instead of talking about how it used to be, Mike? Let's analyze the problem by using the example of that requirement for the steel mill that you gave to Melanie. Melanie, what did you see happening then?

Melanie: As I saw it, there were ten really hot projects that were taking up all my time. I have been working more than fifty hours a week for the past two months and I've been trying to satisfy all the requirements placed on me, but they seem to be increasing, not decreasing, and the overload is getting to be too much. I can understand that everybody wants his or her job out first, but that is getting to be impossible. I need more structure in the jobs assigned to me, I've got to have some authority across division lines because my projects cross out of this research and development department into other divisions of the company, and I believe that there should be some coordination and training meetings with those other divisions in order to acquaint them with our needs for prompt response.

As far as Mike's request is concerned, I believe that the questions I asked were necessary to prevent a lot of wasted work if my own definitions were wrong or if I misunderstood the problem. I haven't always been able to follow exactly what marketing wanted and this is one way to be sure that the problem is clearly defined. Also, since I cannot control the other people who have to work with me on the project, I wanted some priority to use in coordinating the requests for their time that I have to submit to their bosses better. I look silly when I have to change schedules every week.

QUESTIONS

1. When Mike started with the company, the R&D group was quite small. How have the leadership requirements changed as the company has grown? What has happened to the R&D group?

2. What kinds of recommendations would you make in terms of developing answers for the repetitive problems of definition of the problem, project leader's authority, and intergroup cooperation?

3. How would you set up the process of implementing your answers to question 2 if you were Bob? To whom would you assign it? How should that person or group proceed to get solutions? How would you know if a satisfactory implementation had occurred?

REVIEW

In previous chapters, we developed a model of the technical organization, analyzed its components, and prescribed how it might be optimized using contingency concepts. We now move directly into perhaps one of the most important sets of contingencies that affect both the organization and you: the contingencies dealing with leadership. I won't try to make the difficult value-laden division that could distinguish between managing (doing the thing right) and leadership (doing the right thing) here because, as I see it, technical management spans both of these aspects. However, it is possible to define both aspects so that they are related to each other in a way that is already familiar to us.

Managing can be understood as dealing with the stable and structured part of the organization, that part concerned with setting up repetitive decisions. This includes development of standardized organizational policies and procedures, setting up the manuals and standards that apply to product design standards, and documenting the relatively fixed rules that apply to Mother Nature. For example, supervising participants' actions in determining, purchasing, and using standard sizes of product raw materials, finishes, and components would be managing under this definition. Another example could be the administrative task of scheduling vacations.

Leadership deals with change and innovation that are mainly concerned with nonrepetitive decisions involving people. An example of leadership could be establishing a new project team, supervising field service for new products as part of quality assurance, and implementing a management-by-objective system for the technical group. Obviously, you need both managing and leadership techniques. Consequently, those terms are treated as equivalents here. We follow the by-now familiar pattern of definition, description, and prescription; then attempt to develop some suggestions for leadership behaviors that seem to be particularly applicable.

DEFINITIONS OF LEADERSHIP

The study of leadership has a long and honorable past. Ancient authors on this subject have included Homer, Marcus Aurelius, Plato, and Caesar. Each had his own viewpoint, and although those viewpoints are not as universally applicable as their authors might suggest, many of their basic concepts still survive (Bass, 1981). The "thoughtful" leaders of Marcus Aurelius, the "men of action" of Caesar, the "shrewd and cunning" leaders of Homer, and the "statesmen-philosophers" of Plato may still be found in modern research literature. Niccolò Machiavelli's treatise on leadership, *The Prince* (1940), which was written during the Renaissance, is still quoted today as a guide for a specific type of leadership behavior.

The importance of this subject should never be underestimated, and we will therefore try to give it the attention it rightly deserves. I consider leadership as the cement that binds the total organization together—the components, the information systems, and human decision making. Without it there is no organization, there can be no cooperative effort. In one sense, it is a process that induces one or more individuals to act in accordance with the wishes of another individual. The intended end result of the process is basically to induce compliance.

How can you best get that compliance? We would usually start our descriptions with the primary actor in the drama: the leader. However, it is extremely difficult to use the same descriptive patterns of the previous chapters because we really cannot separate the two major parts of the leadership concept: the leader and the situation. Describing the leader while keeping all other things fixed is an internal contradiction, since by definition the situation includes the followers, and followers are inextricably intertwined with the leader. A leader with no one to lead is an impossibility. That seems rather obvious, but this impossibility did not appear to be insurmountable to early researchers in leadership theory.

These researchers assumed that a leader can lead no matter what the situation, and that he or she would always be recognizable by superior attributes. The dilemma did not exist for them; it was just assumed away. One of these earlier theories was called *trait theory*. It explicitly stated that good leaders have inherent characteristics that support successful leadership behavior in any situation. According to this theory, we should be able to analyze those characteristics, learn what is required for each potential leader, teach that person what he or she needs to know, and there we have it— another successful leader. This kind of thinking is universalistic and is similar to the universal motivation ideas that we discussed in Chapter 3. It has the same obvious problems they do.

But since trait theory and the other universal leadership theories are parts of the foundations of this management literature, we should review them before going on: so we will. Then we shall move quickly into the more realistic and useful person-situation relationships. Some of those relationships assume a relatively stable situation and some do not. Since we know that there are parts of technical operations that are intended to remain quite predictable and constant (such as functionally organized, mechanistic groups) and parts that are expected to change very rapidly (such as project-organized, organic groups), we will try to develop leadership hypotheses to account for both of these contingencies.

DESCRIPTION OF UNIVERSAL MODELS

Trait Theory

As I mentioned before, one theory that is part of the older and most traditional group of theories describing leadership is trait theory. A trait is the "differences between the directly observable behavior or charcteristics of two or more individuals on a defined dimension" (Mischel, 1968, p. 5). Since those differences are subject to interpretation, it appears that the trait being observed must be some type of construct or an abstraction of the observer. There is an impressive literature list that identifies "supposedly essential characteristics of the successful leader, over a hundred, in fact, even after the elimination of obvious duplication and overlap of terms" (McGregor, 1960, p. 180).

These traits are assumed to be the generalized and enduring parts of the leader's personality that cause the leadership behaviors. It is easy to understand how important the determination of these essential leadership characteristics is as a research project. If it were really possible to determine a combination of stable mental characteristics for successful leadership, it would then be possible to predict who will and who will not be a successful leader. It might even be possible to find out what leadership traits are missing in a particular person and train that person, eliminating the deficiencies.

These ideas implicitly assume that the leader generally determines the success patterns of the group being led. Typical traits or characteristics such as choice of associates, biography, height, weight, physique, appearance, speech fluency, intelligence, scholarship, knowledge, insight, judgment, originality, adaptability, introversion-extroversion, dominance, initiative, integrity, responsibility, self-confidence, social skills, and a myriad other attributes were measured or correlated. The correlations were not very high, but some typical findings were that "The average person who occupies a position of leadership exceeds the average members of his group in the following respects: (1) intelligence; (2) scholarship; (3) dependability in exercising responsibility; (4) activity and social participation; and (5) socioeconomic status" (Bass, 1981, p. 65).

The writers of the classical school, such as Fayol, Taylor, and Urwick, extended the definitions of traits to include the thinking activities performed by the manager on the job. They felt that typical activities such as "planning, organizing, and controlling" determined a leader's success in the organization. These activities were supposed to be both basic and universal. If the leader did them well, he or she succeeded; if not, he or she failed. This approach defined behaviors (and inferred traits) and was supposed to demonstrate how every organization could be managed effectively.

As an outgrowth of this classical, relatively fixed approach to managing but modified to suit more modern ideas, McGregor (1960) proposed his well-known theory X and theory Y leadership concepts. His concepts define the leadership traits as resulting from the relatively fixed belief patterns of the leader. These belief patterns were split into two separated sets of beliefs: those supporting theory X and those supporting theory Y.

The believer in theory X assumed that subordinates consisted of people who were passive, not inclined to work, and resistant to organizational needs.

The theory X leader's resultant behavior was consistently directive, with close supervision of subordinates. The theory Y leader, on the other hand, assumed that the group being led already was motivated and each person in it would want to work as much as he or she would want to play. The worker had an internal desire for more responsibility. That leader's behavior was consultative and intended to assist the workers or the individuals in the group to achieve their organizational objectives.

The theory Y beliefs were valued more highly, since they seemed to match the human-relations school of management, which was popular at the time. There were many training sessions in which managers were supposedly indoctrinated with behaviors intended to support the beliefs of theory Y, such as the need for listening to subordinates, gaining their participation, and attempting to develop a team approach regardless of the particular situation.

There is an obvious problem with trait theory. There is no objective way to separate the traits being measured from the evaluation processes of the person doing the measuring.

> Traits are used first simply as adverbs describing behavior (e.g., he behaves anxiously), but this soon is generalized to describe the person (he is anxious) and then abstracted to "he has anxiety.". . . We quickly emerge with the tautology, "He behaves anxiously because he has the trait of anxiety." This is the danger of trait-theoretical explanations (Mischel, 1968, p. 42).

This problem, by itself, is sufficient to cast doubt on the theory's validity and consequent usefulness: One of its basic assumptions is not supported—that of universal application regardless of the situation. The simplistic definition of behavior as solely a function of the internal state of the leader with no provisions for the stimuli of and interaction with the situation is a more analytical way of putting it. But there are always stimulus changes in the situation. If they interact with the leader's personal attributes, this interaction can also cause behavior. That interaction (which is supposedly predictable and fixed) is the basis for *social behavior theories* in the leadership literature. These theories also suggest an approach that is universal, but they consider the relationship between the situation and the leader's behavior, too.

Social Behavior Theory

Social behavior theory is typically illustrated in the organic-mechanistic classification we discussed in previous chapters. If we define a leader's response to fast situational change as always being quick and flexible, as opposed to always being rigid and unchanging, we have the expected behaviors needed in organic structures versus those of mechanistic structures (Burns & Stalker, 1961). This theory maintains that leadership also takes into account the values and expectations of those being led (Likert, 1961). The leader must present behaviors that are perceived by the followers to be supportive, involve them in decision making, and increase their influence in defining and completing organizational tasks. These behaviors are expected to build group cohesiveness and the inferred motivation to produce by supporting the individual's freedom and the taking of initiatives.

Many of the ideas in this social behavior theory have been included in a managerial grid (Blake & Mouton, 1964) that measures leadership behavior with two perpendicular axes: *concern for people* and *concern for production* (see Figure 7-1). As an illustration, a leader may exhibit behavior that seems to be high in concern for people and low in production concerns. This would then result in a country-club type of organizational climate where everyone was relatively happy but production was low. (This climate would confirm the idea that there is little relationship between morale and productivity.) In contrast, we could have a leader who is supposedly high in concern for production and low in people concerns. This would result in a rather directive, mechanistic, production-oriented organizational climate. Accordingly, the optimum leader is one who is a 9,9 (both axes are measured from 0 to 9, with 9 being the highest score). That is the leader who has major concerns for both production and people at the same time.

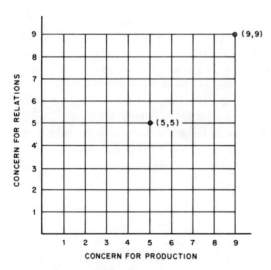

Figure 7-1   Grid theory.

ADVANTAGES OF UNIVERSAL MODELS

These universal models have the same advantage of all universal models: They promise the "one best way" to manage. Trait theory is appealing because traits are supposedly part of the "package" of the best leader. Traits that are not in the package can probably be awakened through adequate training and experience, since everyone can be a good leader if he or she is taught how to be (according to this theory). In effect, a good leader is always effective, no matter where he or she goes, and those who are not can be trained.

Social behavior theory is similar to trait theory in its predictions of universality. It is different when it suggests that there is an interaction between the environment and the still relatively fixed attributes of the leader. And, of course, it is similar in suggesting that those attributes can be modified through training to fit some predetermined and fixed situationally optimal condition. Both the situation (or the organizational climate) and the leader are predictable, and the interaction between them is fixed. With this theory, the leader will invariably succeed if trained to respond correctly to that environment.

DISADVANTAGES OF UNIVERSAL MODELS

The problems in trait definition severely limit the proposed universality of trait theory. Social behavior theory is also limited, because all situations are not alike. The interaction of the leader with a *particular organizational situation* is not considered. All situations are considered to be optimizable and all potential leaders trainable. But we know that all situations are not the same. For example, organizational situations change over time. Other changes affect their technology of production (Woodward, 1965), their mode of decision making (Thompson, 1967), and the perceived uncertainty (Lawrence & Lorsch, 1967). All situations are not optimizable using the same ground rules, and if all leaders were trainable to fit this theoretical optimal situation, we would not have companies going bankrupt after many years of success. Their leaders wouldn't let it happen, because if they succeeded before, they always succeed—but it does happen, so there are problems with these theories.

We can find discrepancies even when using some of this theory's predictions. The theory would generally predict that all leaders should have a high concern for production with a similar concern for people, and yet such a leader in a production-oriented, mechanistic situation would probably fail (i.e., a 9,9 leader in the Blake and Mouton grid theory). Directive leaders with a high concern for production and a low concern for people would be needed in this example. On the other hand, the advanced research and development group in the same company could not tolerate this production-oriented brand of leadership; for it, the theory Y leader or one concerned primarily with people would be best.

There are places for a drill-sergeant, theory X type of leadership behavior and there are places for a consultative-collegial, theory Y leader. What is even more interesting, when the organizational situation changes, if we don't have the ability to change the leader, we may need both in the same person: someone who could be either X or Y, depending on the contingencies. We can see that these theories are limited and can be applied only when the organizational variables will support them. There are other more pragmatic models, however, that make no claim to universal applications. As usual, they are more complex and therefore more applicable to real situations.

DESCRIPTION OF SITUATIONAL MODELS

The situational models are concerned with the changing interaction between the group being led and the leader. The group may be affected initially by the externals of technology, information systems, and/or environmental change, but the primary concern is always with the results of those changes on the basic group-leader interaction in whatever form it takes. The interaction is divided into two parts: the formal part, based on the structural confirmation by the total organization (of which the group is a part), and the informal part, based on the acceptance of the leader's direction by the group. It's the familiar combination of legal power and personal influence as was noted by Blau (1967):

> Compliance can be enforced with sufficient power, but approval cannot be forced regardless of how great the power. Yet the effectiveness and stability of leadership depend on the social approval of subordinates. . . . Effective authority, whether in formal organizations or outside, requires both power and legitimating approval, but the one is more problematical for the informal leader, and the other, for the formal leader in an organization. (pp. 201, 210)

The idea that the approval of the group itself affected leadership success was accepted by the classical theorists, but as a potential source of conflict which was to be minimized if possible. They maintained that formal approval by the organization was all that was necessary for adequate organizational leadership. The subordinates' effect upon the leader was to be minimized and, if possible, eliminated, since it conflicted with the needs of the organization as communicated by the leader. Those needs coincided with the needs of the leader who represented the organization. The Hawthorne experiments confirmed this through the suggestion that the social or informal evaluations by the workers of their management should be manipulated in order to obtain agreement between organizational goals and those of the workers. When this agreement took place, the workers accepted the leadership of their supervisors.

However, there were those who recognized that this could not always be done unilaterally by management. Workers (and especially "knowledge workers") think for themselves, and when they perceive a difference between their own needs and those of management, their informal social organization often becomes strong enough to assist in mitigating those management needs. In that case, authority "rests upon the acceptance or consent of individuals. . . . The decision as to whether an order has authority or not lies with the persons to whom it is addressed, and does not reside in persons of authority or those who issue the orders" (Barnard, 1938, p. 163).

You can see why the armed services would not voluntarily adopt this suggestion in many of their organizations. In combat organizations, the authority must lie with the symbols of office or the uniform, not the wearer, because there are often rapid situational changes and subordinates' compliance must be an almost automatic response. However, in modern industrial organizations, the independence of the knowledge worker suggests that these ideas of automatic obedience are not quite valid and the worker can accept or reject the leader in charge. As one engineer told me, "When they sent out the memo announcing my promotion to chief, it was like getting a hunting license for lions. Now all I have to do is get the lions to cooperate or else I could get badly hurt." In this case, getting the lions to cooperate was defined as obtaining the approval of the group or, in more formal terms, achieving a positive interaction between group and leader behaviors.

Path-Goal Theory

How this interaction of group and leader behaviors was to be gained was the subject of much research. One important question in that research was, "Why should the group accept the leader?" One potential answer was the path-goal theory (House, 1971). This theory suggested that the interaction was based on the behaviors exhibited by the leader that aroused subordinates to perform and to achieve satisfaction from the job to be done. The leader did this by clarifying the goals of subordinates and defining the paths to achieve them. Perhaps more importantly, he also controlled the rewards that subordinates valued and cleared a path for them to achieve those rewards.

This brief description of path-goal theory shows that it has roots in many theories of motivation and organizational structure design. The motivational roots include the appeal to the self-interest of the subordinates since the leader controls the subordinate's rewards and the structural roots, the ability to organize the structure that clears a path for the achievement of these management-valued goals. The theory's effectiveness lies in the specific definition of the methods by which leaders can achieve the compliance of their subordinates. Those methods are clarifying goals and defining paths to achieve them.

Contingency Theory

For technical operations, however, it seems to me that one of the better theories fitting the changing situations involving both leader behavior and group interactions is the general contingency leadership theory (Fiedler, 1967). It describes measurable aspects of the leader's personality and relates those aspects to unique characteristics of the situation. It predicts which leadership personalities will succeed in which types of organizational environment. The measurement of the leader is operationally determined by a test called the Least Preferred Coworker Test (LPC). The leader is requested to complete a paper-and-pencil test checking off the characteristics of someone she or he knows or did know in the work environment whom she or he would define as her or his least preferred coworker. For example:

> Now think of the person *with whom you can work least well* (emphasis is mine, not the author's). He may be someone you work with now, or he may be someone you knew in the past. He does not have to be the person you like least well, but should be the person with whom you had the most difficulty getting a job done. Describe this person as he appears to you.

Pleasant	Unpleasant
Friendly	Unfriendly
Rejecting	Accepting
Helpful	Frustrating

etc. (Fiedler, 1967, pp. 268–269)

These four items are typical of the total of sixteen items in all and the leader checks the appropriate answer for each of those items. The favorable pole of each item is rated as 8 and the unfavorable as 1. I have added a numeral score above to clarify this test. These scores don't appear on the actual LPC test but are added by the test scorer after the appropriate check mark is made. The sum of the item scores constitutes a person's LPC score. A high LPC score denotes a person who is relationship-oriented and a low LPC score, one who is task-oriented.

. . . a person who describes the least-preferred coworker in a relatively favorable manner tends to be permissive, human relations oriented and consid-
erate of the feelings of his men. But a person who describes his least preferred
coworker in an unfavorable manner, who has what we have come to call a low
LPC rating, tends to be managing, task-controlling, and less concerned with
the human relations aspects of the job (Fiedler, 1976, p. 485).

Research on LPC scores suggests that they are relatively stable over time
and that low-LPC people receive satisfaction from successful completion of
tasks, while high-LPC score people are concerned more with successful
interpersonal relationships. While you might consider this to be a possible
extension of trait theory, it is much more than that. Measuring a global
characteristic such as LPC is more applicable to a particular person than
measuring pieces of that person, such as traits, and then attempting to
reassemble those pieces back into the whole person. The test reliability and
validity for the whole is greater than the sum of the tests of the parts.

The definition of the situation was also operationally defined by questionnaires resulting in three measurements: leader-member relations, task
structure, and position power. The leader-member relations measurement
determines how well the group and the leader interact—how willing the
group is to follow the leader's guidance and direction. The task structure
measurement is how well the task is spelled out in a step-by-step methodology. Is it well-defined or is it nebulous and poorly defined? Finally, the
position power is the measurement of the power inherent in the position
itself aside from any influence that the particular leader might bring. An
analogy would be the high-position power of a general in the army, regardless
of the personal influence of whoever happens to have that job.

The research shows a correlation between the LPC score of the group
leader and the group performance on the y axis and the three determinants
of the situation on the x axis. (See Figure 7-2; Fiedler, 1967, p. 146) It was
found that low-LPC leaders (task-oriented) performed better and managed
more effective groups when the quality of leader-member relations, the task
structure, and the position power were either very favorable or unfavorable
for the leader. The measurement used was the correlation between LPC
score and group performance in specific situations. The high-LPC leaders
(relations-oriented) were more effective when neither of these extremes
appeared. In other words,

A positive correlation (falling above midline) shows that the permissive, nondirective and human-relations oriented leaders performed best; a negative
correlation (below the midline) shows that the task controlling, managing leader
performed best. For instance, leaders of effective groups in situation categories
1 and 2 had LPC [versus] group performance correlations of -0.40 to -0.80,

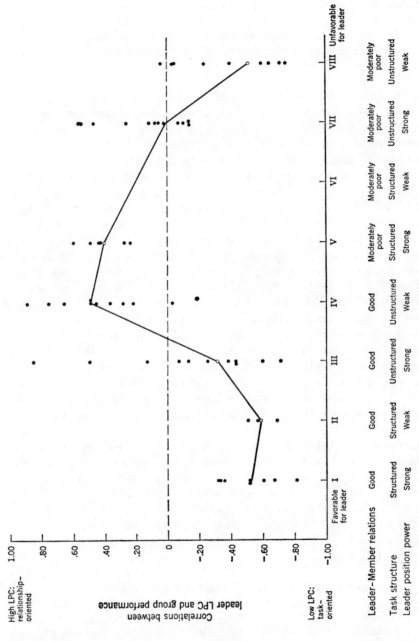

Figure 7-2 Leader LPC and group performance correlations.

with the average between -0.50 and -0.60; whereas leaders of effective groups in situation categories 4 and 5 had LPC [versus] group performance correlations of 0.20 to 0.80, with the average between 0.40 and 0.50 (Fiedler, 1976, p. 488).

The implications for leadership prediction and control become clearer with this approach. There are suggestions for almost all kinds of leaders in Fiedler's model—those who are at the extreme of being totally task-oriented (which might be similar to the concern for production of Blake and Mouton) or those who are totally relations-oriented (such as concern for people). But there are also implications in his models that there are potential leaders who can operate successfully between these extremes. Even the two extreme leaders with either task or relations orientation can exhibit temporary behavior that is supposedly atypical. The task-oriented production foreman who would be concerned about the family problems of one of the workers would be exhibiting atypical behavior, but that behavior could still represent the particular manager adequately, since this measurement indicates an *overall* tendency and doesn't pretend to describe all possible behaviors. The LPC scores indicate the dominant mode of behavior; it is not intended to be exclusive. More recent research findings (Kennedy, 1982) apply to middle-LPC managers, who seem to be the best all-around managers, according to these data.

> . . . the results . . . provide strong support for the hypothesis that middle LPC leaders perform well in all leadership situations . . . these results suggest that the middle LPC leader who . . . is least concerned with the task and the opinions of others, appears to be most capable of performing leadership tasks in an effective manner, regardless of the situations . . . more flexible, not overly constrained by any one goal orientation, and therefore better able to employ the behaviors that will maximize performance. (pp. 7–9)

This has some aspects of trait theory, since it suggests that there is one type of leader who will perform well in all situations. There are two differences from trait theory, however. The measurement of the leader is global, not in traits, and performance was determined against the kinds of situations noted in this contingency theory. This research, however, is only indicative, since the researcher is describing only limited experimental situations. We still don't have sufficient replications that are needed for effective reliance on these initial data in our own organizational situations. Fiedler's work has been replicated many times.

Now, assuming that we try to use these data in building our own theories, we should start with some typical questions:

1. How do the experimental situations compare with those that we perceive in our own organizations?

2. Do the extremes of our situations apply—i.e., does task orientation describe our technical organizations better than descriptions of relationship orientations or vice versa?

I'm sure that there are others you have considered. This contingency model, as noted before, has been supported through replication many times, so there is a higher confidence in using its recommendations. These are offered later in this chapter.

A final model within the situational framework is that proposed by Hersey and Blanchard (1969). They suggest that the model of the leader group interaction should be responsive to, as they call it, the maturity of the organization. They also have a graphical representation of the theory (see Figure 7-3), but they are concerned with a "life cycle" approach.

> According to Life Cycle Theory, as the level of maturity of one's followers continues to increase, appropriate leader behavior not only requires less and less structure (task) but also less and less socio-emotional support (relationships). . . . Maturity is defined . . . by the relative independence, ability to

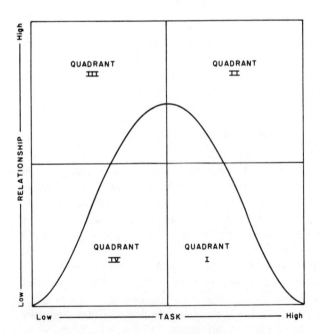

Figure 7-3 Life-cycle theory of leadership.

take responsibility and achievement motivation of an individual or a group. (Hersey & Blanchard, 1969, p. 29)

The components of maturity typically include level of education and amount of experience. As an interesting aside, maturity is not related to chronological age. The researchers consider chronological age almost irrelevant. They are only concerned with psychological age, and there may be no relationship between the two. The life cycle theory suggests that the leader's behavior should change as the organization matures. It should move from high-task and low-relationship behaviors (quadrant I in Figure 7-3), to high-task and high-relationship behaviors (quadrant II), through high relationships and low tasks (quadrant III), to low relationships and low tasks (quadrant IV) as the followers move from immaturity to maturity. Life cycle theory seems to be analogous to the parent-child relationship and is similarly expected to exhibit the slow, gradual developmental process that is evolutionary, not revolutionary, in nature.

ADVANTAGES OF SITUATIONAL MODELS

The situational models have attempted to solve the measurement problems of trait theory. Within the limits of accuracy of the social sciences, they seem to have made a step in that direction by considering both the person and the situation. The major research problems of construct and predictive validity and test result reliability seem to have been met; additionally, the data have been replicated many times. Fiedler's situational constructs do not agree exactly with Hersey and Blanchard's but there seems to be sufficient coincidence to indicate that there may be larger contingency theories yet to be formulated within which these models will fit.

Therefore, although situational theory has not solved all the problems in defining leadership-group interactions, it seems to be an important step forward in describing and explaining more of the intricate relationships between the leader and the group. It is complex theory, but sometimes complex theory is required to explain complex phenomena. It assists in predicting how the behavioral characteristics of a particular leader will interact with the particular characteristics of a group and/or situation and it points to better solutions than we have been able to use before. It begins to provide the general theory into which many of the smaller theories can be fitted. For example,

> Leaders differ in their concern for the group's goals and the means to achieve them. Those with strong concern are seen as task oriented (Fiedler, 1967), concerned with production (Blake & Mouton, 1964), in need of achievement (McClelland, 1961) and production oriented (Katz et al., 1950). . . . Such lead-

ers are likely to keep their distance psychologically from their followers and to be more cold and aloof. . . . When coupled with an inability to trust subordinates, such concern for production is likely to manifest itself in close controlling supervision (McGregor, 1960). . . .

Leaders also differ in their concern about the group members in the extent to which they pursue a human relations approach and try to maintain friendly, supportive relations with followers (Katz et al., 1950), concern for people (Blake & Mouton, 1964), and in need of affiliation (McClelland, 1961). Usually associated with a relations orientation is a sense of trust in subordinates, less felt need to control them, and more general than close supervision (McGregor, 1960). (As noted in Bass, 1981, p. 331)

DISADVANTAGES OF SITUATIONAL THEORIES

These theories are helpful within the technical management functions, but in my opinion they have not matured to the point where they can handle different descriptions of the appropriate leader behaviors when there are major environmental changes that we have discussed before. These changes affect different parts of the organization in different ways. Changes in technology (e.g., moving from a typewriter to a computer terminal) or in the level of economic uncertainty (e.g., crude oil prices have tripled within one year and we make petrochemical-based plastic parts) are not handled well so far and these can occur more quickly causing great situational changes in the technical operations.

It seems to me that the Hersey-Blanchard model approach tends to handle this problem of rapid change better than most, but it proposes that the total organization itself changes and matures, with required consequent changes in the leader's behaviors as it does so. However, technical organizations are not monolithic; they can operate in different degrees of maturity at the same time. They can operate simultaneously both in more certainty or maturity, as in the mass production department, and in less certainty or immaturity, as in the advanced product design department. Mass production (which might be defined as "maturity") requires high task orientation and low human relationships orientation while unit production (the organic operations of "less maturity") requires the opposite.

And now we must consider the effect of change upon the leader. It takes a long time to change anyone's behaviors permanently. We are all products of both our environment and our heritage, and changing human beings as quickly as the organizational situation changes is a difficult task indeed because of these semihidden mental processes that we all carry with us. It seems to me that individual change can occur with trauma (either physical or mental), therapy, or changing of *organizationally required* behaviors. Trauma is not socially acceptable in most technical organizations, and therapy is not usually considered to be part of management's prerogatives. Therefore, changing required behaviors is the only practical path left.

That means, for example, that when the organizational culture rewards a leader who wears short-sleeve shirts and no ties to work, those who show up in a double-breasted suit with matching shirt, tie, and shoes will probably not be rewarded. In a reasonable time period, the nonconformist leader either changes his clothing styles (which may change how he thinks) or leaves the situation. Nonconformity, by the way, is not necessarily a one-way street; it can support freedom or constraints (or anything in between). Therefore, although change in the organization's environment may require different leadership approaches *within* the technical organization, it is only possible to obtain this by correctly defining the new leadership behaviors. That's neither easy to define nor to control. There are further consequences for different parts of the technical organization based on Lawrence and Lorsch's ideas. While I believe that change on an overall basis is important to consider in technical organizations, it is also important to remember that the level of uncertainty *within* the total organization may differ. According to the findings of Lawrence and Lorsch (1967), the research group in the organizations they gathered data from had higher levels of uncertainty than the manufacturing group (i.e., the technical economic area). This leads to a similar conclusion we reached using the Hersey-Blanchard model.

Leadership behavior patterns must change, but now they are based upon differences within the overall total organization as a function of perceptions of environmental uncertainty. Using the ideas in this theory, behavior results from changed perceptions of organizational participants. In other words, thinking precedes behavior. That is a much more theoretical approach and provides you, as the technical manager, with no management tools to accomplish a change in the appropriate leader's behavior. In my opinion, this is not as applicable. I suggest, therefore, that if you can change behavior, thinking will follow.

This research defines and measures several kinds of variables that affect leadership better than many of the other theories that we have reviewed. However, it is basically descriptive. We must now consider prescriptions that can apply more closely to technical functions.

PRESCRIPTIONS: TECHNICAL OPERATIONS LEADERSHIP

These prescriptions are specifically intended to match the situations in technical functions. They assume that if they can define the situation, you will improve the consequent management behaviors. They should be used in this sequence: gather appropriate data, analyze them to determine what the problem is (i.e., diagnosis), and then modify those general prescriptions that are applicable to fit your diagnosis based on your data. This method provides personal theory (and prescriptions or hypotheses that flow from your theory) that is grounded in the empirical data from the organization. The interactive, changeable theory is a type of "grounded" theory (Bass, 1981, p. 26) that is often used by managers and leaders. It isn't the kind that theorists in the physical sciences might accept, since it has few experimental controls and is not really replicable, but it has been very useful in the social sciences and can be very useful to you, since it applies to your situation. Situations are part of a social science system.

According to the concepts proposed earlier in the chapter on structure, technical operations can be generally divided into two related situations: functional and project operations. If we are to use leadership models that are applicable, we have to develop two models adapted to these two different situations. However, we understand that these two models are only proto-types, and your real situation may require extensive modifications. These models are also based partly upon my background and experience in technical organizations. They may, therefore, not match your data and diagnoses but they should provide some starting points from your own analysis. We start with functional operations.

Functional Leadership

DEFINING THE GROUP

I would assume that most people in technical jobs have been through some type of structured training process. This provides a degree of "maturity" (as defined in the Hersey and Blanchard model). They have probably learned to discipline their thinking and behavior (if you don't learn that, you just don't graduate from most technical curricula), and they are generally aware of the contributions that they are making to the firm. It is possible that they have the independence of the traditional journeyman-technician. Generally, there is some degree of group cohesiveness or group identity. They think more like each other than they do like members of other functional groups in the same organization (Lawrence & Lorsch, 1967).

DEFINING THE SITUATION

There is more uncertainty in their tasks than in those of most of the rest of the organization. It may not be at the highest level for the entire organization (and that could also happen in some marketing or in advanced R&D functions, for example) but it is higher than the organizational average. There are some tasks, however, that are repetitive, and these tend to lower overall uncertainty. Those tasks typically deal with production support, maintenance, standards, quality control, and evolutionary changes and improvements in products and processes.

DEFINING THE LEADER

By definition, most of us who have managed technical operations have been trained throughout our academic experiences to operate in a task- rather than a relations-oriented mode. The basic undergraduate courses in mathematics, physics, chemistry, etc. demand that the student provide only one answer, and logic is the method through which that answer is obtained. The instructor isn't really interested in a creative approach to that answer to his calculus problem; most of the time she or he is looking for a specific answer on the test and wants it provided only one way. If this is the training path that you have followed in your academic period(s), we can assume that you, as the technical manager, approach the processes of leadership in a functional organization with an initial disadvantage. You have been trained in a task environment, but the functional situation generally appears to need a relations-oriented approach. The following models seem to show this.

Using Fiedler: When the leader initially assumes a leadership position over an operating functional group, the leader-member relations could be moderately poor at first, the task well-structured, and the leader's position power strong. This would be equivalent to category 5 (see the chart in Figure 7-2) and require a high LPC or relations-oriented leader. However, when leader-member relationships improve with time or design problems suddenly occur and there are changes in the organizational environment, with the tasks becoming less structured, a different leadership pattern is required.

According to Fiedler, this changed situation would then require a task orientation in the leader. But this change is temporary in functional situations and *if* the relations-oriented leader recognizes it as a temporary change *and* is capable of temporarily assuming a task role, that leader is again matched with the situation in an optimum leadership pattern. This appears to be possible, just as it is possible for task-oriented leaders to become *temporarily* relations-oriented if required. Fiedler's general model seems (to me) to be satisfactory when applied specifically to functional technical organizations. Here, leaders should be relations-oriented.

Using Blake and Mouton: The initial approach for a newly appointed manager should be concern for relationships until a positive human interaction is established. That seems to me to be the first goal in managing a functional group, since the manager will probably be with the group for quite a while. The next step is concern for production until a satisfactory production output is reached, and then a concern for both, or, as the authors describe it, a 9,9 approach. The path of optimum leadership behavior is almost sinusoidal, with a dampening influence until the leader gets to the 9,9 position. This model is also applicable, but it doesn't provide for rapid changes in the situation or in the uncertainty as Fiedler's does. It is concerned with only two variables: relationships and production.

Using Hersey and Blanchard: Initially the group would require a high relationship, moderate task orientation if we were to follow our prior scenario. With time, "maturity" increases, thereby requiring less relationship, less task, and more self-direction for group members. With increased uncertainty, there would be more emphasis on task behaviors. This model is the most difficult to apply, since it assumes a constant positive set of behaviors in the leader and a close diagnostic interpretation of the maturity of the followers. It provides fewer allowances for externally caused changes—just for those related to the people in the group being managed.

SUMMARY OF FUNCTIONAL LEADERSHIP PRESCRIPTIONS

As I interpret these models, the behaviors most appropriate for long-term success in leading functional technical operations would be supportive, relations-oriented behaviors with occasional short-term changes into a task type of behavior intended to solve specific short-term problems.

Project Leadership

DEFINING THE GROUP

While we can assume that the same training has been imposed on the people who work in projects as in functions, we cannot assume that they were very proficient in project operations before this project team was organized. That is apparent, since project operations may be relatively new and the specific project is very new by definition. In other words, even if people are quite familiar with project operations in general, they have never interacted in *this* specific project organization before, and each project is new by definition. Members of a project group are more independent, have lower group cohesiveness, and are probably more different from each other than functional employees (Lawrence & Lorsch, 1967).

DEFINING THE SITUATION

There is high uncertainty at the beginning of a project with respect to both project task unknowns and human interactions. If the declining S-shaped curve for uncertainty versus time (see Figure 4-7) is reasonable, I suggest that most uncertainties probably tend to decrease at a point that is about 25 to 30 percent into the project. Then, after the middle of the project passes and project task unknowns continue to decrease as design and production problems are solved, uncertainty (as perceived by participants) again begins to rise for human interactions, because the end of the project is approaching.

Since we are always concerned with human interactions, the uncertainty (as viewed by project participants) would probably initially be high in projects, drop to moderate at about midpoint, and then very slowly climb as the project end approaches and the people begin to wonder about the next project (or if there will even be one). The declining S curve of uncertainty versus time (Figure 4-7) refers to the total project uncertainty—that of reaching all the project goals *and* not only that of the participants' perceptions of their own uncertainties, which is slightly different as the project ends. The total project's uncertainty curve, which means meeting the project's goals, continues to decrease; the participants' rises slightly (I believe).

DEFINING THE LEADER

Using the three models noted above, the suggested leadership behavior recommendations are quite different.

Using Fiedler: Since all the conditions at the start of a project usually seem to be negative for the leader—i.e., poor leader-member relations, task structure, and position power—a low-LPC leader would be appropriate. The expected task orientation of the technically trained person is necessary and an asset: to organize, define, and, in general, get the project moving. This would probably fall into category 8 of the Fiedler graph shown before (Figure 7-2). As project task problems are solved and uncertainty decreases in the middle of the project, the task structure increases, requiring a change to relationships-oriented behaviors. Then, as the project begins to end, and if leader-member relations have improved, there is again a need for a task orientation.

Using Blake and Mouton: There is an initial need for concern for production. If that is satisfied and the project moves along, the general recommendations that Blake and Mouton make for a 9,9 become useful in the middle; concern for relationships should happen toward the end. These predictions conflict with those in Fiedler's model.

Using Hersey and Blanchard: There would be an initial high task behavior, then high relationship, and, as the project began to end, back to high task. This coincides with Fiedler, as I have defined it.

SUMMARY OF PROJECT LEADERSHIP PRESCRIPTIONS

According to my general interpretation of these models, and fitting them to my diagnosis of project operations, I would suggest that as an initial management hypothesis the leadership behavior patterns should start out as task oriented, change to those that emphasize relationship in the project middle, and then back to task during the final phases. As I've said before, that's not easy to do. Deliberately changing one's behavior to fit a changing situation requires unusual personal discipline and patience, which are usually in short supply when uncertainty begins to rise. That would, however, be my prescription for the optimal leadership patterns in managing projects.

There is an interesting extrapolation to this prescription that would apply if you were managing several projects, all of which were in different stages of uncertainty (i.e., some of them beginning—task—some of them in the middle—relations—and some of them at the ending—task again). I think that I've described a multiple personality in one person, and that, surely, is not the usual description of a technical manager! What it seems to mean is that whatever dominant type you are, task- or relations-oriented, there are occasions when you have to be able to use other behaviors, at least for a short time. So it might be useful to consider what is required when facing different stages of your projects. That is really the basis of my prescriptive hypothesis.

Since these recommendations have not been tested except through my own experience, there should be a caveat or warning given to you here. I suggest that before using these or other recommendations, you evaluate thoroughly the applicability to your situation. In other words, since there has been no in-depth replicative research or testing done, these general recommendations for effective leadership behaviors in functional and in project operations are based primarily upon nonreplicable personal experience in the field and extensive anecdotal empirical confirming data gathered from others. If the data, the assumptions, and the descriptions of the people and the situation do not hold, the diagnoses and prescriptions will fail.

Several other considerations should be kept in mind:

1. Any relationships between leaders and subordinates are built over time and involve multiple exchanges between leaders and subordinates. Those exchanges include both tangibles, such as rewards and compensations, and nontangibles, such as commendations and approvals. They affect the relationships, which change with time and affect both the leader and the subordinates. Behaviors, in most technical organizations, are difficult to permanently change, even though that's the suggested management route, and as it is virtually impossible for a leader to change from task to relationship quickly, it is just as difficult for groups to build internally cohesive and predictable behaviors that approve a leader quickly.

2. There is no assumption in the above prescription of leader behaviors that an either-or behavior is required. Task-oriented leaders seem to be capable of relations-oriented behaviors to achieve their ends, just

as relations-oriented leaders are capable of acting in task-oriented ways. The one research finding on middle-LPC leaders (Kennedy, 1982), indicates that a middle-of-the-roader could also function well. My opinion is that middle-LPC managers would be able to handle the higher uncertainty levels less well than the LPC extremes (i.e., either high or low). As human beings, we are capable of many different behaviors, and we can change those behaviors for fairly short periods of time fairly easily. It is the more or less permanent behavioral changes that are very difficult and keep the psychiatrists busy.

3. These prescriptions, in general, match two fundamentally different but related organizational models—one directed toward nonrepetitive problem solving, innovation, or unique decision making (relationships and organic structures) and the other toward disciplined performance or production (tasks and mechanistic structures).

4. According to one source, "rather than standing apart, the situation perceived to exist may be [the leader's (my comments in parentheses)] creation. It is now possible to see that trait (or universal) and situational approaches merely emphasize parts of a process which are by no means inseparable" (Hollander & Julian, 1976, p. 476). That reinforces and repeats what has been stated before in this book. The situation-leader (or person) interaction is a gestalt or total process that cannot easily be dissociated for analytical purposes. The process may even be a creation of the observer. Those could be major factors in the inability of both universal and situational theories to explain all the observed behaviors. We have separated some of the parts here merely for instructional purposes, and have pointed out that situational theory is more applicable to technical operations. It can never be the entire answer, since that answer depends upon you—and you are somewhat of a variable.

AFTER DIAGNOSIS, THE TRAINING

It is very difficult to change the behavior of leaders or potential leaders permanently.

A person's leader-style . . . reflects the individual's basic motivational and need structure. At best, it takes one, two, or three years of intensive psychotherapy to effect lasting changes in personality structure. It is difficult to see how we can change in more than a few cases an equally important set of core values in a few hours of lectures and role playing or even in the course of a more intensive training program of one or two weeks. (Fiedler, 1967, p. 248)

It seems to be much easier to change the organizational situation than the leader's style. The job can always be changed by organizational rules. As an example, we have the newly appointed project engineer who has had his or her job changed drastically. The engineer has just been given the authority to hire or fire people on the project, after serving in a staff position with none of this power. His or her position power has been increased, and according to most of the models, the leader requirements have become more relationship-oriented.

Then, considering what we said before about projects, the engineer's behavior should be changed over the project life; initially it should be task-oriented (project beginning), then relationship-oriented (project middle), and finally revert back to being task-oriented (project end). Therefore, if the people on the project can feel confidence about new projects coming up when this one ends, even the project end could become relationships-oriented. According to these new contingencies, the project manager is supposed to behave like a functional manager after the initial organizing or task phases of the project beginnings are completed. Changing behaviors in organizations is usually attempted on a formal basis through training. After the analysis of the difficulties of changing behaviors, does training really work?

There are reports from all the situational researchers that we have discussed—Fiedler; Hersey and Blanchard; and Blake and Mouton—of successes in training managers to fit the requirements of the situation as they have defined it. Fiedler, Chemers, and Mahar (1976) concluded that there were supportive results for participants going through the four- to six-hour self-paced programmed instruction workbook titled *Leader Match*. Hersey and his associates provide industrial training for many organizations, and there are anecdotal reports of success as a result of it.

In a more controlled research study, there was an interesting report about using the Hersey and Blanchard system for leadership training of adult women (Hart, 1975). The report stated that the trainees perceived themselves as better able to take control over their lives, become more active, and make better decisions. Blake et al. (1964) concluded that there was an increase in positive managerial attitudes and an improvement in departmental productivity as a result of grid training to achieve a 9,9 framework.

One major problem is reported with all this leadership training. It is nicely quoted as follows: "There is evidence that leadership training programs result in some behavioral and attitudinal changes, although these changes are dependent upon a supportive organizational climate" (Schein & Bennis, 1965). A supportive climate includes the ability to practice what has been learned on the job, and to receive feedback or corrected reinforcement for these different practices. Perhaps the most important kind of reinforcement is that received from one's immediate supervisor. Researchers (Haire, 1948) have often made the point that for management or leadership training to be effective, the entire management group should be trained. It is nonproductive and even possibly destructive for lower levels of management to be trained in a supervisory style that is incompatible with that of upper management.

This was further supported by some recent research on the practicality of committing funds for individual management development that includes leadership training. A major conclusion was that the primary focus should be on promoting the development and institutionalization of specific organizational practices before committing resources for individual development purposes, since the environment of the organization itself constrains what leaders can do (Dreilinger et al., 1982, p. 70).

In one organizaton with which I am familiar, the lower levels of leaders were trained to be supportive, creative, and relations-oriented when dealing with their technical personnel. The positive effects of this training stopped short when a memo arrived from the engineering vice president stating that he wanted as much creativity and innovation as possible, but that any changes that did not agree with existing product lines or personnel procedures had to be approved beforehand by him. In other words, beginning immediately, you are directed to be more creative, but only within the limitations that I have established.

To put this in more appropriate research-oriented language, the results of training are unsatisfactory when they collide with a top management group that is insensitive to the suggestions for change the training was intended to initiate (Sykes, 1962). Therefore, it seems reasonable that training to achieve better leadership behaviors can be effective provided two general criteria are satisfied:

1. The best job situation is developed as perceived by the organizational participants.

2. Top management either goes through the same training or commits itself to supporting the results of that training.

In Chapter 8 we will deal with training again, but from a different viewpoint: that of achieving total organizational change rather than individual change. As with many of the other subjects that we have covered, training can be analyzed from several positions.

SUMMARY

This chapter has reviewed some of the pertinent leadership theories that seem to apply to technical organizations. It has attempted to describe them considering both the situations and the leadership patterns that should fit the two principal operations of most technical departments: functions and projects. Guidelines for organizational and personal diagnosis have been suggested and general prescriptions developed. The pluses and minuses of training individual people in leadership behaviors have been covered. As usual, the application of this material is up to you. A straightforward procedure that I have often found helpful in developing my own theory is:

1. Try to define what presently exists.

2. Determine what should exist.

3. Determine how to go from what exists to what should exist.

For example, you might want to quantify the *what is* and the *what should be* subjectively, so that you'll have some guidelines on what to do first, then second, etc., to attain your goal. Using a Likert-type scale (i.e., a subjective value from 1 to 10, with 10 being the best) might give you the following results:

Typical definitions	What is	What should be
1. Projects are well-defined initially.	7	10
2. Reports are delivered promptly.	5	9
3. Problems are uncovered immediately.	4	7

This could mean that problem 2 is most important. (The score difference between 5 and 9 is 4, greater than for problem 1 or 3.) One possible way to improve would be to redesign the management information system to get faster data. You may have your own interpretation, but the idea is there. This method is repeated for you in Chapter 9.

I suggest that this procedure is less costly to put on paper than one that is attempted without any kind of a blueprint or plan. Define your theory, set up your hypotheses on paper, then try them in the organization. Just the process of writing things down helps the behavior-cognition interactive process, which is more simply called *learning*.

SUGGESTED ANSWERS TO CASE QUESTIONS

1. The leadership requirements seem to have moved from a task orientation to a relationship one (Fiedler), from an immature or young group to a mature one (Hersey & Blanchard), and from a strict concern for production to concerns for both production and people (Blake & Mouton). The leadership require-

ments involve more coordination of others and coaching tasks and less doing the work. The R&D group has moved from a functional to a project orientation, since the tasks are no longer concentrated in one area, but are spread across departmental boundaries.

2. **a.** *Defining the problem:* Problems should be documented in some type of standard format, perhaps by major customer or industry, because requirements change across different product lines.

   b. *Project leader's authority:* The project leader should have the ability to select and relinquish the personnel needed to do the projects assigned. This is one of the better times to install project controls, such as the ability to *open and close* financial aspects of projects.

   c. *Intergroup cooperation:* One mechanism would be to install dual reporting. Another would be to schedule total project design reviews at predetermined intervals during each project life. These reviews would be formal meetings during which project problems and progress could be discussed.

3. Implementation of change requires communication and expertise.

   a. *Communication:* What will the proposed design mean to participants and how will it benefit them? Ask the people affected for answers in order to determine what the proposed change means to them.

   b. *Expertise:* The "experts" in this case would be Melanie and Mike. They should be selected as an ad hoc design team to develop proposals for the new system. What would happen next and how would they get their answers? The success of a system implementation is determined the same way that any forecasted achievement is measured: according to predetermined standards. What would a possible measurement be that would mean the successful achievement of a new scheduling system? How could it be measured?

If this had happened in your company, what would have occurred in Bob's office? At the staff meeting?

Do you have alternative answers?

REFERENCES

Barnard, Chester I. *The functions of an executive*. Boston: Harvard Univ. Press, 1938.

Bass, Bernard M. *Stogdill's handbook of leadership*. New York: Free Press, 1981.

Blake, R. R., & Mouton, J. S. *The managerial grid*. Houston, Tex.: Gulf Publishing, 1964.

Blake, R. R., Mouton, J. S., Barnes, J. S., & Greiner L. E. Breakthrough in organizational development. *Harvard Business Review*, 1964, 42, 133–155.

Blau, Peter M. *Exchange and power in social life*. New York: Wiley, 1967.

Burns, T., & Stalker, G. M. *The management of innovation*. Chicago, Ill.: Quadrangle Books, 1961.

Dreilinger, Craig, McElheny, Richard, Robinson, Bruce, & Rice, Dan. Beyond the myth of leadership—style training—planned organizational change. *Training and Development Journal*, October 1982, *36*(10), 70–74.

Fiedler, F. E. *A theory of leadership effectiveness*. New York: McGraw-Hill, 1967.

Fiedler, F. E. Styles of leadership. In E. P. Hollander & R. G. Hunt (Eds.), *Current perspectives in social psychology* (4th ed.). New York: Oxford Univ. Press, 1976.

Fiedler, F. E., Chemers, M. M., & Mahar L. *Improving leadership effectiveness: the leader match concept*. New York: Wiley, 1976.

Glaser, B. G., & Strauss, A. L. *The discovery of grounded theory*. Chicago, Ill.: Aldine Press, 1967.

Haire, M. Some problems in industrial training. *Journal of Social Issues*, 1948, *4*, 41–47.

Hart, L. B. Training women to be effective leaders. *Dissertation Abstracts International*, 1975, 35, 1977.

Hersey, P., & Blanchard, K. H. Life cycle theory of leadership. *Training and Development Journal*, 1969, *23*, 26–34.

Hollander, Edwin P., & Julian, James W. Contemporary trends in the analysis of leadership process. In E. P. Hollander & R. G. Hunt (Eds.), *Current perspectives in social psychology* (4th ed.). New York: Oxford Univ. Press, 1976.

House, R. J. A path-goal theory of leader effectiveness. *Administrative Science Quarterly*, 1971, *16*, 321–338.

Katz, D., Maccoby, N., & Morse, N. C. *Productivity, supervision and morale in an office situation*. Ann Arbor, Mich.: Univ. of Mich. Institute for Social Research, 1950.

Kennedy, John J., Jr. Middle LPC leaders and the contingency model of leadership effectiveness. *Organizational Behavior and Human Performance*, August 1982, *30*(1), 1–14.

Lawrence, Paul, & Lorsch, J. *Organization and environment*. Boston: Harvard Univ. Div. of Research, Grad. School of Business, 1967.

Likert, R. *New patterns of management*. New York: McGraw-Hill, 1961.

McClelland, D. C. *The achieving society*. Princeton, N.J.: Van Nostrand, 1961.

McGregor, D. *The human side of enterprise*. New York: McGraw-Hill, 1960.

Machiavelli, Niccolò. *The prince and the discourses*. New York: Modern Library, 1940.

Mischel, Walter. *Personality and assessment*. New York: Wiley, 1968.

Schein, E. H., & Bennis W. G. *Personal and organizational change through group methods: the laboratory approach*. New York: Wiley, 1965.

Sykes, A. J. M. The effect of a supervisory training course in changing supervisors' perceptions of the role of management. *Human Relations*, 1962, *15*, 227–243.

Thompson, James D. *Organizations in action*. New York: McGraw-Hill, 1967.

Woodward, Joan *Industrial organization: theory and practice*. London, England: Oxford Univ. Press, 1965.

FURTHER READINGS

Andrews, K. R. *The effectiveness of university management development programs*. Boston, Mass.: Harvard Univ. Grad. School of Bus. Admin., 1966.

Bennis, W. G. Leadership theory and administrative behavior: the problems of authority. *Administrative Science Quarterly*, 1959, *4*, 259–301.

Burns, J. M. *Leadership*. New York: Harper & Row, 1978.

Fiedler, F. E. The contingency model and the dynamics of the leadership process. In L. Berkowitz (Ed.), *Advances in experimental social psychology* (Vol. 11). New York: Academic Press, 1978.

Hersey, P., & Blanchard K. H. Leadership style: attitudes and behaviors. *Training and Development Journal*, 1982, *36*, 50–52.

Hilgard, Ernest R., & Bower, Gordon H. *Theories of learning* (3rd ed.). New York: Appleton Century Crofts, 1966.

Hollander, Edwin P., & Hunt, Raymond G. (Eds.). *Current perspectives in social psychology* (4th ed.). New York: Oxford Univ. Press, 1976.

Korman, A. K. Contingency approaches to leadership: an overview. In J. G. Hunt & L. L. Larsen (Eds.), *Contingency approaches to leadership*. Carbondale, Ill.: Southern Illinois Univ. Press, 1974.

McGregor, D. *Leadership and motivation*. Cambridge, Mass.: MIT Press, 1966.

Seashore, S. E. *Group cohesiveness in the industrial work group*. Ann Arbor, Mich.: Univ. of Mich. Institute for Social Research, 1954.

8
GETTING THINGS CHANGED

Case Study
THE CASE OF HEISENBERG THE UNCERTAIN

CAST

Hank Heisenberg: Chief project engineer

Mike Johnson: Project accounting

Lief Gilder: Vice president, engineering

Marvin Swerley: Chief draftsman

Dr. Peter Hazbee: A consultant retained by the engineering division of the company.

Caring Steel Company was a major producer of advanced state-of-the-art aircraft radars. The company had always been able to design and build the best equipment that the industry could offer, but it had always operated under defense cost-reimbursement contracts. When it started a commercial division to supply civilian airports with similar equipment, it consistently lost money. Lief Gilder had called Hank Heisenberg, Mike Johnson, and Marvin Swerley to a meeting in his office to try to find out why the performance of the commercial products was so poor.

The time was 8:30 A.M. in Lief's office.

Lief: I've asked Mike to review our cost figures for the last three contracts that we've had in our commercial division. The numbers are terrible. Mike, you put them together. What do you see?

Mike: Well, in each case the estimates were reviewed by the project team and then top management and were finally accepted by the customer. I must admit that the price had to be shaved a bit during final negotiations, and the project team never met its schedules. The preliminary design seemed to go well, and the initial design review occurred on schedule, but the development costs went right through the roof and the rework on the field installation was about three times what we estimated it would be.

Marvin: I can explain some of the overrun costs in development, because my group expended a lot of those funds in redrawing the designs several times. The engineers just couldn't make up their minds. They kept changing the design, and you know what that does to costs.

Hank: Well, I really can't explain it. We manage those commercial projects exactly the same way we do our defense projects. We've built that general kind of equipment several times before, so it's not really unusual for us. I just don't know why we haven't been making our budgets. Of course, when we worked with the defense department, any changes they wanted, they paid for. These commercial airports are different. They think that they can get anything added on that they want without paying for it.

Lief: Well, we've got to get to the bottom of this. I've asked a consultant to come in to help us. He'll be talking to you all fairly soon. Maybe he'll come up with some suggestions. This meeting is over.

Several weeks went by. During that time, Dr. Peter Hazbee had a chance to visit all of them and discuss the problem. He then inspected engineering operations, drafting, support functions, project management, and the manufacturing departments. There was another meeting in Lief's office several weeks later, at which Dr. Hazbee offered some suggestions.

Peter: I'd rather make my report orally so that we all can offer contributions and suggestions and in any other way comment upon what I have to say. First of all, I'll give you the obvious conclusion, and that is that the nature of the business changes when you work with commercial clients. The government will pay for a change in scope if it wants another thing added to the design. Commercial customers expect you to deliver the latest state of the art as a regular business practice. If you won't, they'll get someone else who will.

The project engineering design group seems to be reluctant and uncertain about writing an impact statement for two reasons, as I see it: When the changes are transmitted from this office, it feels that it's a top-down direction and the group might appear disloyal if it didn't meet its original budget. If it comes from the customer, the group feels that if it charges for it the customer will complain to this office and the group will get blamed. There are a lot of other situational factors, but the people in this company not only know what they are, they even have some ideas about how to solve them.

Lief: Well, fellows, do you agree with these findings? Why doesn't anybody say anything?

Hank: Lief, I run my projects according to company policies and I don't really know if there are any solutions in the company, as Dr. Hazbee says. Consultants are supposed to come up with answers. Where are they? I haven't seen a written report or anything.

Lief: Well, Peter, what do you have to say?

Peter: A written report would cover exactly the same things that I'm telling you, and whether it's in writing or not, it's the ideas that are important. I suggest that an ad hoc committee be appointed to evaluate the problem thoroughly, develop solutions, and even implement them.

Lief: Well, I wish that I could be on it, but I'm too busy. Why don't you fellows work with Peter here and become the committee? Set it up just the way that Peter described it. This meeting is over.

Several months later, the committee had accomplished the following list of items.

1. It had sent out questionnaires to the technical group asking its members to list the five most important problems in order of importance.

2. It then sent out another questionnaire with problems listed according to the data it received and asked for suggestions on how to solve them.

3. All the suggestions were scored and a design plan to correct the worst three problems was set up. That was circulated to the people who had responded to the questionnaires. There was further feedback and the committee started to implement their suggestions.

4. They kept the entire company advised through periodic progress reports and requests for feedback.

One month before the design corrections were to be implemented, Lief called Dr. Hazbee into a meeting in his office.

Lief: Peter, I don't mind telling you that all of this committee stuff has the organization in an uproar. That's all everybody is talking about, and

I'm not sure that I want to agree to the committee's suggestions. It wants to have project management issue an impact statement almost every day, it seems. And this idea of the project team having to approve a price after I've negotiated it is just not workable.

Peter: Lief, everybody is talking about it because they're interested in it and can see how they'll all benefit from the proposed changes. If you don't agree with the committee's suggestions, why don't you do something about it? And why shouldn't the project people approve the final price? After all, if they don't feel that they can make it for the price that your salespeople have agreed to, you'd better know it in the beginning. And Heisenberg, your project manager, always seems to be uncertain about his team being able to get the job done because of changes that are imposed by top management and the customer, without any equivalent increase in time or funds.

Lief: Well, I'm going to sleep on it. I'll let you know what I'd like to do next week.

QUESTIONS

1. What do you think Lief's alternatives are? What do you think that he will do?

2. Do you believe that Hank really is uncertain? Why do you think that he wanted the consultant's report in writing? What tools does he have or can create to help him get the job done?

3. Why did Peter say that he believed that the people in the organization knew what the problems were and could offer suggestions for their solution? The original meeting in Lief's office didn't provide any answers. Has this ever happened in your organization?

MAJOR CHANGE: HOW TO INITIATE OR RESPOND TO IT

The last chapter was about leadership concepts to use during the regular business of the day. We found that even highly complex projects intended to result in completely new products or services have some semirepetitive elements that are predictable if we use a contingency-based framework to define them. Therefore, the manager's job of making decisions by absorbing uncertainty does not include dealing with a completely new project every time, just parts of it.

But other kinds of projects extend beyond the usual boundaries of the technical department and can be larger and more complex than those usually handled by technical managers. These different and very unusual kinds of projects are not concerned with a very limited end goal, product, or service produced by the technical group and intended to satisfy some customer or client, but with developing a *new* methodology or plan that is intended to affect the entire technical organization. There are fewer predictable parts to these projects. These projects are almost entirely unique and are attempts to solve major internal problems. They are more difficult to manage than those concerned with new products or services because they involve changing the organization itself—one of the most important tasks that any manager faces. This task requires different kinds of leadership techniques. These are needed to respond to or to initiate major change.

These projects are started because some of the repetitive, ongoing management decisions have become either inadequate or inappropriate. There is a perception that something has happened in the environment, the organizational structure, the technology, or the people and a "new" situation exists with more than the usual uncertainty in it. It extends over the whole technical organization, not just in a part of it, as before. The standard sequence of fitting as much of it as possible into repetitive decision matrixes, then working on the few parts that don't fit by absorbing uncertainty, no longer operates satisfactorily. Too much is new. The interaction of this new problem with the organizational model can no longer be contained within past practices. There are new contingencies to consider, and the leadership concepts that worked well in more stable situations (even those of project operations) may not work well anymore. Something else must be done, and that something affects the design of the total organization, across the limits of the technical department.

As we said before, minor change for individuals and small groups is almost an everyday occurrence. It is generally defined as no major alterations among the relative positions or sizes of the model components noted in Chapter 2 (people, technology, and structure, plus the information systems surrounding them). However, when a change does affect the relationship or the size of these components, the magnitude and the frequency of internal interactions increase greatly and the complexities of the situation seem to become geometrically larger. The leadership tasks are correspondingly increased.

As an example, consider starting a project on the development of a special series of high-pressure, corrosion-resistant widgets. There might, of course, be some minor changes in the organizational components when the project gets going. (I'm assuming that these types of operations have been done before in your situation.) They might include the assignment of several designers, engineers, technicians, etc., to the project manager on a temporary basis (either directly or in a matrix form), which would change the situation. This is defined as a minor change because the development of the widget line is not usually expected to affect the organization itself permanently. Of course, it might (since anything could), but it's not intended to.

On the other hand, if the project to be started is to redesign the engineering department from a functional basis (design, production support, maintenance engineering) to a product basis (widget products and blodget products) in a permanent arrangement, most of the organizational model would have to be restructured. The shapes and relative sizes of the organizational components might be permanently changed. (Some of the possible changes are discussed in Chapter 4.) This would definitely be a major change and should happen less often. Of course, this change classification is not that rigid. Some organizations might even have both major and minor changes all the time. (I was once in an aerospace company that changed its organization chart so many times that the inside joke was that the company was going to put it on the computer on a real-time basis. Another saying was, "If my boss calls and I'm not at my phone, be sure to get his number. I want to be able to call the right person back!")

Perhaps a better definition of *minor* versus *major* is the degree of repetition in the change that is perceived by organizational people and in the leader's ability to control the situation. I suggest that repetition in the change means "This is just another widget design to add to our regular line of widgets," and *not* "This is another change in the whole organization structure, since they now want quality control to report to production." For further discussions here, let us assume that the degree of change will be major (however you define it for your organization) and therefore the leader's ability to control the change will be our primary concern. We will discuss how to optimize the leader's control of major change—that change involving a new or different organization.

In the prior chapter, we defined leadership as an interactive process and that means that there must be members or other people in some type of relationship with the leader. That relationship is not fixed; it varies. Multidirectional change and response of leadership behaviors with varying contingencies is thus an integral part of the change process. Even in the minor change that was discussed in the prior chapter, we became familiar with the example of the take-charge, task-oriented leader in a mechanistic, production-oriented organization who found that his or her heretofore successful leadership pattern failed when he or she was reassigned to the research and development group assigned to develop new products. Creativity seems to be a fragile flower that responds best when supervised least. It seems to do well when exposed to a supportive, relationship-oriented leadership set of behaviors.

Conversely, the personally supportive, relationship-oriented leader who was superb at managing subordinates in the research laboratories while producing the new and improved widget failed when transferred into the production department. Production usually depends upon consistent, repetitive, and disciplined actions, and creativity may result in lowered quality and output. Typically, this type of transfer of a successful leader from one environment to another as part of a major change process is one type of change that probably could exceed the leader's control ability. She or he must use techniques that are entirely new and are different from past techniques if he or she is to succeed. Since there is always an interaction between leaders and subordinates, I suggest that we begin our analysis of potentially useful successful change techniques that you as the leader can use with understanding how the subordinates perceive the leader when the situation does change. We will start by discussing some of the research on those perceptions.

As noted in the prior chapter, the way in which the leader is perceived by his subordinates is one of the most important factors in his or her success in managing technical operations. Perception affects the way subordinates work. For example, under conditions with ambiguous tasks, unclear policies, and less need for personal autonomy, such as in a critical or confused situation, the directive leader is usually received well, and there is a positive correlation between this behavior that he or she exhibits and the personal satisfaction of subordinates (House and Mitchell, 1974). However, when the task becomes clear and policies become quite specific and are interpreted as being personally restrictive, this same leadership directness usually becomes a hindrance rather than a help to subordinates. The prior chapter suggests that this happens in leading smaller projects that don't have an impact on the total organization.

When we deal with major change, there is a corresponding increase in situational ambiguity, and we would expect the leader to be much more direct because of it: even more task-oriented. There always seems to be less clarity both in tasks and in the organizational structure during any change process that involves more than very small groups of people. When we deal with familiar people around us, most of us can fairly well define the problems that we have and can relate to them because of past experience. When dealing with entirely new kinds of things and different people whom we know less well, we're not as sure about new problems that occur.

Therefore, the technical leader who must oversee the design and implementation of a major change is in the same position (although greatly amplified) as the project manager at the beginning of an *entirely new* project: He should be a very task-oriented and directive leader. That seems, however, to conflict with the idea that the psychological needs of technical subordinates or knowledge workers doing creative or novel tasks generally favor the supportive leader for them to gain maximum satisfaction at work. In effect, major change seems to require directive leadership initially and major creativity seems to require supportive leadership initially.

The resolution of these two conflicting recommendations lies in an implicit assumption that we have made about the knowledge level of the leader: that she or he knows what to do with respect to the tasks that have to be supervised. If she or he does have the answers, the leader is directive ("OK, while this group hasn't worked together before, we'll start the design of the new widgets similarly to the project that our competitors produced last year on the new blodgets, and we'll see how our plans can be adjusted to fit this new situation. Therefore this is what I suggest that we do. . . .") If the leader doesn't even have the beginning answers for whatever reasons, she or he has to be supportive ("OK, while this group hasn't worked together before, the task of designing the new widgets has been defined for you. How shall we organize ourselves and what should the sequence of design be?")

We are familiar by now with these two different approaches, but most of the time these kinds of answers are not that obvious. Therefore, I suggest caution as a first choice, using the supportive mode initially, if possible, since the support of subordinates is always a major situational variable when major change is to be accomplished. We know that people interpret change differently. Therefore, the experienced leader will always begin the change–design-implementation process carefully, in a supportive leadership posture. The attitudes and acceptance of the subordinates are important design variables, and since the leader can rarely predict with certainty how organizational change is understood, I believe that the participative, supportive approach is recommended as a first choice. But people's interpretations of change are not entirely an unknown area, even if it is not completely predictable. There are some data on it and those are the next items to be considered.

THE KNOWLEDGE WORKER: ATTITUDES AND ACCEPTANCE

It is perhaps belaboring the obvious to point out that everyone is always in some type of change process. Similarly, the organization and the economic environment to which it reacts are not static; there is some constant motion among the parts of the organizational model. Therefore, as noted before, the technical leader is responding simultaneously to many different contingencies and so are the subordinates. But there is a suggestion that there is a very limited range in the change that subordinates will allow the leader to impose or manage before major problems begin to happen (Barnard, 1938). The subordinate will allow the leader to control certain of the subordinate's behaviors. As long as the proposed new change doesn't require the subordinate to exceed that range of behaviors that the subordinate has established, the subordinate will *allow* the leader to control the situation. According to this concept, the subordinate evaluates the proposed change and is probably saying (unheard), "I am in control of what I allow you to do to affect my behavior and, for a price (money, position, status, social conditions), I will allow you to direct some of the things that I do. This permission can be revoked at any time." When the proposed organizational change is within the limits set (or understood by) the subordinate that affect his or her own behaviors, the change is probably evaluated as a "minor" one: otherwise it is probably "major."

This is particularly applicable to knowledge workers in the technical organization. Therefore, the effective technical leader understands this unheard message and develops "minor" changes such as in new product development in a task-oriented way, but "major" changes such as organizational reorganizations in a relations-oriented way. Therefore, leadership direction either is applied only in a limited way, a way that either does not violate the limits set by the subordinate extensively or in a way that is intended to revise those limits by raising the price the leader and/or the organization pays to the subordinate to achieve his or her cooperation. When a major change is to be designed and implemented, it usually violates predetermined limits, and supportive leadership is necessary as part of the "price" to be paid. This is not a fixed process, since leadership is not institutionalized but is personalized. This does not, of course, minimize using other rewards such as money, increased job titles, etc., as another part of the "price" to be paid, but we are concerned with technical leadership at this moment.

The design and implementation of a successful change depend upon an (unstated?) agreement between the leader and the subordinate. When the leader has to exceed some preexisting limit, as defined by *the person affected*, that leader is dealing with an automatic definition of "new or major" change. (This is another one of those operationalized definitions.) Therefore, in this example, the leader has to consider how to modify the behavioral limits set by the subordinate or the person experiencing the change as part of the leader's design of that change process.

Some considerations could include, how does the subordinate evaluate the change? If it's evaluated as positive, the personal limits may automatically be lifted. If it's negative, something should be done to obtain that person's personal commitment, since the end result of the change depends upon it. As noted before, commitment is usually tied to some price (however defined) that the organization must be prepared to pay. In turn, that payment seems to be related to some reward that the person perceives he or she will receive for supporting the proposed change. The key to major change is therefore finally tied into personal motivation (see Chapter 3).

As noted before, this important contingency of the subordinates' behaviors is affected by how they interpret the proposed change. These interpretations can sometimes be predicted during the implementation of changes that are viewed by subordinates as being intermediate in their effects. One very important kind of intermediate sized, leader-initiated change is called *strategic planning*. Technical organizations are often directly involved, since they are often called upon to provide the innovation upon which much of the overall organizational strategic planning is based. This change certainly has more effect on the overall organization than project-based change, and less effect than restructuring the way a company operates.

STRATEGIC CHANGE

Strategic changes in most operating organizations that I have seen are rarely managed in the standardized way that many textbooks seem to suggest. Just as in organization design as we discovered in Chapter 4, there is no single theory or paradigm that works for all. By this time, we probably are all familiar with one well-established and open management pattern that suggests:

1. Establish goals.

2. Forecast resources and needs to reach those goals.

3. Set up measurements of progress or achievement.

4. Develop detailed plans to coordinate the activities of the various organizational groups.

5. Reevaluate achievements on a timely basis and go back to step 1 to start the process all over again.

It's the scientific method dressed up in planning terminology. It is all very clear and direct, but it never really happens that way. This method is a top-down, unilaterally accepted direction that would follow the classical theory of management. It assumes that subordinates' behavioral limits will not be violated. They will evaluate the process as almost a regular one, that is, somewhat repetitive over time. Developing major change for a technical organization involves more than this. That organization is not an army that happens to be involved in technical operations; it is a changing group of highly individual knowledge workers (and that includes managers as well as subordinates) involved in a shifting, complex process, possibly including first supportive, and then directive, leadership strategies.

Without the commitment of the team, change quickly goes wrong, because no plan can provide for all contingencies. (One way to bring an organization to a grinding halt is for subordinates to do exactly what they are told to do *and no more*.) Therefore, strategic planning (probably the most familiar and largest intermediate change) requires a somewhat different and more adaptive leadership process than is suggested by consistently applied relations-oriented models. According to Quinn (1980), leaders can begin the process using nondirective, relations-type approaches. The stages are similar to a muddling through or "cut and try" system. They are:

1. *Creating awareness*. By alerting the informal networks to intended changes and obtaining feedback from them, leaders can circumvent the tendency to shield themselves from potentially negative responses too soon. The ideas can be circulated as trial balloons in order to elicit informal suggestions that would either support or mitigate future implementation.

2. *Generating alternatives*. The trial balloons may not fly well, but maybe someone has an idea that modifies those balloons into something that will. Just the process of evaluation and suggestion fed back by some subordinates to the leader increases the potential for cooptation of those subordinates who are against the ideas or else provides modifications intended to improve the original ideas proposed.

3. *Broadening the support for change*. This may be done by allowing a reasonable gestation period for the ideas, setting up study or advisory committees that include both those who are for and those who are against the ideas (but mostly those who are for them), assigning "champions" to develop detailed plans, and making sure that discussions about the ideas are conducted under "no lose" ground rules. This means that all ideas are accepted and evaluated as positive, with the discarded ideas being considered only slightly less acceptable than the others. There is not supposed to be any personal stigma attached to having suggested an idea that is not accepted. At this stage, the relations-oriented process tends to minimize the "hardening" of subordinates' behavioral limitations.

This is proposed as an existential, step-by-step approach that is expected to modify the personal psychological limits of persons affected by the proposed change. It allows the leaders and subordinates to define the proposed change (for and to themselves), contribute to it and tentatively predict how their own behaviors will be modified by the change. It includes both the familiar behaviors, formal and informal evaluations, and those of power politics. This procedure is intended to be conceived broadly and to be under constant refinement. It is never supposed to be completely finished, and it usually means a great deal of organizational, in addition to personal, ambiguity affecting large numbers of organizational participants.

While this existential procedure can be useful as one alternative for changing the limits of personal or psychological acceptance of change, it doesn't help us to determine the total direction of that proposed change, nor to implement it. That is defined by the difference between the existing organization-environment fit (as it is presently understood) and some expected optimum fit toward which the organization should be moved.

In other words, when we can determine both the end point and where we are at present, we then can understand both the differences between them (i.e., the amount and the direction of the required change) and can begin to plan and direct the existential process that not only changes psychological acceptance but also works to minimize this total difference. Since the process of defining where we are can be similar to defining where we should be, we can start the definition process with either point. Since we can usually be more objective about defining where we want to be rather than where we presently are, I suggest that the first step in designing and implementing this kind of change should be defining our end point, optimum, or best situation. Goal forecasting is easier, and less threatening personally, than defining the present unsatisfactory situation.

THE OPTIMUM FIT: WHAT DOES IT REALLY MEAN?

Many factors can be used to define and measure the optimum. We have covered some of them in other chapters. For example, Chapter 4 analyzed several improved designs for the technical organizational structure. Since structure was defined as repetitive behavior patterns in both the formal and informal systems, there were several designs available to you, as the organizational designer, that were to be used depending upon the amount and rate of change to which the structure had to adjust. With slower change, the functional organizational structure was suggested to best handle "immortal" or continuing operations such as product support, drafting, and design standards. These operations have a fairly low, but continuing, level of problem solving, with very little overall change. With more rapid change, but limited to products or services developments, project organizations would be best since they were for "mortal" or relatively short-lived operations.

Technology was another variable component defining type of "fit." In Chapter 5, it was defined as the methodologies used by the organization to transform inputs into outputs. Typically, those methodologies could be manufacturing processes, decision-making processes, or others, such as information processing. The transformation process in manufacturing generally involved organic structures for unit and process production and mechanistic structures for mass production.

Other chapters dealt with the environment, information systems representing repetitive decisions, and, finally, some of the tools of leadership to manage it all. The one conclusion was that all the optimums were contingency (or situationally) based and that there was no "one best way." There are many optimums; not only one, and there are many end points or goals that you can define. As the amount of change needed to reach those defined goals increases or, in other words, as the difference between "today" and "what should be" increases, the change methods change. In the structural example, it goes from functions to projects. Similar changes are needed for technology, as noted before. The "optimum" depends on where you want to go and if you can get there from here. But how should one select the best optimum *at the moment?* What are the contingencies that tie all these variables together? What is an end point or optimum design that we can at least start with, understanding that when we reach it, we may have another goal or end point at our new "optimum"?

Starting with People

One optimum based on the person-organizational interaction is described in Lorsch (1977). That model involves the relationship between the organizational tasks and the individual sense-of-competence motivation. This model is very heavily influenced by human needs and expectancies, and since those types of variables are key ones in technical organizations, the model could be a very appropriate beginning to starting on the road to our temporary optimum design. The variables are defined by Lorsch:

—An individual's sense of competence is a self-reinforcing reward. As an individual performs a job successfully, feelings of competence encourage continued efforts to do the job well.

—Different tasks seem to be attractive to persons with different psychological makeups. For example, research scientists who work on uncertain, complex and long-range tasks prefer to work alone, with freedom from supervision, and enjoy highly ambiguous and complex tasks. In contrast, factory managers, whose jobs are more certain, predictable and short range, prefer more directive leadership, closer relationships with colleagues and less ambiguity (Lorsch, 1977).*

The organizational structure that is designed acknowledges the differences among people. Table 8-1 can be used for development of ordinal measurement of your optimum. It has been slightly modified (Lorsch, 1977). Using these data can help in defining the organization's end point. Obviously, any discrepancies between that proposed end point and the present situation would be a major reason for a strategic plan of organizational change. While these data are illustrative of a specific research point of view, they are only qualitative and involve the general fit between the needs of the economic environment and the requirements of the individual and the organizational culture.

The assumption that the organizational culture does fit most of the eco-

*Reprinted by permission of the publisher, from *Organizational Dynamics*, Autumn 1977, © 1977 by AMACOM, a division of American Management Associations, N.Y. All rights reserved.

TABLE 8-1 DEVELOPMENT OF ORDINAL MEASUREMENT OF OPTIMUM

Design dimension	R&D lab	Manufacturing plant
Structure		
Spans of control	Wide	Narrow
Job definitions	General and broad	Specific and detailed
Measurement	Less frequent	Very frequent
Planning	General, goals	Specific, detailed
Rewards	Money, professional recognition, careers	Money, management, careers
Selection	Technical qualifications, relations leader	Process and cost analysis, directive leader
Training	Professional conferences	Human skills, quantitative technology

nomic demands of the environment is not always an entirely valid assumption. Misfits do occur and may take a long time to cause failure. But, there are occasions when the misfit is obvious enough for management to perceive it and take corrective action. That corrective action might even require changing organizational culture, which is a major, lengthy operation, well outside the central scope of this book, even though we dealt with a small portion earlier in this chapter concerned with methods for changing subordinates' psychological limits. We therefore deal with it as a supplementary, rather than a central, concern.

ORGANIZATIONAL CULTURE

The culture of a firm is defined as the unique and readily observable behaviors that bind its members together and separate them from other organizations. That culture may be good or bad, functional or dysfunctional, or anything else, depending upon how we choose to evaluate it. Although generally acceptable, economic evaluations are not the only ones used. Companies may be losing money because of extensive R&D or market development activities and still be viable for long periods of time. If we choose to evaluate it using an anthropological framework (O'Toole, 1979), we can use the term *cultural relativism*, which means that there is no single absolute standard against which to measure an organizational culture, but there are relative standards that can change over time.

For example, there may be an accepted code of behavior within a specific firm that determines all interpersonal contact and is a part of the culture. People may be encouraged to use everyone's first names, even when addressing the president of the company, but everyone is also aware that this freedom doesn't apply when meeting socially after work. That culture has then a "relative" standard of behavior. Similarly, you're supposed to ask about each other's family and health, but not to respond to those questions in great detail. These patterns are learned quickly, and while the company will hire the maverick, he or she is quickly worn down into the accepted modes or is ejected (voluntarily or not) from the company. Although all types of people may be selected (and even this might be questioned in some organizations), only certain behaviors are rewarded. Culture may not wholly determine personality, but it greatly influences which personality patterns will succeed.

For example, one of the major Fortune 100 manufacturing companies was very centralized, with most of the important decisions being made at the top levels. All the top executives were members of the same golf club, and it was their practice to discuss business and make decisions while dining or playing. That club excluded women, blacks, Catholics, and Jews from membership. In recent times the company found it harder to recruit and retain top young managers and the board of directors could not understand why the quality of those younger executives it did retain was not up to the standards of its competition (based upon O'Toole, 1979, p. 20). That organizational culture gradually contributed to an organizational misfit of the company with the needs of the environment. The change required for a new optimum fit was certainly beyond the capabilities of any internal technical manager. In other words, these were gradual but important changes in its economic and social environment. Those environments had moved on past the familiar landmarks that existed when the company was started, and old answers no longer applied.

You can see how changing the organization's total culture would be beyond our scope here, since it involves extensive management actions and an extended time period (sometimes years) to accomplish, but there are some general ideas that might apply in the very limited circumstances surrounding the technical group. (I'm not suggesting giving up. I'm suggesting handling the part of the problem that we, as technical managers, might be able to solve.) Ideas to accomplish limited change would certainly start with some type of analysis of those desirable and different optimum structures and/or cultures, then define the existing organizational structures and sanctions that either support behaviors we want or discourage various kinds of behaviors that are no longer useful.

For example, if the limited intent is to develop new or innovative products, an organizational structure's insistence that time sheets be completed every week by a certain time of the day and that lengthy plans be approved by multiple levels of management before funds are spent would probably be one of those dysfunctional cultural sanctions and be among the first things to be changed. Achieving even this relatively trivial change in the limited cultural circumstances in and around the technical group could be difficult, but it can be approached most effectively by changing the *systems* within which people work, *not* by changing the *people* in them. Changing people cannot be done easily, if at all, and is probably not equitable, if we wish to moralize for a minute. In other words, "any approach that singles out people for change when it is just the institution that is determining their behavior is patently unjust" (O'Toole, 1979, p. 27).

It therefore follows that if you want to consider changing some of the miniculture around the technical groups, that might be a reasonable task, but that process is best accomplished as an open process, since cooperation of the knowledge workers is essential after you have achieved the easier task of changing the work systems. This process cannot be manipulative, as in Skinnerian conditioning. Obviously, human beings are a great deal more complex than pigeons, and even managers who believe in operant conditioning recognize that there are many forces upon human beings at work that don't necessarily begin, end, or even apply, in that work environment. Even if one assumes the extreme position that people can be manipulated, it's very difficult to do, since the manipulators don't have complete control over the experimental subjects.

Another point against complete acceptance of Skinnerian conditioning is that any change in the miniculture is almost always pluralistic. You can't change one thing and expect everything else to remain constant. Moreover, the proposed change cannot be monolithic, since it is intended to alter the way things *are done now* to some *other way* of doing things. Many things happen during the day, and they can all be affected by (and affect) the proposed change which must, by definition, be pluralistic. Change cannot be intended primarily to modify the participants themselves; they will still generally be the same loving, nasty, kind, hostile, selfish, or generous individuals before and after the limited change in the miniculture of the technical group.

However, there is always some interaction between the culture in which we work and our personalities, and a change in that culture always tends to have some effect on us. That effect is neither entirely predictable by management nor management's primary concern. When the change occurs, the reaction may range from complete acceptance to complete rejection. (The people could decide that "We're changing the limits within which we'll allow the organization to control our behavior.") When those limits are opened, people allow more of themselves to be controlled by the culture, and when the limits are closed down, people may leave the organization (either physically—i.e., I quit—or mentally—i.e., I just do my job and not one bit more). Culture and personality are interactive, regardless of the management intent behind the change, and the interaction always has some unpredictable parts.

Therefore, to decrease this human unpredictability, the general rule that

I suggest is that this kind of change should be, if possible, participative. As technically trained managers, we are not in general an emotionally adaptive group. In most cases, the things that we know and the ways in which we do them are familiar and comfortable. Changing them, by definition, is an uncomfortable process. Opposition can come from an entirely human response for self-protection and one's way of life. It is a major design consideration in any change process. On the other hand, participation and commitment also exists, as well as unpredictability during a change in the miniculture. The problem that you face as the leader is to determine if that commitment and/or unpredictability is for you or against you, and that is generally decided by the self-interest of the people concerned. And that, in turn, depends to some extent on *how* you present the proposed change, as well as *what* the proposed change is.

HOW TO GET THERE: EQUIFINALITY

When you have completed the process of defining the optimum organization (i.e., what would I like to have happen?) and a similar definition of where your organization is now (i.e., what is happening now?), the next task would be to develop some method for getting there from here. Just as there is no one optimum (the optimums could be different, depending on the people, technology, and structure) or the total culture, there is no one way to get there. This idea of the existence of multiple ways to reach some organizational goal is called *equifinality*. It represents that there are many possible processes of growth and development; not a *specific* one. It suggests that organizations are not wholly constrained by their initial structure, but can adapt without losing their basic form and can continually redesign themselves in more than one way.

The growth pattern from the small, single proprietorship into the multinational conglomerate can occur through the merger and acquisition path or through internal growth and diversification. These growth or change alternatives or combinations of these and other ways are equally likely according to the concept of equifinality. That seems reasonable when one considers the different ways in which many different kinds of organizations grow or are modified. Usually, the tendency is toward greater organizational complexity and size as a response to a larger range of environmental conditions.

We have assumed that the organization is an open system, accepting inputs, modifying those inputs through the technology chosen, and producing outputs. The concept of equifinality indicates that there are many ways for organizational change to take place and the one you choose should be related to your estimation of the ease of implementation which is the same thing as saying, the cost to get it done. This estimation must include an evaluation of the potential paths to the goal selected and the possible advantages or disadvantages to the people who have to implement the path selected. Remember that estimates and forecasts do not predict the future. They are only present evaluations of the relative costs and benefits that result from selecting an optimum path from those presently perceived by the estimator or forecaster. Therefore, forecasts sometimes don't come true. There are always obstacles to change that seem to be based on the individual's (i.e., the leader's and peers' or subordinates') behaviors and attitudes.

Assuming that equifinality is a reasonable concept, the selection of the path to effect change often seems to be some sort of wish fulfillment on the part of the leaders who desire the change. One leader may try to modify the structure ("What we need is to decentralize."); another, to change the technology ("We really need to computerize."); and the least productive, to change people ("We need to have everyone go through sensitivity training.").

The common misconception here is that the leader's way to make the change is the best way. It's almost an act of faith. I have noticed that it is particularly prevalent among leaders who have had no record of any previous disappointment in the work environment (the brilliant wonder who turned that losing division around in six months) and among inexperienced consultants ("Well, we did it before and it worked with Amalgamated Iron Works, why shouldn't it be just as effective with this company, the Universal Toy Factory?"). It demonstrates a naive approach to solving problems that denies the undeniable complexity of human beings.

Manipulative personal relationships designed to change people's feelings about security, morale, and personal sensitivity usually result in less than optimum change because of two implicit assumptions which are not true:

1. *Leader controls*. The assumption that the leader alone can design and develop an optimum plan.

2. *Uniform treatment*. The assumption that all those who interact with the change can be treated uniformly, because they will all regard the proposal the same way.

Of course, neither assumption is valid. The attempt to change people using only one path for the achievement of some organizational goal will invariably be less than successful. Using the "indirect methods" of sounding out potential opposition, gaining converts, changing the plan as new ideas are brought forward, and generally attempting to connect the revised (and undoubtedly improved) plan with the particular individual's self-interest is about the limit that the leader can expect. These indirect methods may seem to take longer but they result in lower implementation costs. There are many roads to a goal, and those that are chosen in an open, participative management environment usually are more successful because of the implicit commitment of those involved. The human costs and the consequent financial costs decrease with commitment.

SOME METHODS FOR BEGINNING IMPLEMENTATION OF CHANGE
Congruence Model

Obtaining the involvement of the organization in the change process can perhaps be started by modifying our by-now familiar scientific method. Some researchers (Nadler & Tushman, 1980) have developed a step-by-step approach, using a well-named congruence model.

1. *Identify symptoms*. List data indicating possible existence of problems.

2. *Specify inputs*. Identify the systems. Determine nature of environment, resources and history. Identify critical aspects of strategy.

3. *Identify outputs*. Identify data that define the nature of outputs at various levels (individual, group-unit, organizational). This should include desired outputs (from strategy), and actual outputs being obtained.

4. *Identify problems*. Identify areas where there are significant and meaningful differences between desired and actual outputs. To the extent possible, identify penalties; that is, specific costs (actual and opportunity costs) associated with each problem.

5. *Describe components of the organization*. Describe basic nature of each . . . component with emphasis on their critical features.

6. *Assess congruence (fits)*. Conduct analysis to determine relative congruence among components. . . .

7. *Generate and identify causes*. Analyze to associate fit with specific problems.

8. *Identify action steps*. Indicate the possible actions to deal with problem causes. (Nadler & Tushman, 1980)*

This may seem to be a very structured approach, but it is really no different from the seemingly undirected and less structured approach of Quinn (1980). The method suggested here doesn't state that this task falls only to the leader. It may (and I strongly urge it should) include the views and design modifications of the knowledge workers affected. The step-by-step approach simply provides a firmer thinking and estimating framework. The final change is always the result of interaction between the framework and the actual implementation.

Organizational Development

This is another process that has been used to effect organizational change. It is defined as:

> a long range effort to improve an organization's problem solving and renewal processes . . . —with special emphasis on the . . . work teams—with the assistance of a change agent, or catalyst, and the use of theory and technology of applied behavioral science, including action research. (French & Bell, 1978, p. 14)

There are several assumptions about people in this type of change methodology. They are:

1. If provided with a supportive and challenging work environment, most people have drives toward personal growth that are positively valued.

2. Most people can make a better contribution to the organization than the organizational environment will allow.

In my opinion, these assumptions are both the major assets and the major liabilities of organizational development. They depend on the implicit idea that there is some uniformity in the way that people regard their work. In some cases that we have discussed (Lawrence & Lorsch, 1967), we have a great similarity *within* technical groups but great *differences* among groups in a direct relationship with the way that they perceive change in their organization's environment. It appears that this research might conflict with these assumptions. Therefore, these assumptions are a major consideration if you intend to use organizational development techniques.

There are other conflicts when using these assumptions. Considering the mechanistic organization which is optimum in mass production organizations (Woodward, 1970), we find that human growth and improved contributions could be dysfunctional there. They would probably disturb the machinelike operation of this organizational model. That kind of organization doesn't respond well to this kind of behavior, and the assumptions, if true, would not support positive change. Of course, it might be said that this kind of organization doesn't provide supportive and challenging work either, but we know from Chapter 3 that the definition of supportive and challenging work is in the mind of the worker, not that of the researcher.

Organizational development, however, has been used successfully as a change technique in many situations in which it has been used. Those situations included positive elements such as open organizational systems, negative elements such as automatic resistance from those affected by any prospective change, and neither positive nor negative, but descriptive elements such as solving nonrepetitive problems that were not of central concern to the organization. Those elements defined as of central concern typically are the responsibility of top management—the survival of the organization, acquisitions of new companies, long-range strategic planning, product-line design, etc.

The *action research* process of effecting change (using the terminology of organizational development) is the process of collecting research information systematically about some organizational, goal-oriented system; feeding the findings back to the people in the organization; altering selected variables from the system based upon the hypothesis chosen and the data collected from the organizational people, taking some appropriate action; and then evaluating the results of that action through the acquisition of more data (based on French & Bell, 1978, p. 88).

Action research is obviously a participative process and seems to work very well when the situation requires destructuring or loosening up of tight organizations. It doesn't work well the other way. That usually requires strong, central direction. Our standard assumption that the knowledge worker of the technical organization usually prefers less directive, relations-oriented leadership supports the use of this action-research technique in implementing major kinds of change in technical groups, especially when that change is expected to apply immediately. It's almost irreversible once started, since movement in an opposite direction from open, participative groups and leaders toward closed, nonparticipative leadership situations usually decreases the "mental limits" of the knowledge worker that he permits the organization (or his boss) to use in controlling his behavior. The movement toward openness raises expectations and tends to increase the mental limits. It's difficult to put the escaped genie back in the bottle or the spilled sugar back in the pot.

Therefore, action research as a tool in organizational development requires the active participative leadership and continuing support of top management and the need for on-the-job application. Since the result will usually be a move toward more organic structures, there may be a decrease from the speed with which management decisions were made in the past. These kinds of organizations seem to make progress on an overall basis less quickly than mechanistic organizational structures since these top management may be making fewer quick decisions for the whole organization. It's no longer necessary for top management to be involved since others are doing it for their own areas without being told what to do.

Summary of Change Methods

We have discussed some of the change methods that may be used to modify the miniculture of the technical organization. Which method is selected depends a great deal upon the leader's forecasts of the "optimum" fit and the problems that he or she perceives *now*. The concept of *equifinality* becomes very applicable here. If the beginnings of the organization do not set it irrevocably upon only one possible path to the future and we understand that there are many potential paths to follow in achieving a goal, the leader is free to choose any method of change that he or she feels will work. The answer to the question, "How do I get there from here?" is straightforward enough. Just determine where you want to be, where you are now, and use any of the available techniques that you can learn about to get you from here to there. Any one of them may work for you.

Some of those techniques that deal with groups include Quinn's interactive strategic planning methods, congruence methods, organizational development, and organizational development's internal tool, action research. We are concerned initially here with changing the way that groups function, since we have more or less concluded that it is both undesirable and almost impossible to change the individual to fit our goal(s). We change the way they function by changing systems first. However, when these systems change the group's methods of working, interaction will make something of a consequent change in the individuals' behaviors.

So one might say that the individual is also changed, but that cannot be the primary aim. The individual is much more complex than any group. This would seem obvious, but a great deal of time and effort is expended in organizations on a point of view that is contrary to everything we have said. That point of view seems to be concerned specifically with attempts to change people themselves. That change process, which we discussed briefly in the last chapter, is called *training*. And since our plan in this book is to move from general to specific recommendations, we will now consider the general training methods that organizations use in attempting to change people's behaviors.

FORMAL TRAINING: WHAT IT IS SUPPOSED TO DO AND WHAT IT REALLY DOES

In many cases, the purpose of training a person is to cause some change in that trainee that will result in a greater return to the organization providing the training. The intent is no different from those noted above when working with groups. But critical reviews of research in the training fields shows that there is no scientific evidence to support claims that training in various individually oriented techniques is generally effective for the organization (Perrow, 1972). Attempts to produce long-lasting and positive change in the organization through typical training methods such as managerial grids, sensitivity training, job enrichment, and even job enlargement did not succeed.

Leadership training and leadership experience also do not seem to contribute uniformly to group or organizational effectiveness. But even though training does not produce uniform results for the organization and there is no *overall* gain to the organization providing the training, the results that are produced seem to have some limited gains because the *individual achieves*. While organizational effectiveness in general may not be improved, the skills, attitudes, behaviors, and technical and administrative know-how of the person is improved.

If training is considered within our contingency framework, the person being trained will probably have considerably greater control and influence over the job and subordinates than someone who is untrained and inexperienced. Leadership training and experience seem primarily to improve the ability of the leader to increase the favorableness of the leadership situation for him or herself. The conclusion to be drawn from this is that although training is intended to benefit the organization, it does so not directly, but indirectly, through benefits for the person. If the person then contributes more effectively (and only he or she will know that), the organization will gain; otherwise, not. Training is just as much a capital investment, in this respect, as the purchase of a machine tool or the improvement of a manufacturing building, but the end benefits are not as predictable for the organization *as a whole*.

The complexity of the people being trained and the interaction of the existing information that they carry into the training situation with the new information learned do not make for easy predictions of positive results. The organization as a whole, therefore, must be willing to accept a lack of predictability about the training end result, since the primary benefit will accrue to the person being trained. He or she will be better equipped to improve the favorableness of the leadership situation *for himself or herself*.

Any improvement in the overall organization will then depend partially upon factors outside the individual's control. That could mean that the optimum result of training would occur when there is a complete coincidence among the organization's goals, the individual's goals and new learning, and an implicit agreement about the best way to achieve these mutual goals. Since this rarely occurs, we have a situation in which the individual benefits the most from training and the organization as a whole benefits only when there is goal and methodology congruence. This assumes, of course, that the training has all the good things any training program is supposed to have: a direction, a method to assist the learner in moving in that direction, and techniques to support positive on-the-job applications and feedback at the conclusion of the training.

INFORMAL TRAINING: COUNSELING

Managers and leaders use other change methods in technical organizations, methods that are intended to deal primarily with the individual. These methods can all be categorized under the one-to-one interpersonal or communications patterns. A part of the management-by-objective technique includes the leader-subordinate interactions in setting specific goals for the subordinate to achieve in some future time period. It also includes personal counseling.

While the intent often is positive, the result often is not. The problems in attempting change through one-on-one discussions are major ones, since the attributes of an honest, supportive discussion between leader and subordinate rarely exist. Even professionally trained counselors must concentrate to provide an environment that will allow for discussion of the problems of the subordinate, for positive inputs to support personal development for the subordinate, and for an unbiased or subordinate-oriented viewpoint; the problems for the manager who attempts this are almost insurmountable. However, counseling (or supportive criticism) can occur occasionally when the subordinate recognizes its value; when self-interest is aroused. Under those conditions, it does change an individual's behavior.

The initial task would be to differentiate between task-related and interpersonal counseling or evaluations. Task-related information should be handled in an open, problem-solving process. While objective decisions may eventually become subjective as the data about product designs, cost, and delivery schedules are sifted and resolved, the process of open discussions (such as in design review meetings) should always be chosen first. Since problem definition is a first step in problem solution, task-related information can be defined initially through open discussion of the relevant facts.

Interpersonal and subjective information-handling methods, however, are different. They are always used on a private, one-on-one basis, since they implicitly contain potential conflict situations that could permanently damage relationships. They involve listening to the other person and understanding what emotions as well as data are being transmitted; e.g., "Let me tell you what you told me. If I can do that and you agree that what I said is what you meant, we will at least be able to understand each other's position. After I have done that, you try it for the things I said and I'll tell you if that's what I really meant."

The *training* that is done in individual counseling is always interactive. The counselor as well as the counseled is affected. In many organizations, the boss is more of a mentor (Schein, 1981), since there is a long-term relationship-oriented leadership situation and the boss trains through cues rather than direct feedback. Direct feedback usually is used in task-related evaluation. Therefore, task-related information is seldom the subject of personal counseling, and can involve more than the boss-subordinate dyad.

Conversely, interpersonal counseling, being unique in that it only concerns two people—the boss and the subordinate—is intended mainly to benefit only those two participants. If the organization benefits, that is a secondary gain.

Summary: Training

Training is usually the process of choice in organizations when the intent is to change the behavior of individuals through providing new information. On a formal basis, the results have not been uniformly effective for the organization because the information that person brings to the training situation interacts with the new information being transmitted in ways that are not always predictable. However, assuming that it is possible that the training provides new and helpful information to the person, the training results in some type of change intended to help that person be better able to control and improve his or her own leadership or working situation. Therefore, training leads to individual improvement and the desired organizational improvement may or may not occur. The organizational improvement depends on coincidence of organizational and personal goals, acceptance of the newly learned information, and positive reinforcement of these new personal behaviors.

On an informal basis, individual counseling can be divided into task-related information, which should always be handled in an open forum since it is based primarily on objective data, and interpersonal information, which should always be handled in a private, one-on-one discussion. The results here are similar to those of formal training since they also depend upon goal coincidence, acceptance, and positive reinforcement. The major differences between formal and informal training are the methods used. The end result is always the same: some change for the individual.

GETTING YOUR IDEAS ACROSS

After considering the general methods to be used in changing the miniculture, groups, and individuals, we come to very specific prescriptions for presenting these change ideas so that they generate minimum opposition. I suggest that an adequate plan and preparation before any proposals are presented win most of the battles. Some of the basic planning is fairly obvious. For example, know your audience. Who is going to receive this plan of yours and who has to approve it?

The effectiveness of any proposal depends initially upon the ability of the proposal receiver to understand it. If you are presenting information to someone who is technically qualified to understand the jargon, by all means, use it. Jargon is a form of oral shorthand that can cut down on supplementary explanation time. But use less complex forms of communication if the receiver is not up to it. Also, you might emphasize those parts of the change proposal that would appeal to the receiver. For example, someone in charge of plant operations would certainly be influenced more positively by a discussion of the potential for a decrease in maintenance costs rather than of an increase in production, assuming that both of these factors are part of your proposal. Conversely, the person in charge of production would have a reverse point of view.

Timing is a vital consideration, so do it now, but only if it's appropriate! Make the presentation as soon as you think you can get a positive reaction. There will be delays enough that are not of your making. Rehearse your presentation; then try a dry run with some cooperative associates to become more familiar with potential problem areas. There was one manager who always had a dry-run staff meeting before presenting his yearly budget to the board of directors. It accomplished two things for him: He was able to eliminate some problems before any board member could bring them up and his own team (which served as the "dry-run board") came to appreciate the complexities of their boss's dealing with the board. The best impromptu speeches are those that have been rehearsed thoroughly. Try listening to yourself reading one of your reports into a dictating machine if you wish to test this idea. What you hear will not be the polished logical input you thought you were providing.

Use visual aids, if necessary. When slides are projected on a screen in front of the room, you can point to appropriate areas and have everyone's attention on the same thing. Three-dimensional models are also effective, since everyone can then visualize what the final result should be like. A word of caution is necessary, though. These visual aids should be simple and easy to grasp. Colors and perspective drawings should be used to make things apparent, not to dazzle the viewer by themselves. The best visual aids quickly become almost invisible as the viewer concentrates on the idea expressed by them, not on their intrinsic value.

No one knows your subject better than you do. Therefore, these techniques are expected to make you more familiar with the act of transmission, rather than with the content of that transmission. Emphasize the receiver's point of view, rehearse your presentation, use visual aids and jargon where appropriate, complete dry runs, do it now, simplify your presentation media, or use any other technique that seems reasonable, but prepare your presentation thoroughly before you make it. Up-front planning is the best kind, since it costs less and prevents problems later on. Decrease the amount of uncertainty as much as you can at the beginning of any proposal and you'll be absorbing less uncertainty later on.

SUMMARY

Effecting intended change in an organization is a complex process that doesn't become easier as we move from large organizations through groups to individuals. The variables are different, that's all. And the actual change rarely seems to match the original intent; it may be better or worse, but it is always different. The process itself modifies the original intent into something else. The successful and experienced leader quickly learns that it is the process that is of major concern, and the goal is often achieved by allowing that process to move along, with other people's contributions.

We have explored both some of the reasons this happens and some of the methods intended to keep that intended change as much on the planned track as we can. One of the major supports for a change process in technical organizations is the need for open, participative discussions and commitment. Since the work is done by knowledge workers (both leaders and subordinates are in this classification) and no one can completely predetermine the amount of effort knowledge workers exert, the need for their support is apparent.

Commitment and support come more easily when there is occasional ambiguity in the leader. It's easier to obtain cooperation when one says, "Perhaps you can think about this a bit more" rather than "You're completely wrong. Redo it!" Laying all the cards on the table may be personally satisfying (especially if you are the one who lays them down), but it may not always be required or even desired if you want others to cooperate. In functional technical operations, directive leadership is not the primary mode; relations orientation is. Fortunately, the needs of the organization and those of technical participants often coincide, and effective techniques for organizational and systems change interact positively between the organization and the people. These suggestions are offered as pragmatic methods to get change accomplished. It has little to do with a claim to making organizations more humanlike but rather with using humanlike techniques to gain some organizational goal. In other words,

> Whether deservedly or not, humanism adapted to the management process has the taint of manipulation. It is difficult to imagine management using techniques like organization development, sensitivity training, or job enrichment out of the pure milk of human kindness. (Scott, 1974)

But in our examples, the "manipulation" is planned and controlled primarily by the organizational participants, *for their own self-interest*. If that "manipulation" is highly differentiated to account either for the different kinds of people in the organization or for the different tasks that each group within the organization has to do, the organization gains the flexibility and adaptability needed to modify its components and support its own further development. The organization (if we can speak of it as a total thing) is not being altruistic. Its aims just happen to coincide with ours—in most technical functions. Those aims are not always completely coincident, however. Each of us, as an individual, has unique aspirations, which no organization can satisfy completely. But there is a higher degree of coincidence in our kinds of work now than ever before, and we have the ability to continue to change the organization to suit more of our needs or the freedom to choose another organization. Which would you like to do?

SUGGESTED ANSWERS TO CASE QUESTIONS

1. Lief has four alternatives: (1) He can join the committee meeting and help in the problem diagnosis and solution. (2) He can accept the committee's work and support it. (3) He can have all the committee's suggestions sent to him for approval prior to implementation. (4) He can dissolve the committee. I would guess that he will choose alternative (3) first; then find out that the quality of the committee's work immediately drops and go to alternative (4). Of course, he will be firmly convinced that it was all a waste of time.
The best solution for the technical group and then, of course, for the organization would be for him to choose alternative (2). That choice is rarely made, but when it is, everyone gains.

2. Hank was probably very aware of the reasons for the project cost overruns, but he is a survivor. His request for the consultant's written report is a typical delaying ploy that allows him to find out how Lief feels about it and act accordingly. With an oral report, Hank had to rely on his own thinking processes. He has the ability to match the cost of each element of the work breakdown schedule against the contract and find out why costs were so high. All he has to do is compare actual to budget and define the largest variances first.

3. Most organizational participants are familiar with the reasons for and the potential solutions to repetitive problems. Often, it is top management who cannot or refuses to listen to them. If management would follow the decision-making flow diagram in Figure 1-2, the solutions to repetitive problems could quickly be placed in a solution matrix for all to see.

REFERENCES

Barnard, C. I. *The functions of the executive.* Cambridge, Mass.: Harvard Univ. Press, 1938.

French, Wendell L., & Bell, Cecil H., Jr. *Organization development.* Englewood Cliffs, N.J.: Prentice-Hall, 1978.

House, Robert J., & Mitchell, Terence R. Path-goal theory of leadership. *Journal of Contemporary Business,* Autumn 1974, 81–97.

Lawrence, Paul R., & Lorsch, Jay W. *Organization and environment*. Boston: Harvard Univ. Press, 1967.

Lorsch, Jay W. Organization design: a situational perspective. *Organizational Dynamics*, Autumn 1977.

Nadler, David A., & Tushman Michael L. A model for diagnosing organizational behavior. *Organizational Dynamics*, Autumn 1980.

O'Toole, James J. Corporate and managerial cultures. In Cary L. Cooper (Ed.), *Behavioral problems in organizations*. Englewood Cliffs, N.J.: Prentice-Hall, 1979, pp. 7–28.

Perrow, Charles. *Complex organizations: a critical essay*. Glenville, Ill.: Scott, Foresman, 1972.

Quinn, James Brian. Managing strategic change. *Sloan Management Review*, Summer 1980, *21*(4), 3–20.

Scott, William G. Organizational theory: a reassessment. *Academy of Management Journal*, June 1974, *17*(2), 242–254.

Woodward, J. *Industrial organization: behavior and control*. London: Oxford Univ. Press, 1970.

FURTHER READINGS

Cummings, Thomas G., & Markus, May Lynne. A socio-technical systems view of organizations. In Cary L. Cooper (Ed.), *Behavioral problems in organizations*. Englewood Cliffs, N.J.: Prentice-Hall, 1979, pp. 59–77.

Farris, George F. The technology supervisor: beyond the Peter principle. *Technology Review*, April 1973, 75(5),

Fry, Fred L. Operant conditioning in organizational settings: of mice and men? In Fred Luthans (Ed.), *Contemporary readings in organizational behavior* (2d ed.), New York: McGraw-Hill, 1977, pp. 377–382.

Galbraith, Jay. *Designing complex organizations*. Reading, Mass.: Addison-Wesley, 1973.

Gluck, Frederick W., & Foster, Richard N. Managing technological change: a box of cigars for Brad. *Harvard Business Review*, September–October 1975.

Hickson, D. J., Pugh, D., & Pheysey, D. Organization, is technology the key? *Personnel Management*, February 1970, 21–28.

Holmen, Milton G. Action research: the solution or the problem? In Cary L. Cooper (Ed.), *Behavioral problems in organizations*. Englewood Cliffs, N.J.: Prentice-Hall, 1979, pp. 203–228.

House, R. W. Some steps and approaches for conducting collective inquiries. *Proceedings of the IEEE international conference on cybernetics and society*, Denver, Colo.: Oct. 8–10, 1979, pp. 908–913.

Leavitt, Harold J. Applied organizational change in industry: structural, technological and humanistic approaches. In James G. March (Ed.), *Handbook of organizations*. Chicago: Rand McNally, 1965, pp. 1144–1170.

Lorsch, Jay W., & Morse, John W. *Organizations and their members: a contingency approach*. New York: Harper & Row, 1974.

Mirvis, Philip H., & Berg, David N. (Eds.). *Failures in organization development and change*. New York: Wiley, 1977.

Morison, Eltin. *Men, machines and modern times*. Cambridge, Mass.: MIT Press, 1966.

Pascale, Richard Tanner. Zen and the art of management. In *On human relations*. New York: Harper & Row, 1979.

Porter, Lyman W., Lawler, Edward E., III, & Hackman, J. Richard. *Behavior in organizations*. New York: McGraw-Hill, 1975.

Rickover, H. G. A humanistic technology. *Mechanical engineering*, November 1982, 44–47.

Salancik, Gerald R. Commitment is too easy. *Organizational Dynamics*, Summer 1977.

Schein, Edgar H. Does Japanese management style have a message for American managers? *Sloan Management Review*, Fall 1981, 77–90.

Skinner, W. The focused factory. *Harvard Business Review*, May-June 1974.

Torgerson, Paul E., & Craig, Robert J. A reexamination of Barnard's theory of organization. *Engineering Management International*, May 1982, *1*(2), 125–130.

Weisbord, Marvin R. *Organizational diagnosis: a workbook of theory and practice*. Reading, Mass.: Addison-Wesley, 1978.

9

SPECULATION ON FUTURE USES

REVIEW

We have used the triangular organizational model outlined first in Chapter 2 as a faithful tool to explore, categorize, and select some modern ideas for your own theories. The by-now familiar components of people, structure, and technology, all held together by leadership and surrounded with information systems, have served us well. But each time that we have attempted to define better ways to manage, there seemed to be more than one way. There were, in fact, many ways, if you recall the ideas about equifinality (you can get there using different paths).

But we seem to have conveniently categorized all the ways into some kind of a bilateral model. The first bilateral model was that of science and art. For example, under the *science* heading we have the logic and consistency of nature, and under the *art* heading there are the emotions and creativity of human beings. The organizational structures could also be divided into a bilateral division of mechanistic and organic, or functional-product prototypes and project prototypes. Technology can be divided into mass production and unit or process production, with equivalent kinds of divisions in the other areas of technology we discussed. Leadership was either task- or relations-oriented or was classified by initiating versus consideration behaviors. And even information systems can be arbitrarily divided into financial and management accounting systems.

NOT LIMITING EQUIFINALITY

I am *not* suggesting that equifinality is limited to either of two paths. It is not. Equifinality really means many paths, but limiting many paths into two general categories simplifies our ability to define, understand, and then begin to build our personal theory. It's a clarifying process. Another reason for the selection of two choices is that limiting all those paths to two general alternatives can happen because the definitions for either side of the bilateral division are not that clear and precise. For example, we will probably never see a pure mechanistic structure or an equally pure relations-oriented behavior pattern in a leader. So these imprecise definitions allow us to make convenient bilateral classification. They give us two general, easy-to-use categories in which to classify all the research that we have reviewed. But remember that this is an artificial kind of dividing process that is only one way to start your understanding of the management ideas we have discussed. (You may want or need to have a model of three, four, or even more divisions as you become more familiar with them.)

BILATERAL MODEL OF THE TECHNICAL FUNCTION

If we can begin with a bilateral division of ideas, it would seem reasonable that the technical function itself might also be bilateral, with one section intended to handle problems primarily in the natural sciences, using logic and consistency, and another in the social sciences, being primarily concerned with creativity and innovation. These sections or parts are not, of course, entirely separated in any organizational structure. They are related because their different kinds of assignments require both general kinds of skills within *the same technical function*.

We already know that science is not absolutely objective, because it is done by human beings called "scientists," and these scientists are subjective to some extent in what they perceive and in what they allow themselves to perceive even though they have trained to follow socially accepted "objective" methods. Therefore, the personal theory of the technical manager must include both the rational tasks of the relatively well defined natural sciences and the less well defined, emotional tasks to produce required innovations. The triangular model of Chapter 2 could therefore really be a section view through an organizational "ball." Figure 9-1 shows the ball cut apart to demonstrate both the continuity of the total technical organization, since it contains both aspects, and the vital differences that should deliberately be included in technical management theories.

Speculatively (since I have no data about this idea), the "ball" model is a starting point for integrating many of the diverse concepts discussed in the book. It provides for many different kinds of ideas about people, structure, technology, etc., but it also indiates that these diverse (differentiated?) components are tied together. When there is unequal weight that is unintentionally given to any component, the model should be able to show this and simultaneously indicate the components that have to be corrected. Since

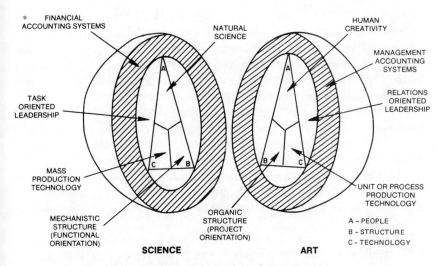

FINANCIAL
ACCOUNTING SYSTEMS

NATURAL
SCIENCE

HUMAN
CREATIVITY

MANAGEMENT
ACCOUNTING
SYSTEMS

TASK
ORIENTED
LEADERSHIP

RELATIONS
ORIENTED
LEADERSHIP

MASS
PRODUCTION
TECHNOLOGY

UNIT OR PROCESS
PRODUCTION
TECHNOLOGY

MECHANISTIC
STRUCTURE
(FUNCTIONAL
ORIENTATION)

ORGANIC
STRUCTURE
(PROJECT
ORIENTATION)

A – PEOPLE
B – STRUCTURE
C – TECHNOLOGY

SCIENCE **ART**

Figure 9-1 The bilateral model of the technical organization.

our model, at this point, is bilateral, the following example considers only two areas.

MEASURING THE AMOUNT

Assume that you have been able to reduce the major descriptions of each component to an ordinal measurement of some kind. For a brief review, ordinal measurements are numbers that can be greater or lesser but not linear; i.e., on a scale from one to ten, our drafting department organizational structure should be an eight (that's good) but is really a four (that's not so good). How can we improve it to be an eight, and what measurements will we use to determine when it reaches that point?

In effect, you measure *what you have* against *what you would like to have* for each component. The difference in these scores (i.e., the score of four, the difference between eight and four for the drafting structure) would tell you which of the components has the highest score and is therefore the primary problem. In other words, it would tell you which component has the greatest distance to travel to reach the goals that you have established for that component. If you made the model area size for a component equal to the score number for that component, the component in the biggest trouble would be the largest one in your drawing of the model (e.g., if the sum of the difference in points for production technology is larger than the point difference for either structure or people, technology has to be brought into line first.) The bilateral split of the organizational ball model between the somewhat arbitrary definitions of science and art refines the subjective measurements that you place on the measurement of each component.

Assume that the point difference for organic structure is six, say, for the research and development group and five for the mechanistic production group. The area for structure on the science side of the model at that point would be smaller than that on the art side. If that were intentional and an acceptable condition, the situation might not have to be changed, although the model wouldn't be a ball anymore. It would be pushed out of that shape. But if it were unintentional or unacceptable, it would indicate to you as the technical manager of both areas that the first item to be corrected is the structure of the research and development group. (Remember that a high score means that there is a larger difference between what you would like to have and what you actually have.) That would be true until a new ball model were set up.

To reiterate, the model of the organization as a bilateral design is not really being suggested here, since it is only a speculation. Other designs might be even more applicable, and *that* point is vitally important. Whatever you choose, it should be understandable to others (because they will want to know how you came to your conclusions) and it should be measurable (for the same reason). No one design can ever have all the answers. This is obvious, because if none of them can account entirely for the vast complexity of even one human being, how could it account for the different interactions that many people in an organization present? Conversely, there are no useless designs. Some are just better than others. They are all incomplete; and that makes you, as the technical manager, invaluable. You can handle all that complexity, build and adapt your personal theory, and do your management job by making decisions.

SPECULATIONS

Speculatively, these ideas incorporate two major themes that have been emphasized throughout this book:

1. *Integration of ideas*. You must develop the intellectual tools that allow you to integrate and use management ideas in your unique situation. These ideas can come from research, from other people's personal experiences, which you have observed, or from trying out your own ideas. The major task is the development of a model, a superordinate theory, or a mechanism of your own for this purpose.

2. *Improving your theories*. The ideas that you adopt must be measurable by whatever methods you choose. These measurements don't have to be exact or even exactly replicable, since they are subjectively based,

but you must have them in order to compare the value of dissimilar ideas.

Without a personal theory for integration of ideas and some method of measurement, we are as savages, living in accordance with magic rites that seemed to have worked once, although we don't know why and are unable to predict when they will happen again. That is not a proper posture for a manager. I hope that I have helped you in improving yours.

Case Study
THE CASE OF SEÑOR PAYROLL
by William E. Barrett*

Larry and I were Junior Engineers in the gas plant, which means that we were clerks. Anything that could be classified as paperwork came to the flat double desk across which we faced each other. The Main Office downtown sent us a bewildering array of orders and rules that were to be put into effect.

Junior Engineers were beneath the notice of everyone except the Mexican laborers at the plant. To them we were the visible form of a distant, unknowable paymaster. We were Señor Payroll.

Those Mexicans were great workmen; the aristocrats among them were the stokers, big men who worked herculean eight-hour shifts in the fierce heat of the retorts. They scooped coal with huge shovels and hurled it with uncanny aim at tiny doors. The coal streamed out from the shovels like black water from a high-pressure nozzle, and never missed the narrow opening. The stokers worked stripped to the waist, and there was pride and dignity in them. Few men could do such work, and they were the few.

The Company paid its men only twice a month, on the fifth and the twentieth. To a Mexican, this was absurd. What man with money will make it last fifteen days? If he hoarded money beyond the spending of three days, he was a miser—and when, Señor, did the blood of Spain flow in the veins of misers? Hence it was the custom for our stokers to appear every third or fourth day to draw the money due to them.

*Reprinted by permission of Harold Ober Associates Incorporated. Copyrighted 1943 University Press, *Southwest Review*, Autumn 1943.

There was a certain elasticity in the Company rules, and Larry and I sent the necessary forms to the main office and received an "advance" against a man's paycheck. Then one day, Downtown favored us with a memorandum:

> There have been too many abuses of the advance-against-wages privilege. Hereafter, no advance against wages will be made to any employee except in a case of genuine emergency.

We had no sooner posted the notice when in came stoker Juan Garcia. He asked for an advance. I pointed to the notice. He spelled it through slowly, then said, "What does this mean, this 'genuine emergency'?"

I explained to him patiently that the Company was kind and sympathetic, but that it was a great nuisance to have to pay wages every few days. If someone was ill or if money was urgently needed for some other good reason, then the Company would make an exception to the rule.

Juan Garcia turned his hat over and over slowly in his big hands, "I do not get my money?"

"Next payday, Juan. On the twentieth."

He went out silently and I felt a little ashamed of myself. I looked across the desk at Larry. He avoided my eyes.

In the next hour two other stokers came in, looked at the notice, had it explained and walked solemnly out; then no more came. What we did not know was that Juan Garcia, Pete Mendoza, and Francisco Gonzalez had spread the word and that every Mexican in the plant was explaining the order to every other Mexican. "To get the money now, the wife must be sick. There must be medicine for the baby."

The next morning Juan Garcia's wife was practically dying, Pete Mendoza's mother would hardly last the day, there was a veritable epidemic among children and, just for variety, there was one sick father. We always suspected that the old man was really sick; no Mexican would otherwise have thought of him. At any rate, nobody paid Larry and me to examine private lives; we made out our forms with an added line describing the "genuine emergency." Our people got paid.

That went on for a week. Then came a new order, curt and to the point:

> Hereafter, employees will be paid ONLY on the fifth and the twentieth of the month. No exceptions will be made except in cases of employees leaving the Company.

The notice went up on the board and we explained its significance gravely. "No, Juan Garcia, we cannot advance your wages. It is too bad about your wife and your cousins and your aunts, but there is a new rule."

Juan Garcia went out and thought it over. He thought out loud with Mendoza and Gonzalez and Ayala; then in the morning, he was back. "I am quitting this company for a different job. You pay me now?"

We argued that it was a good company and that it loved its employees like children, but in the end we paid off, because Juan Garcia quit. And so did Gonzalez, Mendoza, Obregon, Alaya, and Ortez, the best stokers, men who could not be replaced.

Larry and I looked at each other; we knew what was coming in about three days. One of our duties was to sit on the hiring line early each morning, engaging transient workers for the handy gangs. Any man was accepted who could walk up and ask for a job without falling down. Never before had we been called upon to hire such skilled virtuosos as stokers for handy gang work, but we were called upon to hire them now.

The day foreman was wringing his hands and asking the Almighty if he was personally supposed to shovel this condemned coal, while there in a stolid patient line were skilled men—Garcia, Mendoza and others—waiting to be hired. We hired them, of course. There was nothing else to do.

Every day we had a line of resigning stokers, and another line of stokers seeking work. Our paperwork became very complicated. At the Main Office they were jumping up and down. The procession of forms showing Juan Garcia's resigning and being hired over and over again was too much for them. Sometimes Downtown had Garcia on the same payroll twice at the same time when someone down there was slow in entering a resignation. Our phone rang early and often.

Tolerantly and patiently we explained: "There's nothing we can do if a man wants to quit, and if there are stokers available when the plant needs stokers, we hire them."

Out of chaos, Downtown issued another order. I read it and whistled. Larry looked at it and said, "It's going to be very quiet here." The order read:

> Hereafter, no employee who resigns may be rehired within a period of thirty days.

Juan Garcia was due for another resignation, and when he came in we showed him the order and explained that standing in line the next day would do him no good if he resigned today. "Thirty days is a long time, Juan."

It was a grave matter and he took time to reflect on it. So did Gonzalez, Mendoza, Ayala, and Ortez. Ultimately, however, they were all back—and all resigned.

We did our best to dissuade them and we were sad about the parting. This time it was for keeps and they shook hands with us solemnly. It was very nice knowing us. Larry and I looked at each other when they were gone and we both knew that neither of us had been pulling for Downtown to win this duel. It was a blue day.

In the morning, however, they were all back in line. With the utmost gravity, Juan Garcia informed me that he was a stoker looking for a job.

"No dice, Juan," I said. "Come back in thirty days. I warned you."

His eyes looked straight into mine without a flicker. "There is some mistake, Señor," he said. "I am Manuel Hernandez. I work as the stoker in Pueblo, in Santa Fe, in many places."

I stared back at him, remembering the sick wife and the babies without medicine, the mother-in-law in the hospital, the many resignations and rehirings. I knew that there was a gas plant in Pueblo, and that there wasn't any in Santa Fe; but who was I to argue with a man about his own name? A stoker is a stoker.

So I hired him. I hired Gonzalez too, who swore that his name was Carrera, and Ayala, who had shamelessly become Smith.

Three days later, the resigning started.

Within a week our payroll read like a history of Latin America. Everyone was on it: Lopez and Obregon, Villa, Diaz, Batista, Gomez, and even San Martin and Bolivar. Finally, Larry and I, growing weary of staring at familiar faces and writing unfamiliar names, went to the Superintendent and told him the whole story. He tried not to grin, and said, "Damned nonsense!"

The next day the orders were taken down. We called our most prominent stokers into the office and pointed to the board. No rules any more.

"The next time we hire you *hombres*," Larry said grimly, "come in under the names you like best, because that's the way you are going to stay on the books."

They looked at us and they looked at the board; then for the first time in the long duel, their teeth flashed white, "*Si, Señores,*" they said.

And so it was.

INDEX

Abilities as attributes, 74–75
Accounting systems, 265–269
Action research process, 343
Adams, J. S., 125, 141
Administrative theory, 45
Algorithms, 43, 44, 55, 64, 78, 85, 96
Analogies, 46
Archer, Stephen H., 42, 43
Argyris, Chris, 125, 129–132, 141, 173
Art, technical management as an, 11–14,
    16, 29, 66, 353, 356
As, D., 111
Aurelius, Marcus, 294

Balance (equilibrium), 72–73
Barash, M. M., 244
Barnard, Chester I., 301, 330
Bass, Bernard M., 294, 296, 308
Beer, Michael, 157
Behavior and cognition:
    changes in, 63–64
    differentiation in process, 170, 176,
        210, 232, 233
    motivation and, 99–100
    in process model, 6, 7, 29, 36, 46
    social behavior theory, 297–300
Bell, Cecil H., 342, 343
Bennis, W. G., 316
Bilateral model of technical function,
    354–356
Blake, R. R., 298, 299, 305, 307, 308, 311,
    313, 316
Blanchard, K. H. 306–310, 312, 313, 316
Blau, Peter M., 300
Blood, Milton R., 240
Boettinger, Henry, 36
Brain, left and right, 39, 47
Brainstorming, 79
Bronowski J., 224
Budgeting, 267–269
Bureaucracy as organizational design,
    159–162
Burns, T., 230, 243, 245, 298

Caesar, Julius, 294
Cafeteria approach, 138–142

Capital invested and total input, 97–99
Certainty, 34, 48
    (See also Uncertainty)
Challenge and response approach, 65
Chandler, Alfred D., 157, 158
Changes:
    beginning, methods for, 341–344
    in behavior and cognition, 63–64
    computer-caused, 240–243
    congruence model of, 341–342
    in economy, from goods-producing to
        service-producing, 9, 16,
    in environment, 3–23, 28, 300
    major, in organizations, effecting, 323–
        352
    acceptance of, by knowledge work-
        ers, 330–332, 349
    equifinality and, 339–341, 344
    proposal effectiveness in, 347–348
    overview of, 3–23, 28, 300
    reasons for, 81
    in resources, 251–252
    systems and, 70–72
    time and, 72–73
    training and, 344–347
Charters in organizations, 194–195
Chemers, M. M., 316
Chestnut, Harold, 37
Classical theory of management, 157,
    162–164, 167–168, 171, 179, 180
Close-down, project, 205–207
Closed-loop systems, 80, 83, 272–289
Coch, L., 118
Cognition and behavior (see Behavior
    and cognition)
Communications systems (see Informa-
    tion systems and processes)
Companies, effects of size of, 28–29
Computers, 82, 83, 95, 237
    and change in organizations, 240–243
    design of, 251
    technology and, 247–251
    word processors, 250–251
    (See also Information systems and pro-
        cesses)
Congruence model of change, 341–342
Contingency theory, 134–135
    leadership theory, general, 302–305

Miller, R. R., 253
Mintzberg, Henry, 13, 184
Mischel, Walter, 296, 297
Mitchell, Terence R., 329
Models:
 bilateral, of technical function, 354–356
 choosing, 36–41
 components of (*see specific entries, for example:* People in organizations; Structure, organizational; Technology)
 congruence, 341–342
 functional, 178–190, 209–210
 observers and, interaction between, 84
 organizational, 5–7, 59–85
 process (*see* Process model)
 project, 190–209
 situational and universal (*see* Situational theories, models, and effects; Universal theories, models, and effects)
 [*See also* Theory(ies)]
Mooney, James D., 163
Morley, Edward, 30, 31, 38
Morrow, Lance, 101
Morse, John J., 173, 177
Morse, Nancy C., 119, 131
Mortimer, Joylin T., 123, 131
Motivation and goals, 99–142, 178, 331
 as an attribute, 74, 75
 behavior and, 99–100
 cafeteria approach, 138–142
 definition of, 99
 design of motivation system, 137–138
 theories of, 94–95, 100–142, 159, 164–168
  situational, 94–95, 102–105, 110–117, 124–132, 141
  universal, 94–95, 103–104, 108–109, 120–124, 141
Mouton, J. S., 298, 299, 305, 307, 308, 311, 313, 316
Multinational organization, 28
Murray, A. H., 124

Nadler, David A., 341
Need(s), 132, 140
 hierarchy of, 121–124, 128
 in technical operations, 39
 for theories, 65–66

Networks, financial, 200–203
Newton, Isaac, and Newtonian mechanics, 30, 31, 38, 157, 209
Nominal measurement, 98
Nonrepetitive and repetitive classification of decisions and problems, 48–55, 69, 79–81, 83, 85, 142
Nougaim, Khalil, 123
Numbers, interval and ratio, 98–99

Observers and model, interaction between, 84
O'Donnell, Cyril, 32
Open-loop systems, 80, 83, 272–289
Operational definition, 8, 68
Optimal fit, meaning of, 334–336, 344
Ordinal measurement, 98, 99
Organic organizational structure, 231–237, 243–245, 297–298
Organization:
 changes in (*see* Changes, major, in organization, effecting)
 definition of, 67–68
 multinational, 28
 (*See also specific entries, for example:* People in organizations)
Organizational development method of change, 342-344
Organizational management (*see specific entries, for example:* Organizational model)
Organizational model, 5–7, 59–85
Organizational structure and design (*see* Design; Structure, organizational)
Osborn, Richard N., 157
O'Toole, James J., 336–339
Outcomes, 125
Outputs and history of technology, 222–224
Outside, in project close-down, 206

Participation, definition of, 110
Path-goal theory of leadership, 301–302
Patriotism as motivation, 103–104
People in organizations, 3–7, 9–11, 15–16, 40, 67, 69, 79, 222–224
 as components of model, 73–75, 84, 91–147
 as decision makers, 96–97, 110
 in design, assumptions about, 188–189